CREATIVE ACTUALIZATION

VALUE INQUIRY BOOK SERIES

VIBS

Volume 224

a volume in
Studies in Pragmatism and Values
SPV
Harvey Cormier, Editor

CREATIVE ACTUALIZATION

A Meliorist Theory of Values

Hugh P. McDonald

Rodopi

Amsterdam - New York, NY 2011

Cover Photo: Hugh P. McDonald

Cover Design: Studio Pollmann

The paper on which this book is printed meets the requirements of "ISO 9706:1994, Information and documentation - Paper for documents - Requirements for permanence".

ISBN: 978-90-420-3253-8
E-Book ISBN: 978-90-420-3254-5

Printed in the Netherlands

To the people of New York State,

sine qua non

CONTENTS

Preface

AN OUTLINE OF THE PHILOSOPHY OF VALUE

Value theory examines the character, nature, content, concept or identical element of value in different realms, spheres, bearers, instances and loci of value. The notion of "value" is coextensive with "good" for most authors in the twentieth century. "Value" as the most generally used philosophic term in the context of axiology replaced "good" around the turn of the century, mostly under the influence of the "Heidelburg" Neo-Kantians, the Austrian "psychological" school and Nietzsche. These authors confined good to moral goodness, a connotation it did not have in the classics except among the Stoics; or to bearers of value. The notion of the axiologists of the late nineteenth century that "good" meant "moral good" was based on the moral value theory of the Stoics (see Diogenes Laertius on the early Stoics.) Other ancient authors did not share the Stoic view that non-moral good was "neutral" in value, whether the Sophists, Plato, Aristotle or the hedonists. Aristotle and Plato explicitly discuss the value of economic goods while the hedonistic theory is one big argument against the moral theory. For most of the authors prior to the twentieth century, "good" covered moral and non-moral senses.

In any case, value theory was supposed to cover non-moral sources of value, for example, in economics, esthetics, knowledge, and so on. Contrary to Heidegger, however, "value" was used occasionally prior to the nineteenth century, for example, in Immanuel Kant and Thomas Hobbes (Hobbes, 1994).[1] The problematic of value arose as a distinct problem for the late nineteenth century because value was *not* conceived as an attribute of being. Specific value contraries were interpreted as normative (standards), such as good and bad, right and wrong, true and false, reality and appearance, beautiful and ugly, and so on. The study of such values was distinct from metaphysics and, in some versions, prior to metaphysics.

Value theories can be schematized as transcendent, objective, relational and subjective. In a sense, all value theories are relational, as they involve some relation of value to a bearer. However, the relation can vary from external to internal. One main group of value theories describes value in terms of external valuation. In turn, the external relation theories can be separated into transcendent, in which ultimate values transcend any value bearer, or of an object to a valuing subject, Transcendent theories involve a relation in which ultimate values have an extrinsic relation to bearers or objects.

"Relational" theories involve an extrinsic relation of a subject to an object, in which value is defined in relation to and for a subject The other two groups stress internal valuation in different degrees: whether and attributive-objective, or as internal relations of the subject. Objective theories involve an intrinsic or compositional relation of substances, things, objects, bodies or other actualities distinct from a valuer; or as the attributes of substances, objects, activities or things. Subjective theories stress an internal relation within the subject. Nevertheless, as the relational theories involve relations of both subject and object to value, they were given the generic label "relational" theories, despite the relations involved in all such theories (Pepper, 1949).

The metaphysics of value is usually closely connected with the model contained in the metaphysics of the philosopher examining value. Raising the question of the "ontological" status of values presumes a universe of beings, just as the "subjectivity" or "objectivity" of values presumes a Cartesian universe. In some cases, they are mutually exclusive: if value is subjective, it cannot be objective. However, some combinations are possible, for example, transcendent standards of value but immanent bearers of value (Max Scheler). In part, the metaphysics of value reflects the history of philosophy, in which philosophies of being and becoming competed with philosophies of the Good in ancient times. Transcendence in value theory was developed by Platonism, which characterized the "form of the Good" as "beyond being," while ontological metaphysical theories tended to treat value as an attribute of being. Descartes initiated the tendency to give this relation an inward or subjective turn. The relation of a quality was retained, but referred to a subject, not an object. Beginning with G. E. Moore in the twentieth century, analytic philosophers articulated theories of value that reflect and accompany the examination of language as the task of philosophy.

Transcendent theories locate ultimate value, or the source of value beyond either subjects or objects in some transcendent realm. One group of transcendent theories refers to values as standards, which instances of value (attempt to) exemplify more or less well. Such standards are independent of and prior to instances. In some theories they are causal factors as well, that is, they are the cause of the being of the instance of value. Platonism argues for ultimate values as standards, in the Form of the Good, and in Neo-Platonic theories which identify the transcendent One and the Good.[2] These theories envision a physics of flux in which values play some sort of stabilizing role. Theological ethical and value theories use a similar model, for example in agapism, or in divine will as the source of all good (John Duns Scotus). Another, closely related group of theories understands the relations of value in terms of ideals that transcend both subject and object, those of the Neo-Kantians and Max Scheler. Scheler's theory provides a transition to objective theories, since values have ontic "bearers," that is, goods, but values also have their own essence, objective relations, and objective reality as transcendent ideals (Scheler, 1916). Scheler's theory is an affective, objective theory:

values are objective as they have their own *a priori* relations, for example, hierarchy, polarity, and so on apart from any subject; and are revealed in feeling. These relations are "essences" or essential to value as such, thus objective and a priori in the Kantian sense. Ironically, a quite different philosophy of flux also conceives ultimate value as a standard. Friedrich Nietzsche's theory conceives of ultimate values as more naturalistic: "Dionysian standards of value" compete with otherworldly, Christian-Platonic standards.[3] All of these views also include instrumental value theories which may not be transcendent, but whose ultimate justification is in relation to transcendent value standards.

Another group of theories envisions values as in some sense foundational instead of as hierarchical standards. Values are the "basis" instead of above or beyond being. This last may be a verbal distinction, as both envision a metaphysical hierarchy from value to reality, but standards tend to culminate in a highest value that justifies the rest, perhaps the origin of the *summum bonum* distinction. "Basis" does not necessarily imply a one in many of the "standards" theory. In any case, there are at least two such theories, Rudolf H. Lotze's teleological naturalism, in which he proposes "that which should be [is] the ground of that which is" (Lotze, 1878, 319), and F. Canning Scott Schiller's humanism which argues for "the ethical basis of metaphysics" (1903, ch. 1),[4] a model echoed by Emmanuel Levinas.

Objective, or ontological, theories argue either that value attaches to the whole object, being or a thing; or is a quality, attribute, or character of being or an object. In some sense value attaches to the object or entity apart from or independently from a subject. An example of value defined in terms of a whole object is objective vitalism, in which value is defined in terms of overall contribution to life (Everett, 1918, Dewey 1920).[5] Albert Schweitzer's arguments for the intrinsic value of life constitute a type of objective vitalism. In a sense, John Dewey's theory of value is vitalist, but only with qualifications. For Dewey, life is an overall value but value can be attributed to some things regardless of whether they directly contribute to life. Clarence I. Lewis' theory is similar in this respect to Dewey's but is relational. Nietzsche has a vitalist theory of values, but value does not attach to life as an "object" but as a standard. He is a vitalist without a Cartesian metaphysics.

Value as an attribute of an entity has naturalist and non-naturalist versions. Aristotle argues for natural value in his equation of the good and the end and his arguments for natural teleology.[6] Stoic naturalism, based on the (species) nature of humans is another form. By contrast, George E. Moore argues that value is a "non-natural property" (Moore, 1903, ch. 1; 1912, ch. 7). Moore's theory of value is holistic and constitutes a unique type of a relational theory, so may be taken as transitional between objectively and subjectively based theories. Idealistic theories of value based on necessary duties, norms or ends of a subject are another type of objective non-naturalism, as is Robert Hartman's "axiometrics."

Relational theories argue that value consists primarily in the relation of a valued object to a valuing subject. They are intermediate between objective theories, in which value attaches to an object in some respect, and is thereby a property of the entity or object in a compositional relation; and purely subjective theories in which value is confined to subjective states. Relational theories can be more active, in the sense that an active part is taken by the subject, or more passive, in the sense that value somehow impresses itself upon the subject. Roughly, this distinction marks the divide between conative and affective theories. Conative theories involve some sort of striving or activity on the part of the subject toward the object of value. They include voluntaristic, conative and interest theories, that is, theories that value is an object of human will, desire or interest. Value involves more than just a subject, as desire is for an object of desire or interest is in an object of interest (Prall, 1921; Perry, 1926, 1954). Several affective theories, by contrast, posit the power of the object to affect the subject in valuable ways. A beautiful object, for example, arouses specific feelings in the subject, which are valued states. Clarence I. Lewis's theory of value is of this type as is Alexis Meinong's.[7]

Finally, relational theories shade off into subjective theories. Often subjective theories are difficult to distinguish from relational theories as they locate value in the same subjective states or posit similar psychological faculties or powers as connected with value. The affective theory is transitional in this respect, with both relational and subjective forms However, subjective theories tend to emphasize the subjective state more than the relation: the locus of value is at least partly independent of any capability of objects to affect such subjective states. What connects subjective theories is the location or connection of value with subjective states, whether attitudes, feelings or emotions. A purely subjective theory would only involve an internal relation of value to a subjective state: good feelings, proper attitudes and the like. They include attitudinal, affective theories proper, egoistic and emotive theories.

Theories of value are independent of the metaphysics of value, but tend to fall into patterns closely related to the metaphysics of value. I will give a quick summary of the main theories of value below, but caution that the list is by no means complete. The theories which I am summarizing either have a long history of treatment in philosophy or are of recent prominence.

The mystic theory is that good is the bliss attending unification with the divine: unity in the One. This theory was articulated in the West by Plotinus and echoed by Aquinas, Eckhardt and others, but is much more common in the East. Related theories are those of divine grace, or divine love, in which good emanates from the benevolent actions of a monotheistic God, not human efforts (predestination and some religious forms of fatalism).

One influential holistic theory is that good lies in the order and harmony of the parts. Although this theory is Platonic in origin, it is not necessarily transcendent, since it fits well with organic philosophies. Leibniz and G.E. Moore articulated variants of this theory. In the East, Kong fu-tse (Confucius) included it in his philosophy. The most impressive contemporary version of the harmony theory is in Dean Robert Neville's *Reconstruction of Thinking*.

Vitalism, the value of and for life, can be idealistic, as it is for some Eastern religions (Jainism and Buddhism). Greek drama raised the question of the value of life as a standard and Schopenhauer revised this issue for the modern world. In turn, Schopenhauer influenced Nietzsche, who in turn, perhaps, was the source of the problem for Camus, since for both life forms a standard of value. Both Nietzsche and the pragmatists, who also use life as an ultimate standard, were influenced by the nineteenth century developments in biology, especially the philosophical implications of Darwinism. Life is the bearer of value in almost all theories in some sense. For example, hedonism has a theory in which the good is identified with pleasure but in turn requires a life of pleasure (Herbert Spencer and John Stuart Mill).

Perfectionism, or what has become known, unfortunately, as the "virtue" theory, comes in two main strains. The first is the happiness theory of Aristotle. Happiness is not identified with pleasure by Aristotle but with "flourishing" or "functioning well" (*eudaimonia*), that is, a state in which a person has reached his or her fullest and highest development. Such perfections of development are virtues or excellences, although something can be virtuous, in the sense of good, without excellence. Good, then, is identified with virtue and what is conducive to virtue (the perfection of any substance). This strain includes the "self-realizationism" of Theodore H. Green as a form of perfection and Abraham Maslow's "self-actualization."[8] Aristotle, unlike Plato, separated moral and intellectual virtue. Intellectualism was Socratic in origin. Virtue or the good is equated with wisdom and wisdom in turn with knowledge. Thus good depends on knowledge: knowledge of good. Since reason is the part of the soul concerned with intellectual matters, this theory is tied to human reason. Education in virtue is the path to perfection of human nature. Some authors have combined perfectionism with conativism: good as the virtue after which we strive or perfection as the object of desire because of its goodness. Baruch Spinoza first articulated this approach and in many respects he was echoed by T. H. Green. "Good" for this theory involves perfecting our nature by reason, habit or striving after excellence. Following Spinoza, perfective conation can form a transcendent theory. Rene Le Senne has argued that the perfections of God are the source and model for values created by humans (La Senne, 1961).

Conative theories are a generic label for several closely related kinds involving relations of a valuer to a valued object. Since these involve psychological capacities, whether desire, demand, will or interest, they form one main branch of the psychological theories. They are collectively referred to as conative theories because they are closely connected, as "will" may be a determination of desire, and interest may also be based on desire. The origin of this theory is with Spinoza. However, different versions of the theory may stress one element more than the others, for example, interest in Perry and Prall. An emphasis on the will is referred to as a voluntaristic theory. "Willing" usually involves a modification of desire in some form and may be transcendent (Scotus, Ockham) or immanent (Lavelle 1972, Sartre).[9] Voluntarism connects Scotus, Schopenhauer and Schweitzer, but the first is probably not a subjectivist in his theory of will, which is more complicated, and Schopenhauer's voluntarism is metaphysical as much as subjective. Desire as the source or basis of value is connected with hedonism not as the character of value, but its source; and similarly to the interest theory. Conation implies a striving for the object of desire. Thus the theory can be either objective or relational, depending upon the stress of the philosopher: value as the *object* of desire after which we strive; or value as the object of *desire*, which requires the relation to the subject. The mode, that is, is distinct from the conative theory as such. The origin of this theory in the extant literature can be traced to Plato's Callicles (*Gorgias*), but in the modern period was developed by Hobbes, Christian Von Ehrenfels, Samuel Alexander, and Stephen Pepper. Taking an interest in something also involves desire and the proponents of the interest theory, Perry and Prall, identify it with desire. A subjective interest is taken in an object and the relation is a valuation or value situation.[10]

Affective theories are another generic label for several closely related kinds also involving psychological capacities, in this case feeling, attitudes, emotion and sentiment. The most widespread is hedonism, the theory that argues that pleasure is the good or the ultimate good. The Cyrenaics, Epicureans and Utilitarians, among others, propound hedonistic theories.[11] The hedonism of Mill is an example. Although pleasure may be connected with desire, as its natural object, hedonism is more rightly classed as a kind of affective theory as pleasure is characterized as a kind of feeling and good or value is identified with the feeling of pleasure or happiness.

The affective theory proper is any theory of value that ties value to feeling, regardless of mode. This theory is important in esthetics and has been advanced by Meinong and some feminists, among others. Another subjective-affective theory that connects value with human feelings is the "moral sentiments" theory of David Hume and Adam Smith.[12] Egoistic theories locate value in the good of an individual ego. Another type is the agapism of Christianity, a modern form of which can be found in Gabriel Marcel. Finally, the most subjective theory of all, emotivism, restricts value to human emotions, that is, "good" in a judgment is the expression of emotions.[13]

Franz Brentano posited the theory of good as having specific "proper attitudes." Attitudinal theories argue that specific attitudes are proper or fitting regarding objects. Although a relation to an object is involved, the locus of value is in the subjective attitude; good is a quality of the attitude as such. As these attitudes are proper or fitting apart from any object and have their own logic, they are more subjective than relational. Brentano analyses value in terms of affective attitudes. Another example, which explicitly appeals to Brentano, is Noah Lemos 1994. A related theory is the "approval" or "esteem" theory of Earl of Shaftesbury Anthony Ashley and of Francis Hutcheson (Laird, 1929).

Several other theories deserve mention. One is the moral theory of the Stoics and Kant. Kant is often classed as a voluntarist, but Kant's theory of good is actually a version of the moral theory, since a *good* will is a moral will.[14] Moral values are the condition of the goodness of other goods, including will, which can be bad. Will is the bearer of value in Kant, especially of absolute value, but is not the value element in a good will, or the principle of goodness in a good will. This modification was an advance on the Stoic version of the theory, in which non-moral goods were considered "neutral," neither good nor bad. The Stoics combined perfectionism with the moral theory, since human nature was identified with the species difference, rationality. A related theory is the personalist theory of Scheler, which follows some trends in Kant. The values of the person become the ultimate standard. Finally, there is the theory that human "needs" are the basis for values and the labor to produce them the creator of value; respectively the psychological and economic versions of the theory.

The modes of valuation within such relations vary. Modes of valuation involve the relation of values both to bearers and to other values. The relation to bearers involves the distinctive mode required by the metaphysics in which value is placed, or the mode required by the value theory itself. Modes are in different relations in different theories of value. In some cases, these modes are similar to the above relations attached to theory of value. In other cases they vary, since a value can be an attribute of different bearers, whether ontological, subjective or other. Such issues are often referred to as the "ontology of values." However, "ontology" is only one mode, since value may attach in modes other than "ontological" and indeed exclude a metaphysic of being altogether.

The first model is transcendence, where the good is an independent principle from being. Relations in this mode are hierarchical, in which the good as a supreme principle regulates subordinate principles. The mode of value is participation or imitation.

The second model is differentiation, where good is a qualitative difference between kinds. Relations are external or extrinsic, between incompatible and incommensurable differences. The mode of value is hierarchical in the random occurrence of such kinds.

The third model is attribution, where the good is an attribute of being, or substance, and not an independent principle. Relations are composite, involving potency and actuality. The good comes to be as an attribute of being. The mode of value is categorization, the referring back (*pros hen*) of value to a substance.

The fourth model is universalization, where good is a common or universal element of different bearers. Relations are implicitly reciprocal between goods. The mode of value is inherence.

The fifth model is externalization, where the good of an object derives from a subject. The relation requires both a subject and an object for the relation. The mode of value is contextual positing, that is, value is situational for a valuing subject that projects value onto an object.

The sixth model is internalization, where good is an attribute of internal states, and not of being. Relations are to such internal states, to an ego or subject or to states of the subject. The mode of value is in a proportion to the maximization of such states, that is their enhancement (for example, pleasure) or diminishment (for example, pain).

Some of these modes are compatible with value as a first philosophy, namely, the first, second and fourth. Five and six are subjective; two, objective. One and three are contraries, as are three and six, and four and five.

The question raised by the modes of value is how are values related? Are they manifestations of one supreme value, instances of a conceptual essence, conditions of other concepts; or only related equivocally or by analogy? Creative actualization as a theory of value must answer this question.

Apart from theories of value, several other distinctions in the philosophy of value should be mentioned. The first is the distinction of a first philosophy of value from a theory of value. A first philosophy of value is a whole philosophy centered in or based on value in some sense. Philosophers who articulate first philosophies of value deliberately create a philosophy that is not based in ontology, the subject or language. An example is the Platonic and Neo-Platonic philosophy, in which the "Form of the Good" is "beyond all being" and gives being to the forms. A theory of value is not equivalent to a first philosophy of value and can be articulated separately from any philosophy of value or even any metaphysics.

The second distinction, which will be the most prominent in this book, is the distinction of evaluation and value. Evaluation differs from value in one sense as method to subject or content. Evaluation is a method of determining, judging or positing value. In some sense, any event, thing, person, action or other actuality can be evaluated: we may ask, "what is the value of x…?" We can also evaluate possibilities for decision or in the abstract, or possibilities in relation to actualities. All such evaluations can also be critical, that is, we can judge the ultimate or relative value of any actuality or potentiality, even scientific theories. Is one theory superior to another, more adequate? For example, Einstein over Newton in physics? Is a new model of a technology truly better or not? Is allowing pornography to be sold openly better or not?

Such issues and questions constitute the problematic for evaluations. Evaluation as a method is the logic attached to value, including norms and rules governing activity.

However, as the term "positing" may indicate, there is some controversy over whether value is created by acts of evaluation (Dewey, 1949; Polin, 1972)[15] that is, that the distinction is either one of cause and effect or is not even valid,[16] as the act of evaluation is the positing or creation of value. I will argue against the last view, since "inherent" value makes no sense if reduced to evaluation. Indeed, many of the paradoxes and problems in the philosophy of value disappear if the distinction of value and evaluation is applied, notably those of G. E. Moore (1903, ch. 1). Moore asks his famous question, by which he argues that the value of anything can be doubted by posing the question "but is it good?" or does it have value? In his attempt to argue for the fallacy of infinite regress, he thereby confuses value and evaluation.

While evaluation may involve a hierarchy of different possibilities, or result in one, inherent value is not hierarchical. Multiple inherent values are compossible in an external relation that does not involve a hierarchical structure. Nor is evaluation as a critical notion equivalent to value, since we can critically evaluate the value of what has inherent value or do so comparatively. Above all, we can also evaluate the value of what is not actual, does not bear inherent value, but is potentially valuable.

Evaluation, however, is not only critical, not merely a method. Evaluation also has to do with what will become actual from competing potentialities. Different choices are evaluated and the ostensive best is creatively actualized in acts of value. We evaluate different ends and one is chosen as superior; we can also evaluate means to the ends. Evaluation does have its own content in the realm of potentialities, whether these prove impossible, possible but not yet achieved, possible or necessary. What is actualized is finite, but evaluation is not limited to actuality; we can always evaluate potential improvements, comparatively superior outcomes and the like. Evaluation may not be infinite, but is not limited to actuality. Thus evaluation constitutes a new problematic for seeking grounds. Such grounds are neither finite nor infinite. We can also speculate that, since we can always pose the question of the value of the actual, evaluation never reaches a foundation, a "first." Nor is it ultimate grounding, since grounding requires actualization. However, the ultimate question is still "What is the value of x…" Such speculations are limited in another respect by practicality, since evaluation of potential consequences is not enough to prove value. Value requires some sort of actualization in the world, grounding as actual of what had been potential. While the value of a good in the world is inherent, it may have its origin in the prior evaluation that its actuality would be an improvement of some kind. Evaluation is not totally distinct from value, then, since evaluations are made prior to value as creative actualization of goods.

Individual authors may also use other terms related to value, for example, "valuation," "valuing," and so on, and these may be interchangeable with evaluation, or have meanings peculiar to particular authors. Evaluation and valuation are also sometimes distinguished. Evaluation is comparative; valuation a judgment of particular value. That a particular good has value would involve a value judgment or valuation; that it has value in comparison with other goods an evaluation. However, I have collapsed these two terms.

The distinction of locus and bearer of value is often made by philosophers (Callicott, 1989).[17] The bearer of value is the instance, example, token or individual unit of or bearing value. Locus is the type or location of value, that is, what sort or what type of unit or what qualitative level value is located in: whether in individuals, objects, standards, and so on. The locus is closely connected with the metaphysics of value, as locus in individual subjects presumes some sort of metaphysics that includes subjects. However, locus and bearer may be distinct for other reasons, for example, the composite location of species in individuals in some philosophies of value, but the locus of value in the species. In this case individuals would bear value but its locus would be in species. Aristotle's theory may be of this sort; some radical environmental theories hinge on it (*ibid.*).

The distinction of instrumental and intrinsic value goes back to Plato. However, its relation to the metaphysics of value should be mentioned. Firstly, a theory of intrinsic value can differ from a general value theory. This point is evident from value theories that describe general value characterizations, contain a theory of value, but either deny intrinsic value or are critical of the notion of intrinsic value, such as, respectively, Nietzsche's and Dewey's. Further, the relation of intrinsic and instrumental value may intersect a metaphysical relation, for example, that of subjects to objects, with each type of value located in different kinds of agents or things. Intrinsic value may be located in a distinct subclass of the more general universe of values, as they are in C. I. Lewis, for example.

Instrumental and intrinsic value relations are independent from either the relations of value to reality, objects, subjects, or to the relation of objects to subjects. This is clear from theories of value that posit or accept subjects and objects but are critical of the notion of intrinsic value. Further, value theories which locate value in objects or subjects may not locate *intrinsic* value in such in the same respect. However, in some theories, they may be coextensive. Relational theories, in particular, often view objects of value as instrumental while the subjects for which they are valuable are intrinsically valuable. Thus instrumental and intrinsic value loci can differ in kind. Clearly, the metaphysical status of value is problematic, in view of the philosophies which attach value to contrary loci, whether transcendent standards, objects, or subjects. However, it may be that the internal relations of value, instrumental to intrinsic, in the aspects covered in chapter one, may be a way around this problem. A closely related distinction is value and the condition of value. A condition of value can be a necessary means but can also be another value. Food as the condition of life is a necessity, but may have enjoyment value.

The relation of intrinsic value to the *summum bonum*, the highest good, varies with different theories but, in general, the highest good is an intrinsic value in the highest degree. The highest good has a more restrictive scope than intrinsic value as it only includes the highest intrinsic value. Intrinsic value may be defined in terms of the highest good and may reflect it in some respect. A few authors conflate the highest good with intrinsic value, for example, hedonists. There may also be a distinction between the highest good and some sort of overall good. The overall good is a good that is realized over time, for example, over a lifetime, but not at the end. Happiness or *eudaimonia* is an overall good in Aristotle as well as a supreme good and an intrinsic good. However, overall good can be distinguished from intrinsic good for some authors. C. I. Lewis argues, for example, that an immediate good may have intrinsic or inherent value, but not contribute to a good life as a whole. For most authors, however, overall good is connected with the *summum bonum* and is a more restricted case of intrinsic value. A final or ultimate good is distinct from a mere end in that it is more inclusive and finishes a chain of ends used as means. Many subsidiary ends may contribute to a final end, which may or may not be intrinsically valuable, a highest good or an overall good. Dewey acknowledges final ends that are not in the same respect intrinsic values (Dewey, 1949);[18] other authors distinguish them in other respects.

Most authors distinguish values from norms, "good" from "ought." What ought to be is some projected norm, duty, standard or other principle that is not necessarily actual. We should bring it about even if we do not. While this may reflect evaluation or be a conclusion of an evaluation, norms are not strictly identical with evaluations, since, among other functions, the latter concern comparison, criticism and possibilities of all kinds, not only normative possibilities. Nor are norms identical with value, since they may not be actualized. Norms are sometimes expressed as commands (prescriptions in Hare) while evaluations are judgments that can include such norms but also other value judgments.

C. I. Lewis also distinguishes intrinsic value from inherent value. Inherent value is the quality in an object that arouses a specific state, quality or other response in a subject. Other authors either adopt one term or the other to mean the same norm, either inherent value or intrinsic value; or use the two terms interchangeably. Following Lewis, we will argue in this book that "inherent" connotes the relation to a bearer, while "intrinsic" value is confined to living agents, human or not, but also connotes a different relation, namely, to "extrinsic." Intrinsic and extrinsic are coextensive with three, normatively distinct value expressions that stand for or connote a similar value relation: the means-end relation, that of instrumental to inherent value, and the reflexive relation of "for the sake of" to "for its own sake." Intrinsic and instrumental are generally coextensive with the other relational norms which are distinguished in value theory as connoting the intrinsic and instrumental value relation, that is, the means to end relation, or that of "for the sake of" to "for its own sake." It might be argued that "for its own sake" implies a relation to itself, and not to instrumental value. More properly, extrinsic connotes a relation to intrinsic, but inherent has value in itself. But extrinsic value can have inherent value if the value is inherent to the means and not totally derivative. An instrument can be inherently valuable as a tool but also derive its value from an external relation to an agent. Such an external relation is the product of evaluation of the use of the tool, but does not affect its inherent value. While this may constitute a problem for the notion of inherent value, it does not affect the point that the other normative aspects imply the relation, especially means to ends. Intrinsic value grounds extrinsic for many value theories, and thereby the relation. In terms of contraries, intrinsic is the value contrary of extrinsic in one respect, and so the relation holds in logical terms. "For its own sake" contrasts with actions performed for the sake of something else and arises in relation to extrinsic values.

Other distinctions of value include value analysis, value fallacies, possibility/actuality, and theory and practice. Value analysis consists in analysis of the value judgments involved in any situation, action, statement, text, or choice. Sometimes the values involved are explicit, sometimes hidden. Value fallacies are fallacies in the logic of value or in relation to value. An example is G. E. Moore's so-called naturalistic fallacy, which is itself fallacious. Moore claims both that "everything" is natural and also that good is a "non-natural" predicate, a contradiction.

Theoretical value belongs to possibility; practical value to actuality. Theoretical value attaches to possibilities that have not been actualized but might be worth actualization. However, theoretical value becomes practical through its "realization," or in our language, its creative actualization. A plane begins as a dream, moves through the stage of experimentation and ends with a functioning machine. Thus the realm of possibility is distinct from that of actuality, a distinction roughly parallel to that of evaluation and value. Possible values require a practical test or at least a practical condition. Theory and practice can also be the criterion of a value orientation or perspective. A work of metaphysics may have practical ends, for example, that of Descartes. By contrast, Aristotle argued for a theoretical orientation for metaphysics, and thereby its theoretical value. A related contrast is the distinction of a moral orientation for values from an esthetic one. An esthetic model can ultimately orient values or morals can provide the basic model.

A final distinction is that between absolute and relative value. Absolute value consists in pure value, that is, value that is not relative to persons, cultures, historical ages or other extrinsic factors. Some authors, such as Hugo

Münsterburg, have implied that absolute value is equivalent to intrinsic or inherent value, since inherent value is valuable in itself, and thus not in relation. Others distinguish intrinsic or inherent value from absolute value. In general, the distinction between absolute and relative value follows the divide between transcendent and objective theories on one side; and relational and subjective theories on the other. While not all subjective theories are relative, there is an affinity, at least, between the differentiation of personal tastes and perspectives, and values that attach to them. Differences in value between persons, cultures and groups are emphasized over identities. Nevertheless, exceptions crop up on both sides. Perfectionist ethics can emphasize degree of value in proportion to virtue and thereby differences in rank or worth (Plato, Nietzsche). Subjective ethics can emphasize the identity of psychological faculties or subjective states between humans (Epicurus, Hume).

Value judgments are statements, propositions or observations of value, worth or evaluation. They usually contain one or more value laden terms, such as good, bad, virtue, vice, right and wrong; comparisons of value; or commendatory expressions. The relation of value judgments to values is also forms a controversial or problematic issue.

Classifying value judgments is problematic, since different approaches may be taken to classification. One would be by kind, another by grammatical usage, another by levels and so on. I will discuss kinds of value, levels, functions, and grammatical usage.

Kinds of value represent different value pursuits, whether in a culture, across cultures or among individuals. These include social values, economic values, intellectual values, religious values, esthetic values, entertainment values, political values, and other kinds. Such classifications of kinds are used in the social sciences both to describe the different cultures, including different activities within them, and to compare them. A ritual meaningless to an outsider may have profound meaning within a culture, and this ritual aims at a valued end or constitutes one. An anthropologist may explain that the ritual has religious and social value within the culture and thereby both explain the ritual with a term, "value," that can be understood by outsiders; and differentiate it as culturally peculiar. The implication of such a classification is that values come in kinds, and such kinds are both distinctive and yet have some elements of identity that makes them all values. An "instance" or bearer of such goods is a token, individual example or the like of such kinds and presume an understanding of goods as an actualization in an instance.

Levels: levels of value are distinctive or peculiar to different social congeries. Ends can also be of categorial, cultural, social or personal value, although ultimately, categorial values serve individuals, whether some or all. Categories of value are kinds of value, but distinct from cultural values as general. Religious values may differ from culture to culture but still be classified as "religious" for comparative or other purposes. Categories of value are the most general kinds of value and thereby cannot be subordinated under another category. Religion cannot be reduced to economics and vice versa. Categorial values are not internal, but connected with action, particularly collective action, involving social cooperation. Building a skyscraper as a valued goal involves the actions of many. Cultural values are generally distinctive to a culture, although adjacent cultures may share specific institutions, such as a common religion. This overlap makes differentiation of a culture difficult, especially if their languages are related dialects. Personal values are those distinctive to individuals and involve personal evaluations of quality. Personal values and goals encompass the realm of the private: wishes, desires, feelings and the inner dimension of a person in general. Personal values are not common, since they differ from person to person and thereby are distinct from the shared values of culture, social groups and the kinds and categories of value.

However, if such personal values become goals, they will result in action in the public space of the world, and be manifest in the world. Moreover, these levels of value can overlap, since a person may value a ritual of his or her culture, or be attracted by an institution or custom defined by kinds or categories, such as music. The music of a culture may particularly appeal to one individual but leave another cold. Social values may differ from those of a culture; they include values of social groups within a culture, whether families, fraternities and sororities, or groups bound by a common interest like chess clubs. Action and language can both reveal such values; action through their manifestation or creative actualization, language through articulation. Finally, universal values are values shared by all cultures. Some might argue that such values are alleged, not actual, but murder, theft, incest, rape and lying within the in group are universal or nearly universal taboos. Moreover, even if they are not, the argument may be that they should be.

Functions: Values have different functions, both in language and in the world, including grounding, instrumentality, standards, norms, commendations and evaluations. Values can be the ground or cause of change, the cause of an agent's action that changes the world, even if in small respects. We can infer from such a function that value is in the world, and the cause of changes in the world. Value is not confined to the feelings, desires or intellect, but is in the world as a dynamic element. Fixing a roof is a valued goal that changes the world. A good job of fixing the roof is not one that causes specific feelings, but one that keeps out the rain. The value of keeping the interior dry is the cause of the action. The means may vary depending on the materials used, whether thatch, tar and gravel, slate, shingles or what not but the actions to accomplish this end have a different function than the roof itself: a different value. Means also ground a goal in the world. Finally, the overall purpose of a job, an activity, even a life is a distinct function: the "meaning" in the sense of significance, importance and purpose. The purpose, the goal and the means are all pursued through action and the same action may accomplish more than one of these functions. Values may also justify actions as well as provide their rationale.

The agent may have a specific standard of value that is aimed at and may or may not meet this standard. Standards of value are evaluative and normative measurements for ends. Standards of quality are impersonal, since the agent is expected to meet the standard; thus they are distinct from personal evaluations of quality. Standards are in the realm of potentiality unless or until they are actualized. Thus they are a type of norm and, like other norms, dwell between the actual and the potential, since norms are both actual as principles and potential as regulative over future potentialities that reduplicate the same principle. Standards, norms and principles dwell in the realm of what ought to be in view of its value, but are not Utopian. They can be creatively actualized and their function is to guide action to achieve what is adjudged worthwhile.

A job that comes up to standard is a good or commendable job and can be evaluated as such. Commendations and other judgments of worth are species of evaluation. Another species is any personal evaluation of quality, in which an individual esteems any good. Still another is a judgment of worth that is impersonal, such as the judgment that a specific kind of food is "good for you." Another is a descriptive sense, where an observer evaluates work as "good." Still another is comparative: one roofing job is better than another. The evaluation involved in such comparisons may also invoke standards. Standards can be used for critical evaluations of worth or quality ("the job is no good, because the roof still leaks"). These functions of value are distinct from norms as directly valuetive. Evaluations may be verbal, but appeal to actual features of goods or norms in the world. A roofing job that is commended will undoubtedly not leak. Both the roof and the leak are in the actual world, not the feelings of an agent.

A final function is the use of language to describe actual value: inherent, intrinsic or instrumental value. Economic "goods" have determinate market value, not only a relation to one person's desire. A "good" tool has value as a characteristic of the tool, not simply as an evaluation. A good tool is one that is not defective: it will generally last and not break during a job. A good tool also functions well in the job it was designed for, whether hammering nails, driving screws, planing surfaces or what not. A hammer may function well and not be personally valued by its owner at all. The inherent value of a good does not mean it cannot be evaluated; evaluation is not in another universe. Something may be judged inherently good only to be compared to something better: a reliable personal computer that is nevertheless dated, slow, or lacks features on newer models.

Grammatical: Grammatical classifications are added here more to refute specific judgments and statements about value in the literature, than because they are helpful. Often an author will argue that one grammatical usage involving value language is a paradigm of some sort. A.J. Ayer and C. Stevenson argued for the priority of value used as an interjection. "Great!" "Excellent!" and other interjections were taken as the *summa bonae*. Dewey argued for the priority of the adjectival use of good or value, as a descriptive term. Moore echoed him, following Aristotle in thinking that value is an attribute of a subject, predicated of a subject. Indeed, a model of language also bewitched Moore: the good as an attribute or "predicate," ignoring the other grammatical uses of value terms and expressions. While some instances can be good or bad in the descriptive, adjectival sense, it does not follow that all uses of value expressions or all value judgments are adjectival, a qualification of 'existent.' An instance can be "a good"; an entirety can be good or bad. Part of a plane will not fly—not necessarily a quality of a part. Value expressions are not like colors, especially where they constitute the ground or origin of a thing: the reason it was created. However, choosing one linguistic or grammatical model over the others is an evaluation, not a statement about actual grammar or a summary of usage. It is a fallacious attempt to reduce value expressions and judgments to one form, mode or metaphysics.

Value terms, on the contrary, can form five different parts of speech: nouns, verbs, adjectives, adverbs and interjections. "Goods" as a noun are a mainstay of practical economic life: "give me the goods," did someone "deliver the goods?" or "when were the goods shipped?" "Goods" can also refer to the totality of economic production: the GNP is all the "goods and services" produced in a specified time frame, usually a year. Similarly, we can speak of "a" value or "of value" and the "value of" some thing, event, and so on. Value is also a verb, as in the expression "to value" "I value," or a particular culture "values" some ritual, virtue, or other value-laden term. "I value" can constitute a critical evaluation or can be a statement of personal valuation regarding some inherent value. There is also the use of something that is "valued for" something else.

Good can indeed be adjectival as well: a "good" race horse, car, house, spouse, person, and so on. In the case of adjectival use, a "valuable" or "valued" horse, car, and so on, can be used almost exactly as a substitute for "good." Also, compare "invaluable" as meaning valuable as a means or secondarily valuable. Again, the adjectival use may constitute a critical or comparative evaluation or a descriptive term: a good apple is without visible flaws, such as worms, that would decrease its appeal. (Reductionists, relativists and nihilists, if consistent, would eat the worm.) Value words can also be adverbial: some action is "well done" or, in more poetic language, "goodly." Finally, value terms can be interjections: "Good!" Generally, this marks the point at which value is finally achieved, problems have been overcome and the end or goal is reached: creative actualization. Good! used as an interjection is not only an expression of emotion, although it can include emotion, for there may be good grounds for a person valuing the good they have achieved such that they summarize it in one value expression.

In terms of the subject-predicate distinction, both value and good can be either parts of the subject or parts of the predicate ("his values," "she values"). As a transitive verb, to value can relate a subject to a predicate as a relation: "she values intellectual life."

Critics or reductionists might argue that such a grammatical plethora is problematic, and the reduction or judgment of priority is an attempt to put the anarchy of linguistic expressions in some sort of order, or to prioritize them. However, all linguistic judgments that express value are ambiguous and thereby require critical evaluation to determine the best interpretation. "Good!" can mean personal satisfaction, achievement of a standard that has been aimed at ("good enough"), the achievement of a novelty, an invention or new process, or an expression of comparative evaluation. "It is good" is, *prima facie*, a judgment of the worth of something being evaluated, not subjective or personal feelings, since is not a value judgment about the feelings but whatever thing, object, event, process or other instance that is evaluated. Judgments of feelings would be expressed as such, namely, "it makes me feel good," "I like it," or the like. Reductionists confuse two very different value judgments. "I like it" is a personal evaluation of quality, not an evaluation of the quality of the object as such or perhaps its instrumental value: good for something. A tool that is "good for" fixing a car is not good because it elicits emotions, nurtures feelings, or the like but because it functions well as a means to an end, regardless of feelings. Finally, are different kinds, functions and levels of value all reducible to feeling? It is evident that they are not, since I may feel I should be able to purchase a four-carat diamond for only a dollar, but be unable to complete my transaction. My feelings may be hurt by the actual economic value of the good.

In any case, and as Dewey also noted, language is not very helpful since language obscures as much as it reveals, and there is no part of speech to which "value" is consistently confined. The diverse uses of value judgments in different parts of speech that I have outlined may appear to some readers like so many penetrating insights into the obvious, but have somehow escaped the "detailed analysis" that some philosophers like to think they have produced. Such self-commendations are more accurately described as hidden value judgments fulfilling an agenda, than the study of the language or grammar of value. Moreover, the inexpressibility of unique value places such values beyond language, although metaphor and the like may be used to attempt to capture it in normative identities (concepts). Concepts are not excluded, although many have confused the concept of value for value. Concepts have instrumental value in some circumstances: they attempt to capture some aspect of the unique as a normative identity.

It should also be clear from the different kinds, functions, levels and uses of value expressions that value cannot be reduced to ethics, another claim made by some analytic philosophers. Indeed, one problem in ethics is that economic goods may retain their value even if they are immorally obtained. Values and ethics intersect, of course, and any comprehensive ethic must address this problem.

Acknowledgements

Chapter one appeared originally in *Contemporary Pragmatism*, December 2006.

My thanks to anonymous readers, who had the thankless task of constructive criticism.

INTRODUCTION

This book is connected to my previous book, *Radical Axiology, a First Philosophy of Values*. In *Radical Axiology* I argued for a first philosophy of evaluation, based upon values. Evaluation comes first in philosophy as a whole, which consists in value inquiry. Radical axiology involves a different inquiry than other candidates for first philosophy. Political philosophy, esthetics, and other areas of philosophy are examined in terms of values, not "being," the "subject," and so on. In this view values are not subjective or objective, not an attribute of being, or a predicate of a grammatical subject, for, in a radical move, value grounds them all.

Creative Actualization will continue this project with a theory of values. Although I conceive this new work as in some sense continuous with the prior book, this book can be read separately. It stands apart from the prior book in that I will examine value as such, rather than evaluation in this one. Although there will be reference to first philosophy and ultimate grounding, this book can also be read as a straightforward value theory. The two projects are independent, since a theory of value can be presented that is not a philosophy of value and a philosophy of value that does not confine itself to a theory of value. Of course for a radical axiology they are related, since ultimate grounding is in the world, that is, requires creative actualization. Creative actualization is a theory of the determinate content of ultimate grounding, since grounding in the world is teleological and involves creative actualization.

"Value" is in one sense much larger than what has been actualized, since it includes potentialities and possibilities not yet creatively actualized. The possible is in the realm of evaluation, although actualities can also be critically evaluated. Thus we can evaluate the value of the possible, and of the actual, either in a critical sense or as embodied past values. The value of the possible can be evaluated without reference to the "existence," that is, the actuality of what is possible. Value encompasses the possible and the actual. It thereby may provide a solution to the problem of the "existence" of the "non-existent," for example empty space. I will cover this relation in the text. Actuality is also wider than "being" and thus more general, for it includes actual language, actual justice, actual societies, culture, and other actualities that cannot be described in terms of "being." Creative actualization marks the point at which all ontology is superceded.

1.

One main thesis of this book is that there has been a serious confusion of value and evaluation in the literature of value (and thereby ethics) during the history of philosophy, in the attempt to reduce value questions to a *summum bonum*, a one value in many. Subjective theories in particular are notorious for arguing both for the intrinsic value of subjective states, and for value as evaluation. This tendency began as the hierarchy of degrees of being as degrees of Good in Plato's *Republic*, the mode of value as the metaphysics of reality. Aristotle subsequently retained the relation in his *Metaphysics* and his *Ethics*, since good was adjudged a transcendental attribute of all being as perfection of substance; and was reduced to "happiness" (*eudaimonia*), the highest good. Subsequent theories have adopted this relational model of good, despite the logic of intrinsic good that Plato and Aristotle also examined.

This tendency follows the logic of systems, in which the many and different fields of philosophy are brought "under" a "one," whether being, substance, matter, experience, reason, language, processes, or any other basic first. The "one" forms an absolute standpoint or foundation from which all principles, phenomena, and differences can be understood. All philosophy proceeds from such an absolute first principle, even those superficially opposed to metaphysics. Similarly, different values are reduced to a one: harmony, life, happiness, pleasure, sentiment, desire, and several others. The bringing under a first, even if the model is foundational in other respects, is hierarchical, for a hierarchy is thereby posited. In value theory, the hierarchy is implied by the well-worn phrase "*summum bonum*," the highest good. The highest good in which subordinate values participate, or from which they "emanate," grounds, justifies, or gives value to the rest. Thus even hedonism, systematically connected or allied with atomism—hardly an idealist system—grounds value in a *summum bonum*, pleasure. Pleasure justifies and gives value to other values as instruments of pleasure. All such hierarchies of value represent value judgments or evaluations that one value can justify the rest. Evaluation, then, rules value theory, in the form of hierarchical relations.

Hierarchy is the previous model both for the structures of philosophy (McDonald, 2004, ch. 1) and, by extension, of value theories, as the one in many of value. The structure of systematic philosophy was hierarchical: "being" was adjudged higher as common to all genera, their ruling identity. The "one" subordinates differences to a one, an identity, the "highest" principle. Similarly, value proceeds downward from a hierarchical first, a *summum bonum*. Differences in value are eliminated, their inherent value lost.

By contrast, creative actualization, which takes both pluralism and inherent value seriously, is an argument for reciprocal differentiation of inherent values, and thereby their plurality. A relation to an incomparably different other is prior to any relation to a justifying first, a *summum bonum*. Thus the structure of any value theory that makes any claim about inherent or intrinsic value at all, including subjective theories, should be reciprocal in structure, not hierarchical. A reciprocal relation is a moral condition for inherent values, plural values, and kinds of value.[1] Creative actualization substitutes intrinsic and inherent value for the hierarchical authority of a *summum bonum*.

Different values do not create a grounding relation but a mutually limiting relation of inherent, thus distinctive values, value pluralism. Values in themselves provide an inherent, and thereby differentiated systematic for traditional questions of philosophy: what is truth, beauty, and so on. These values are good in themselves and provide ultimate referents for different fields and activities. Value is not a one-in-many, a *summum bonum*, but reciprocally different. The value of a value does not refer to some ultimate value but is auto-nomic. Value pluralism also implies that values are historically open: any value can become an end in itself, even if in evaluation some values are judged instrumental. Values are insubordinate.

To pursue the value of different values, moral limits should be observed. Morals do not provide a ground for the values but a reciprocal limit. They are not the source of value either as their "foundation," or as a *summum bonum*. Morals are regulative over creative actualization as a limit to value, while agents in turn creatively actualize moral values. Moral values regulate action, and moral norms and principles regulate association. Without morals to limit the kinds or categories of value reciprocally, coercion becomes the ground of endeavor, and categorial value is destroyed, or subordinated to alien values. A first philosophy of values does not reduce to morals as a first philosophy, but morals as regulative within an overall value inquiry. By limiting inherent values, morals differentiate reciprocally.

The other connection of value theory to first philosophy and ultimate grounding is the practical orientation of radical axiology. The relation of value as a first philosophy to value as a practical philosophy parallels the relation of creative actualization as a practical affair with evaluation. Radical axiology is practical in orientation, based primarily on what was considered "practical" in the sense opposed to "pure theory." Previous philosophies that have been adjudged practical in orientation are actually oriented toward theory. Marxism represents the predominance of theory, permeated as it is by the Hegelian dialectic; while even pragmatism is tied to a philosophy of science as a model, that is knowledge apart from social norms and ethics. However, radical axiology is a philosophy of justice, not only a theory of justice or of practical philosophy.

2.

Radical axiology as a first philosophy evaluates all possible differences in value by bringing differences into a relation. A system of values means that "value" is of sufficient scope to bring all values into relation, if not to encompass all differences under a "one" or "indifferent." Thus different relations may be involved such as quantity and quality, but such incompossibles can be brought into a relation. "All polar bears are white" quantifies the quality "white."[2]

How then can value be defined? How are all the different values connected as values? In one sense, to define values would be to limit values unduly. Some values remain to be discovered, as new differences are determined, which involves creative actualization. Our epoch in history, the epoch in which humans have arisen to cover the planet, has been especially fertile in such discoveries. New categories of value are new creations of worth, something hitherto undetermined. What binds such novelties to all the different values created hitherto is their creative actualization.

To define is in the interest of being: fixation of an essence that can be known: metaphysical epistemology. Moreover, the relation of identity and difference is problematic, a problem I will explore in this text. "Good" in its philosophic purport has been used as a reductive term of identity in the tradition. Plato's "Form of the Good," the origin of *summum bonum* theories of value, is tied to intellectual apprehension as its basis. Value, however, is the most general term for differentiation: plurality. Even the kinds of value represent irreducibly different categories of good: beauty, knowledge, prosperity, intimacy, and so on. Differentiation follows from the logic of inherent value: inherent value differentiates each instance of value as having its own distinct value, since inherent values do not derive their value, or refer to some *summum bonum*, or they would not be inherently valuable. The value is inherent, differentiating the value from the inherent value of another value. Culture and other social factors in history further differentiate values. Personal values round out this differentiation. Value as the ultimate term of differentiation is tied to value as the ground of choice, since different goods were differentiated initially by evaluations of worth. Even time is differentiated by such events of choice, since time is marked by novelties, historical time. The differentiation of the novel and superior provides the basis for breaking old cycles. Value, then, is the ground of change, a superior goal providing the end for which agents undertake action.

Creative actualization provides both a theory of value and a definition, however. All potential values must be creatively actualized and thus a working definition is provided: value is creative actualization. This also provides an identity for all modes of valuation, namely, that they have been creatively actualized. What all value terms have in common is creative actualization. Philosophy, art, science, politics, and economic life are all creative in different ways. Even personal values, such as happiness or other feeling states must be creatively actualized.

While creative actualization provides an identity to different modes and acts of value, the definition of value is not a one in many—values as inherent are plural. Creative actualization does not posit a *summum bonum*. Definition in the mode of norms, not values: a different mode of actuality. Norms and values overlap, since norms can be creatively actualized. But norms and values are not identical, since norms can be standards that should be but have not been actualized. Moreover, each creative actualization of the same norm is different in some respect.

The novelty of a first principle is a mark of the creative side of the project of philosophy. Each new philosophy marks a greater or lesser break with its predecessors as elements of a tradition, the contemporaneous philosophic culture, are revalued. The creativity of philosophy, like all creativity, has a teleological directedness as a product or outcome. A philosopher cannot fail to insert some personal values into a system. But such values serve as more than whimsical starting points for articulation and critical evaluation of values. Such values may be fundamental to the logic of a whole group of activities, as in the relation of freedom to creativity of all sorts.

3.

Moral progress is the thesis that greater and greater numbers, and kinds have been or should be given moral considerability: from a few nobles in the Magna Carta to all adults in contemporary times. Moral progress consists in the progressive expansion and extension of moral considerability to those previously without it. At the forefront of value theory in our age is the unacknowledged revolution in values that environmental ethicists have brought about. While some philosophers persist in the project of subjective grounding and thereby remain anthropocentric, mired in the moral past, radical environmental ethicists like Aldo Leopold, J. Baird Callicott, and Holmes Rolston III have expanded moral considerability from humans to other species, and the environment itself. Their efforts constitute a moral revolution comparable to relativity and quantum mechanics in science, even as philosophers who cling to the Cartesian framework fail to acknowledge it. For their work has begun a revolution in values away from the confines of human subjectivity to include all of life. Inclusion of the environment and other species in moral consideration represents a vast expansion in moral progress, despite opposition from anthropocentric conservatives. The rise of environmental ethics is not simply a new "branch" of applied ethics, in this reading, but a complete "revaluation of all values." Humans are no longer at the center of the moral universe, anymore than the physical universe. Value has been recentered in the environment as a whole. The pioneers in environmental ethics have thereby achieved a revaluation and transvaluation of all values in their attempt to expand moral considerability, to achieve moral progress. Ethics can no longer be considered a subject in the humanities.

The most interesting recent work in axiology has come out of this tradition—one might say the only interesting, and close to the only original work in theory of value. Despite attempts by anthropocentric philosophers to ignore ecocentrism, attempts that come, ironically and inconsistently, from the same quarters that have argued for abandonment of dualism in metaphysics and epistemology, environmental consciousness will ultimately prevail. Just as past attempts at blocking progress, of non-whites, women, and workers failed, so will the attempt to ignore or marginalize environmental philosophy. The present work may be seen as a contribution to the effort begun by the ecocentric philosophers such as Callicott, and Rolston. I hope that I have provided a comprehensive theory of values that supplements their work, and further helps to bring about the transvaluation they began from human chauvinism to genuine ethical universality in axiology.

I would remind the anthropocentrists of Mahatma Gandhi's criterion of progress: that progress is to be measured in non-violence, that is, the reduction of violence, including progress in the treatment of other species. Moral progress does not consist in increasing human consumption, since this consumption must always be at the expense of the wild. Moral progress cannot ever consist of the rape of the planet for increases in human "quality of life." On the contrary, the transvaluation of values must result in a wholesale reduction of human impact, and in particular a reduction of the massive violence against other species made possible through technology: clear cutting, slaughter houses, destruction of rain forests, elimination of Biodiversity, and other human perversions. I would also remind them that humans depend far more on their environment than the environment on humans, that humans are capable of incredible self-destruction (war), and that destruction of the environment would be the ultimate self-destruction, the ultimate folly. The anthropocentric subject is in and dependent upon the environment, indeed essentially interconnected with it, as Dewey argued. The good of the subject requires a suitable, healthy environment as a universal and absolute condition. Wisdom, the goal of philosophy as the love of wisdom, lies in the direction of ecocentric ethics.

4.

In chapter one I will introduce creative actualization as a theory of value. Action is the actual agency of value; goals evaluated worthwhile, the ground of action. Since actions slowly transform the world for the better, they provide purpose and meaning. Evaluation, then, precedes action but is grounded by action, in the reciprocal relation of teleological action. I will include an examination of evaluation and norms.

Actualization in the world is a rejection of transcendent or absolute values. The subordination of all different values under a supreme identity, a *summum bonum*, constitutes a reduction of the kinds of goods to one: to happiness, to duty, to feeling, to God as the supreme good, divine will, to ideals, and so on. Such a subordination is a denial of inherent value. I hope to show in chapter two that the logic of inherent value argues against such a hierarchy, and constitutes a confusion of evaluation and value.

The goal of chapter three is a non-foundational axiological ethics. "Grounding" of ethics lies in moral evaluation of possible goals that require creative actualization. Such acts creatively actualize improvement. They are not given since they are a goal that must be creatively actualized in the future. Their condition is moral evaluation and the result is the creative actualization of moral values or moral limits.

In chapter four I use the norms and values in the first three chapters to establish the moral conditions of life, including human life. These include the world, the environment, and the elements of the environment necessary to life, such as air, water, and soil. These elements along with the world are the condition of any values whatsoever, and thus practically required or morally obligatory. We are obligated to conserve them. Moral values are regulative over actions as well as association. Human alteration of the environment should conserve it and work within the constraints of it. Thus a new philosophical anthropology is called for, consistent with both creative actualization and the place of humans in the environment.

In chapters five through seven, I will undertake a critical evaluation of previous value theories with respect to their particular content. Chapter five will cover non-subjective theories, including idealist theories, perfectionism, intellectualism, needs, labor, and practicality. Chapter six will cover subjective theories, including a detailed look at particular theories based on feeling (hedonism, emotivism, attitudes, agapism, and so on). Chapter seven will cover relational theories surrounding desire (including interest, will, liking, demand, and so on).

I will conclude with a look at how philosophy might look if creative actualization as a theory of value could impact philosophy, since philosophy is also hierarchical, also incorporates a "highest truth." Western metaphysics subordinates all distinctions under the same, an identity of what is different, as beings, subjects and objects, or concepts of language. A thing is treated as a case of a universal identity, not as itself, but as different than itself. Subordination to a first is subordination to something different. The value autonomy implied by inherent value could ground a very different view, where the value of differences in a world of plural value is celebrated, not suppressed.

One

CREATIVE ACTUALIZATION

> Valuation is an ontologically creative activity in that what was possible is made actual through the concrete cognitive-affective process by which any given individual 'decides' that an activity, state, mood or concept has value. —G.J. Stack[1]

> Further, the final cause is an end, and that sort of end which is not for the sake of something else, but for whose sake everything else is; so that if there is to be a last term of this sort the process will not be infinite; but if there is no such term, there will be no final cause, but those who maintain the infinite series eliminate the Good without knowing it. Yet no one would try to do anything if he were not going to come to a limit; nor would there be reason in the world; the reasonable man, at least, always acts for a purpose, and this is a limit; for the end is a limit. —Aristotle

In this chapter I will present a basically new theory of value. I arrived at this theory over the course of a long period because of my growing conviction that, despite many subtle psychological analyses of value that had been set forth in philosophy and the literature of value, these approaches actually had little or nothing to do with value. At most the psychological factors and motives may be one, and only one condition of valuation, and only in the human agent. Usually, however, even if they are a prelude to value, they contribute nothing and tell us nothing about value except, perhaps, the relentless egoism of Western philosophy and culture. Value theory, I am persuaded, should be free of "psychologizing," just as logic was liberated from it about a century ago and on similar grounds. Values are also independent of the metaphysics of substance and attribute, of subject and object, and of any other metaphysical commitments. In order to argue this point, this book will necessarily delve into the relation of value theory to such commitments. However, this chapter will concentrate on creative actualization as a value theory as such.

I will first present creative actualization as a theory of value.[2] I will then bring in the elements involved in creative actualization such as teleology, action, and meaning. Since creative actualization is the creative actualization of goods in the world, "world" as an element of this activity will be discussed next. Since the end of creative actualization of goods is improvement in some sense, the amelioration involved in changing the world through creative actualization of goods will be examined. I will conclude with a brief section on evidence of value: what is the evidence for the value of a bearer?

1. Creative Actualization as a Theory of Value

Let us begin with consideration of one variety of a good, namely, a plane, a flying machine. Examination of such a machine has an eidetic-analytic aspect, a norm that distinguishes aircraft from other kinds. However, it also has a genetic-historical aspect, that is, that planes did not always "exist," at one time they "were not." Prior to the successes of the Wright Brothers, flying machines were simply an idea, a possibility, which they and other visionaries made into a project. In 1895, planes "are not." In 1915, planes "are." Attempts to create flying machines prior to 1903 were unsuccessful, that is, the machines were defective and would not fly: they were "no good," giving birth to the well-worn phrase "it won't fly." Planes were desired; it was felt that they would bring pleasure of a sort; the will was there; the interest; the correct attitudes, feelings, and emotions. What was not actual prior to Kitty Hawk was a good flying machine, a plane that functions as it should, that actually flies. The goodness of a plane was not contingent or based on any of these psychological factors. It was entirely independent of them. Desiring a plane or feeling it would be valuable alone could not produce a good plane. Actual flying machines were produced through practice, action, trial and error, and the like. These were end-governed, that is, the goal was a flying machine, not satisfaction of desire or feeling. Again, desire, feeling, and other psychological factors may well have been present prior to the actual creation of a working plane. However, the goodness of a working aircraft was in no way contingent upon them. Those who failed may have had such psychological motives as much as the Wright Brothers. What they did not have was a plane that could fly. They did not complete their project, the goal, since the norm of flight was not actualized through creativity.

In general, value is connected with such a genesis or actualization of a project at a historical point in time. This point separates actualization of a good as an achievement from its mere potentiality. Prior to 1903 planes were a possibility; after that point in time they were an actuality. This movement from potential to actual is the creation of a novel good, a novelty. Thus the world is at least subtly changed, since the world before the time at which the good was created did not contain such goods. We now live in a world of flight; in 1880 they did not. While time continues as before, history is altered, since a novelty has altered the world, whether for better or worse, more or less. History is thereby differentiated as the kind and number of goods is increased. Value emerges at such a point of historical differentiation, marking the pragmatic actualization of an end that, prior to that point, had only been aimed at. The result is good, since the end is actually achieved: the result is true to the aim in some respect. The plan is achieved successfully. The good is feasible and functional: a working flying machine.

The attempt to make values either essences or nominal is a mistake; it is only valid in the context of future to present temporality. With respect to history or genesis, value is the historical point of differentiation. Since planes did not "exist" prior to their emergence as a value or good, value cannot be connected to "being" in any sense. Value is not ontological, contrary to the above quote, as value is the ground of change. Value marks the historical point at which history is altered or changed by the emergence of novelty, which differentiates the moment from the past in greater or lesser ways. "Being" can never be a basis for value as value marks historical differentiation. What "was not" now "is." This change is not a coming-to-be or becoming as planes "were not." There was no being-existence of planes prior to the change marked by creative actualization of planes as a good, as functioning. Planes are historical novelties, not recurrences of ontological patterns in a becoming. Thus they cannot be determined by past conditions.

Value as the historical point of differentiation is the differentiation of an individual value as a good at a distinct point in time. Temporal, as opposed to generic or specific (normative) differentiation is truly novel: the basis for genuine differentiation. To individuate in time is creative actualization of a "good." Creation of a new good, whether completely novel or a novel individual, changes the world. Value is the ground of such changes, since it is the aim of the activity that creatively actualizes the good and marks the point at which successful accomplishment emerges. Thus value is in no way transcendent. If we achieve something in time a "good" of some sort has emerged. The nail did not bend this time but was successfully driven home, the note was hit on pitch, the theory successfully predicted the new planet, the invention works. Value cannot be analyzed in terms of being as the ground of such changes, but the criteria is completely non-subjective since the practical success of planned change is a good, the mark of value as the historical point of change, novelty, and differentiation. Value makes a difference in the world as a historical change.

Since all such goods change the world in some respect, making it different, value is the most general term of differentiation, the ultimate ground of change. Good is maximized by increasing successes, by the increase of creatively actualized goods in the world, whether planes, successful theories, art works or other modes and categories of good. Even though these changes may only increase value one good at a time, the world is ostensibly better, since the quantity of value is increased. The future is ameliorated with improvement over time. Thus logically, value is the ultimate term of differentiation, that is, the true principle of individuation, since value is both the ground and the mark of creative actualization of individuals as distinct goods. Since they were not in "existence" prior to their actualization, their creative actualization as individual goods is genuine differentiation, historical change. Teleologically, values complete action, that is, bring it to an end. Action is not endless, but finite: it comes to an end when the good is creatively actualized. Action is finalized, when and if the project actually achieves worth. As I will argue below, this gives the action both its ground of possibility as an action and its meaning. Action is beneficial, since its end increases good. Normatively, values provide grounding for rules, principles, ideals, standards, and other potentialities. They are creatively actualized as generic or specific individuals. Norms are grounded as creatively actualized goods: "ought" manifested. Thus rules are not for their own sake but are literally grounded in their creative actualization as a project with an end as success. Temporally, or more strictly historically, values open the future to the present by aiming at a differentiated world in the future. Thus a life or ethic of improved future times can be pro-jected as a goal of action. The future is altered for the better pragmatically through teleological or normative actions and activities.

Creative actualization is consistent with the value theory of historical pragmatism, especially John Dewey's. Value is primarily attached to activity and secondarily to objects for Dewey, as activity is the basis or cause for "bringing about" new goods and keeping old ones in existence through acts of "caring for" and the like. Dewey has an essentially active theory of value in acts of prizing, caring for, and artistic creation. Desire and interest are defined in terms of behavior or conation and thus come under this active characterization ("Theory of Valuation," 1949)

Creative actualization involves an ongoing project in which a valued end is pro-jected into the future as worthwhile, a potential improvement, and is completed as a good through action. Value as creative actualization unites worth, action, end, and achieved good as one activity. A potential change is evaluated worthwhile and then creatively actualized as a good in the world through action or activity. Thus creative actualization is in the world, a project actualized in the world. Of all possibilities, those that are creatively actualized are both feasible, and have been chosen over rivals: evaluated as worth creative actualization. There is therein a critical element in creative actualization as much as a practical one. Creative actualization marks the bridge between the speculative and the practical as much as between the present and future or norms and world. Creative actualization of a good is practical and not merely speculative, as the example of the plane indicates. Creative actualization marks the point at which speculation is brought into world as a good, changing the present unalterably.

Idealism taken as a whole is correct in maintaining that the field of values is not limited to the "existing," but larger, since it includes possibilities, norms, and standards. Realists are correct that possibilities have no worth without actualization. Actualizing possibilities is creative, because it is not ontological-existential: it goes beyond the "real." Actualization, that is, bringing about novelties, including norms that ought to be creatively actualized, is value in historically distinct contexts.[3] What is required for value is the creation of such potentialities by grounding in the world as goods, and compossibility with other distinct goods.[4] Value as actualized and *not* a mere wish or desire is essential to the connection and interaction of the possible to the actual. The results of the project of creative actualization of possibilities are actual in the world. They are inherent goods in the world instantiated as a good. Value is actualized by action and activity as good. The worthwhile is grounded in world or is useless.

An example is provided by an artwork. An artwork is a construct in the world, regardless of the feelings involved. Artists creatively actualize external materials in the world into something new and unique. Skill in accomplishing this project is the "art" in one sense, that is, actions in the world involving manipulation of such external materials. Such skill is not based on subjective or psychological states, but practice and to some degree talent. The work of art is creatively actualized without respect to how an observer feels about it or emotes about it. Its beauty ties both the artist and the observer to world as the work creatively actualizes potential beauty into the world as a unique work. A non-artist may have beautiful feelings that she values highly, and may even desire to produce beautiful art. But the feelings and desires are not paintings or sculptures, not creative actualizations.

What is good is that we can create beauty, bring about new works of beauty. We can also creatively actualize designs. Also good is that we can create cures for diseases and actualize a healthy life style. It is also mostly for the good that we can create wealth and the goods that endure, preserving their value. That we can create goods while preserving other goods is also good. That we can feel is good and that we can sense beauty is also good. That we can hypothesize the truth of principles is good and that they can be confirmed through the creative actualization of knowledge as an activity governed by the end of confirmation or discovery is good. That we can communicate is good. That we can live in peace is good. All these goods are grounded in world, as they constitute our world, the world of our own historical time.

The "ontological" at best merely captures such value as the enduring, but since value may fade the value of "ontological" as a concept is questionable. What is good is that we do not live in a world of being qua being: of fixed, destined unchanging stasis without change, novelty or improvement. That we can keep making pottery, keep growing corn year in and year out—that we can bring new instances to actuality—is not, as Martin Heidegger and Aristotle would have it, the ontological element in change, but the value of method as such. The method has value as a method: inherently.[5] It also reveals the continuing potentiality of the world as good. Values are discoveries of practical possibilities, which do not have being as such but mark changes if creatively actualized. Discoveries yet to be made in the future have no existence, no being. They may yet have value.

Outside of philosophy, such creative actualizations, manifesting value, are referred to as "goods." Goods reflect the creative actualization of good instances; instances that have been creatively actualized are "goods." Only where ontological prejudices have prevailed, or the metaphysics of subjects and objects, are goods evaluated as good "things" or good "objects." Economists and the world of commerce refer to created wealth as "goods" as in "shipping the goods," not "things" as units or instances. The downside is the economism of using this way of referring to individual instances. However, this nominative use of good is not confined to economic life. Even economic objects are not "material things" or "material objects" except in a materialist metaphysic. To characterize them as such is a slander on their creation and the imaginative vision that foresaw their possibility. They have been creatively actualized and are valued. The work that went into their creative actualization was the goal of betterment, for a better way of life, since it was over and above biological necessity. It was not determined by that necessity as a novelty. If value is creative actualization, then the goods of the world that have value and to which value is assigned are also created.

Nor are economic goods the only goods. Economic goods delimit a precise field of goods, namely, those with commercial potential. This point answers the question of the "objectivity" of value. However, the point of this section is that "goods" make no reference to objects or subjects, the Cartesian epistemology, which has spilled over into value theory. Only goods are actual, not "objects." But is the use of "goods" in the nominative a surreptitious return to an ontological problematic, with a new form of being? I would respond that goods are precisely not entities on the grounds that were discussed above, that is, that they originally "are not" but then later are creatively actualized. Goods do not "exist" but then are subsequently creatively actualized as novelties that change the world. As I noted, future inventions have no being or existence. They may well have value. Goods, however, are the result of the process of creative actualization, not identical with it.

Similarly, "actualization" is a superior characterization to "existence" or "realization," since "actualization" includes the genesis of the good. This includes artifacts in the actual world of goods, the world of human transformations and change. Goods don't "exist" as if from nowhere, they are creatively actualized, either as discovery or duplication. "Existence" fails to indicate the creativity involved in goods, as if they always were. "Existence" is not critical, evaluated as worthwhile or established by action. "Ghosts" are actually in the mind, but not actual in the world. Mirages "exist" but are not actual. Existence is tied to perception, but perception should be grounded in action for value, for example, in experimentation. Knowledge ought to be established as actual through confirmation. "Realization" as the *in rem* of things is only one mode of actualization and fails to include its genesis: the creation of the "real." Worth is actualized as a good. The real is fixed in things, that is, things that have been creatively actualized. Thus things are only one mode of actualization, which fails to indicate their genesis.

Another question is whether creative actualization is too tied to goods as a whole instead of specific qualities of goods. Is creative actualization the actualization of good qualities or of goods? Is it of a whole or "attributes?" Creative actualization can involve either: a plane is no good without all of its parts, since it won't fly. Thus a good plane is a more holistic notion. However, improvements in the quality of this or that element might be creatively actualized as well, for example, the comfort of the passengers, safety, or fuel efficiency. The goal can be the creative actualization of either the whole or simply a valued quality of the whole. Quality is primarily an evaluative term or a standard in the context of value; creative actualization of superior quality or a "higher quality" good is nevertheless inherent in the good, if it is achieved. Further, there are other modes of good than simply tangible goods. A prediction is a good if successful, for example, the prediction of the bending of light by Einstein. Predictions may also involve tangibles, that is, actualities such as the planet Neptune. The planet "was not," by analogy with planes, prior to this prediction. It was creatively actualized *as a planet* by the successful prediction, since its actuality as a planet was not known prior to this point. The actuality of objects of theory is also a creative actualization of a sort. I would argue against Charles S. Peirce and Karl Popper that successful predictions include more than their being "not falsified" or "not refuted." For an actual instance is predicted that is more positive.

I noted above that quality can be inherent in the good. Creative actualization means that a standard or ideal can be actualized: the creative actualization of a norm. A part or a whole can be made "to standard" as having successfully fulfilled the standard as an instance. The value is then inherent in the instance. If creative actualization can be "inherent" does this make of it an attribute, similarly to a quality that is up to standard? Inherent shares with intrinsic a revaluation in a theory of value as creative actualization: valuable in themselves in some respect and thus not in an extrinsic relation. The standard can be actualized in an instance that need not refer beyond itself in an extrinsic relation for this mode of good, as inherent in it. A good is creatively actualized as such, not in any relation, once it has been completed. If a quality, principle or standard is creatively actualized in an instance, the quality or norm is inherent in the instance. A plane that flies has the principle of flight; one that does not is no good.

Again, the question arises whether inherent value is "objective." Since Radical Axiology involves a rejection of the Cartesian metaphysics, it also rejects objectivity as either a standard or even as actual. A quality of either a part or the whole, I would argue, can be inherent, (a term borrowed from Clarence I. Lewis and not equivalent to intrinsic) without being "objective." There is no *pros hen* relation to a legitimating first, whether of value or substance/subject.

Making may involve repetition of previous creative actualizations, that is, reproduction or copying a good. Since creative actualization involves creative actualization in instances that can be duplicated in further goods, creative actualization can be a reduplicative process. Copies reveal the normative dimension of creative actualization. In this context, the question is whether a good is a composite in a sense, that is, the individuation of a norm as a quality or as a whole in an instance of a normative good. Is duplication the creative actualization of a kind as an instance? The goal of duplication is creative actualization of norms in the form of principles, which regulate both the production and the good. This principle is important for goods, as goods can be increased once they have been creatively actualized successfully: more new planes. Once flight has been achieved it can be achieved again. Thus principles are grounded in world and normativity seems to be worldly, not simply imaginative. I would argue that the good does indeed individuate a normative principle, but as an instance. Creative actualization grounds norms in world as goods.

It might be objected that creative actualization may well be the condition of value but is not the mark of value. I would respond that creative actualization is one process with different elements. The first is evaluation of the worth of a project. Thus evaluation is distinct from value, since evaluation requires creative actualization for grounding in world, and precedes such grounding. The goal is another element: the end, or purpose. This goal is the ground in a sense as the ground of action but not as result or consequence.[6] Evaluation of this goal as worthwhile precedes action and is creatively actualized in or through action. Creative actualization involves acting for such a worthwhile goal in an activity, often of a type, for example, economic, artistic, and so on. The result, if successful or feasible is creative actualization of a good. The action is what is valuable, since it has achieved its goals. Indeed, some actions may be valuable in themselves, and not result in distinct goods. Creative actualization, creating values from the world may be an art, a skill or craft, even a process of thinking. Value creation may be within a category or personal. But it involves a separation, a distinction that includes wresting what has value from out of the potentialities of the world. Goods result but are creatively actualized from a past in which they were not actual.

In summary, value as creative actualization includes evaluation of worthwhile goals, action and activity in pursuit of such goals, and the creative actualization of goods. However, value should not be confused with evaluation, as it has often been in the tradition. I will discuss this point below.

Creative actualization of a good can involve a new kind that was not actual prior to the historical point of creation, that is, actualizing a novel possibility such as a flying machine. This can also include creatively actualizing norms and ideals as instances, from ought to is, that is, grounding them in a world. Predictions are of this type, since confirming hypotheses is confirmation of a norm in the form of an actual principle or law. The overall result of such goods is the creation of a new world, which includes new instances and kinds, differentiating time with historical change. The quality of world is ostensibly improved, since more good has been added to world. The world is ostensibly better off. Value includes the degree to which a goal is actualized; it must function well or it has not been fully actualized. If an end is not fully actualized then the possible good is defective or incomplete: it won't fly. The attempt is not up to standard. (The question of whether the world is actually better off is a question for critical and moral evaluation. I will examine this point in chapter three.) In general, the world is better off if intrinsic and inherent goods are compossible, that is, one good is not creatively actualized at the expense of another: our own survival should not be at the expense or life of another.

Creative actualization includes actuality; actualization, action, and activity have a similar root. The tendency of recent philosophy is to separate these by analysis almost into different worlds. (Activity of course is both a wider term than action and may include the end. An activity may not be "active," that is, resting is an activity but not an action in the normal sense. Activities may be defined in terms of their end while action is merely regulated in terms of its end, unless, following Aristotle, it constitutes an end in itself. I will discuss action in more detail in a later section of this chapter.) I would argue that action and activity create the actual world, and thus are interconnected in significant ways, which supercede the minute differences in importance. The quantitative unity of an action should not be overlooked: that value governs from beginning to end and thus constitutes a unity. Action and activity are the means to effect change. They are grounded in value as creative actualization, the ground of action and activity, which give them an end and regulate how the action will proceed over time. Actualization creates the actual world of change, in which action is combined with actual in the world. Animals and plants are part of this world of change since they are involved in the transformation of the world, creating the biosphere and acting for ends.[7] They act to survive, to preserve, and in some cases to enhance their own life.

Activity requires particular conditions that are universal and hold for all activities. These include a specific time, a historical context; and a location or place. However, for some activities not just any time or place will do but only a specific time or place. Observing an eclipse cannot be done at any time or place but only at some times and places. Eating a meal is more flexible. If value as creative actualization involves activity it may require a specific setting and time. Other aspects of activities may not hold for all activities, but may be required for some or for distinct types. These requirements include culture, and the elements of culture such as institutions, beliefs (ideas), language, and customs. A particular religious practice may be culturally mediated; money requires institutions, and education, language.

Value in general means that we can actualize the possible. We can enact our ends, norms, standards, and other worthwhile possibilities. By acting according to duty, we are actualizing the duty as an instance. By invention, we are creatively actualizing a worthwhile possibility. Thus what "ought to be" meets a world in value: it is actualized as an end. Norms are possible in the world since if they ought to be accomplished, and are actualizable as an end of action, they can be creatively actualized: a worthwhile action. The worthwhile may include both actualized goods, a good tool for example; and unactualized, for example, standards that are possible but not yet accomplished.

Dreams are not actual but can be creatively actualized; the actual is distinguished by value criteria. That is they have been successfully achieved as goods. Actualization involves meeting the world in action and actualizing the possible. The activity of bringing about novel goods, creative actualization, involves bringing to actuality what had only been potentially or possibly worthwhile. The possible may or may not have occurred to someone as an idea; no doubt some possibilities lie hidden in the mists of the future, while others, like hidden treasures, may never occur to anyone. In any case, possibilities are not always potentialities, since the latter implies a potentiality of some actuality, some previously actualized good; while a possibility is more novel. Bringing out potentialities of an actuality might include improvements in quality in some actual good, while creative actualization of possibilities is more like the invention of flying machines. Invention is indeed novel, not merely an improvement. However, both are modes of creative actualization, since improvements in quality must also be creatively actualized.

The question also arises of whether creative actualization as an actualization of a potency is concerned with the action or the good created by the action. Is creative actualization in the act or a good created by the act? Of course if the action is itself good, its creative actualization is inherently valuable. However, the action is generally connected with means, that is, done for the sake of creative actualization as its ground. "A" good is the product or result of creative actualization, a creatively actualized good. Value occurs when the end of action is achieved as success. This may or may not include a product, which as a good, has creative actualization as its ground. Creative actualization may or may not include creative actualization of the duration of the good. In the case of durable goods it does, but in the case of performances it does not. A musician may hit the correct note, but would be incorrect to hold it longer than is suitable to the piece: doing so would not be good but defective. However, as some kinds of goods decay their value may be reduced: they become threadbare and are written off or "amortized." Their actual value fades over time: the basis or grounding in the actual. Art works may be an exception, since antiquity may be a factor in the worth of a piece, even if it is incomplete, broken or depauperate. A faded but genuine medieval tapestry may be worth one hundred shiny, bright, contemporary copies. Goods that are creatively actualized by actions are included in value, since they are the result of value. Value is attached primarily to action, rather than the goods produced by action, as the ground of action.

If actualization constitutes the active and actual element in value, the dynamic and worldly element, creativity constitutes the novel, melioristic element. Creativity involves acting for a new goal and thus action and *telos*, an end. A new good creatively actualized adds to the total value in the world. As I noted above this improves the world in some respect, subject to wise critical evaluation. If it is compossible with other inherent goods it is melioristic. I will argue in chapter three, compossibility creates reciprocity, the reciprocity of prior actualizations with novel goods. This is a norm for creative actualization.

Creativity is neither subjective nor objective but more basic: a creative differentiation in world has been accomplished. When something new has been created, the novelty is not merely "subjective," since it has been actualized in the world. Nor is it entirely "objective" since, like an artwork, it reflects the creator to some degree, not just the world.[8] The act of creation is central in transferring the worthwhile in imagination to the actual. The creative transforms the world as historical differentiation. Value as creative actualization is in one sense the meeting of "subject" and "object" in the making of world as its objective. However, creation in creative actualization is the ground of the object and thus more basic. The subject, granting the actuality of a "sub-jectum" as underlying imagination, is in-formed with values by acculturation and development, thus grounded in value.[9] Yet imagination may be one spring of the novel as idea. Creative actualization of worthwhile projects of imagination constitutes value, that is, grounding of creative images in world as actual goods. Creativity is thus fundamental and as important as labor, which it regulates. People can be creative in small ways as well as large, for example in creating a new kind or variety of a daily meal and actualizing it. Creativity in this sense connects to all activity since successful completion of the activity is a creative actualization. Creativity is a significant and basic element of all activities, personal and impersonal, and supercedes the subject-object distinction.

Immanuel Kant's theory of practice as tied to the "empirical" misses what is novel about creative actualization: what is invented is new (Kant, 1785, 1788). The novel cannot be based on past experience; only repetition of the novel can. Bringing about a novelty often involves trial and error, since there is no experience to guide creation. Craftsmanship is the perfecting and repetition of an initial creative actualization but is itself creative in perfecting and in dealing with new circumstances.

Creation is not creation *ex nihilo*, since elements of world are transformed by creation into novelties. Computers are new, but not created out of nothing. Yet in a sense creativity involves novelty that is "out of nothing," for although computers have physical components, computers as such are a novelty—a new item in the repertory of the feasible. The actuality or good is new as a creatively actualized vision. Novelty always disrupts the stasis, the "natural order." Value can indeed be a "non-natural predicate."[10] Value as the ground of the new, the creation of novelty, is the basis for a superior world, and as meliorist is disruptive of the "natural order," the fixed world that is the hypothesis of philosophies of being. Value is temporally and historically dynamic and as such values can be upsetting. As novel, value is a check upon the stultification of values inherited from the tradition.

I would argue that some goods are inherent with minimal human activity (the Grand Canyon) while others are created (paintings). Creativity is meliorist but the actual is not limited to the novel. The Grand Canyon has distinct inherent value as actual: the beauty is in it and not illusory. The Grand Canyon is different than the surrounding area: both distinct and unique. The Grand Canyon was also created, but its "creation" is by the chance action of the Colorado River on geo-strata. This is only a metaphorical sense of creation, since the Canyon was not created by design or as an end. However, the actuality of the Grand Canyon is good, that it happened. The world contains some good as it is and without which life would be poorer. Creative actualization can enhance beauty in the activity of the fine arts.

Creative actualization, as was noted above, is a discovery of the novel. However, some creative actualizations are more novel than others, implying degrees of novelty. A factory that turns out identical parts creates new goods, but they are new only in one dimension, whereas their original invention was a greater novelty. Accurate predictions of new kinds are a greater novelty than a new act of viewing them. The Wright Brothers broke a new barrier that is more novel than later imitations or improvements, since the idea was actualized creatively for the first time. These different senses of novelty can be referred to generically as original and contextual. Contextual novelty also includes "jury-rigging" temporary solutions to contextual problems. Contextual novelty should not be confused with creative actualization, which can include contextual novelty. Contextual novelty may well result in new goods, or preservation of older ones, but they are not as novel as entirely new creations, that is, the first flying machine. Indeed the two can be in conflict (genetic engineering as a threat to life).

Such novelties are actualized in the world, however, and constitute a discovery about the possibilities inherent in world. The discovery of new values is both historical-temporal and creative, since new values contribute to the change in the world that is historical. The emergence of both new species and technologies has changed the world-scape on the basis of discovered worth. New value determinations have meant a new world. The world is transformed by such novelties. Values "hidden" by the future have been revealed through creativity. Some values remain to be created, no doubt, as new differences are actualized. Discovery of such novelties reveals a whole set of aspects about the world unknown to that point: value is revelatory. New goods as feasibilities reveal something of world previously undisclosed. Our epoch of history includes the discovery of values, including different choices about which values to actualize. New categories of value are new creations and determinations of worth, something hitherto undetermined. These goods are successful in many respects, but they are good more than as merely successful. Value reveals the world in terms of its potentialities: it widens our understanding of the world as capable of new compossibilities. Value also brings new value to the world, since the new value is worth creating. Differences of value that are discovered in turn create new possibilities. Our world considered qualitatively is the latest compossible set of creative actualizations or goods.

Value as creative actualization creates the possibility of amelioration. Destiny is not fixed: the novel is not determined in principle, and thus we are free in this sense. The world can be made a better place. Creative actualization of qualities of a creatively actualized whole is such an amelioration. As I noted above, this could be a quality of a part in some cases, for example, the fineness of color in a gem. Thus a lemon tree may be good, without defects, but another may have a superior form, fruit, and so on, however quality is defined. Thus its qualities or improvements are a further good, even better. However, creative actualization is generally of a whole good, and its parts function within that whole. Flight is not a property of a part of a plane but of the whole: the whole flies. Creative actualization of flight involves the whole. Thus creative actualization concerns amelioration in general in the primary sense although it can involve improvements of distinct parts. Creative actualization of goods is in general ameliorative.

What is the test for value? At what point has value been reached, that is, what is the criterion for creative actualization, the event of value? There is a decisive test for values, namely, does it work? Failures of action are judged by this standard, creating expressions such as "it doesn't work," or, "it won't fly." Thus "it's no good," that is, the item or attempt is defective in some way, whether entirely or in part: not up to standard, broken, in need of repair, disfunctional, and so on. This test is an event in time, that is, it happens at a point in time at which the good passes the test or proves defective. Ultimately such tests involve action, that is, testing culminating in passing or failing. Thus value may not be perceivable at any point in time. Value may be less a "property" of actual goods, than whether such goods fulfill their function well or badly, that is, goodness as relative to purpose. A good functions well for the purpose for which it was designed. It can consistently be used for its end, in activities of creative actualization or value. A plane that is maintained flies properly again and again. In turn, purposes are related to worth or value, not things. They are grounded in value as creative actualization.

Creative actualization can mean "well-functioning," that is, functioning or working well. An invention such as a computer "works," that is, the machine is not defective but works well or properly. Value has been reached when the novelty does work, at which point the activity is finished, ended, and complete in itself, a whole. This may mean a part as well, however, as when lacking a particular part prevents flight. Replacement of the part allows flight again and this ends the activity of repair. The plane has gone from incomplete to completed and from defective to valuable. "Well-functioning" is not the only sense of creative actualization, but can be one aspect. Minimally, it applies to organs, by extension to parts of an organism, and, by analogy, to machines. Functioning well is less applicable to putting norms (other than mechanical principles) into practice, for example, in morals. A painting is complete when it is finished, when all its parts are completed and it forms a finished whole. But we may not regard this in terms of functioning well and may even treasure incomplete works of art. Again, a prediction can be successful but not a case of functioning well, although we may infer that the theory from which it is derived functions well for use in predicting, for example, the theory of gravitation and the prediction of Neptune. The theory functions as it should and norms have been successfully actualized as creative predictions.

Functioning well involves the creative actualization of an idea or norm as a good, that is, one that functions as it should. Thus functioning makes a criterion of when creative actualization has been reached, at which point value has accrued. Oughts and ideas become actual goods when they function according to plan or as they should. Goods, then, are derivative and grounded in norms and ideals that have been creatively actualized as functional. Action here mediates norms and world by their creative actualization, another mode of value. In the case of a flying machine the test is simple and clear cut: the machine flies or it doesn't. In some other cases the criterion may be more complex or ambiguous, for example, a new computer program or even more especially the completion of a painting. Pollock could have added more paint, but had a sense of what was enough. Even representational paintings may be overdone. Mozart was accused of using "too many notes," although I for one would not want a single note less.

Success is another test or criterion. If some good is actualized it has value both in the sense of successful actualization, one that is not defective, and that the good is a feasible creation. I noted above that a prediction is a good one if successful. Good as creative actualization includes the judgment that an outcome has been successfully achieved, by contrast with inconclusive or bad ones. Adversities have been overcome with favorable results. Success is also a matter of degree. A new quality of flight may be partially achieved, but require more testing and development to achieve complete success. Further, degrees of success may not be equivalent to degrees of good. A good off of a mass-production assembly line may be as much of a good as a complete novelty, but the latter may be more of a *success*. Each note hit properly is a good note, although the composition as a whole is more of a success.

Creative actualization is more than just success or achievement as such, since it may involve invention or creation of novelties. Success may simply involve contextual novelty, not the greater sort involved in invention. I may successfully take the next step in a daily walk but this is usually not a great accomplishment. Creative actualization in the greater sense involves bringing the non-actual, what does not "exist" to actuality. New possibilities are opened up by an initial actualization, for example, quick passage from one distant place to another by flying machines. In turn, previous actualizations were at one time new goods.

Ultimately this history creates grounds for evaluation of propriety about what will work and what won't, judgments about feasibility, which is a derivative sense of value. Metal is more suitable in kettles for boiling water than paper. "Paper kettles" would fall apart when filled with water and burn when put over a flame. Paper is no good for use as a kettle since it does not work for the purpose of boiling water, although it may be excellent for printing or design. Creative actualization of paper kettles can be evaluated as a waste of time and resources that could be put to better use elsewhere, regardless of how much we feel or desire that they should work. Creative actualization is actualization in the world, and thus of all the possibilities, what has been actualized is feasible on some level. However, something may function temporarily that is not generally a suitable substitute for a good. Propriety is elastic to some degree.

Closely related to functioning, success, and propriety is practicality. Practicality is a test for creative actualization, since actualizing what is not actual makes practical a theoretical or speculative possibility, an idea. The idea no longer has merely speculative worth, but has been creatively actualized in the world. In a sense some creation is "no good" if it is not practical—useless, unless it has been proven worthwhile in practice. The practical is the mark of grounding in the world, having a basis in actuality as well as functioning well. Value cannot be defined as practical, however, since practicality is a test for specific kinds or modes of value, instead of a mark of value as such. Practicality is the criterion of worth or of future values creatively actualized into a present good or practical actuality. Creative actualization of a future possibility is its practical actuality. An idea may have theoretical value within a system of ideas, a theoretical whole, without having practical application, however.

Functioning, success, feasibility, and practicality are all in contradistinction to what has not been creatively actualized as a good, namely, what is defective, a bad plane, work of art, idea, and so on. Creative actualization is good especially in the sense of not defective. It can be actualized as something that works as it should, functions practically, instantiates a standard or principle, and so on. Planes that fly are not defective and creative actualization is good in the sense of not defective. Flight can be brought about: it is not mere wishful thinking. But more, flight is not defective, a bungled job or operation. A distinct test for creative actualization, then, is that there are no defects in the operation of the good, and this standard includes, but is distinct from, defects in functioning, success, feasibility, and so on. The good is, as a whole, not defective; a non-defective instance can be creatively actualized. In sum, the test or criterion for creative actualization is that the good, the act, or the other bearer of value is not defective. It must function as it should, successfully and practically or it is not worthwhile. It must be suitable for the function it plays, although this sense of value is derivative.

A new good has been successfully created and these new, successful goods, that is, planes that can actually fly, are a new *kind* of good. What is the relation of a kind of good, a "good of its kind," and creative actualization? A novel success in this sense is a novel kind of functioning. Planes that actually fly are a new kind of good in a distinct mode: actual flight. Since flight in general can now be achieved, flight can be a normative or repetitive good. An identity of norm and instance, of idea and actuality has been creatively actualized. Applied principles are simply a species of good of its kind, since if they are properly applied they result in a well-functioning version or bearer of the instance, a creative actualization of the principle. Standards are another species, since the standard can be used as a norm for a great many individuals that "come up to standard." If they are creatively actualized they have value as the instance, if not they are defective. Standards are still transcendent, since even if they are never achieved, they remain as ideas. They are constants in a changing world precisely because of their transcendence from the changes creatively actualized as valuable. Unactualizable standards, for example, flying unassisted, are Utopian. Actualizable standards may have worth, subject to critical evaluation. However, value is not transcendent, as the creative actualization of the standard: the point at which someone achieves the standard. The point here is that norms, principles, standards, and the like are, if creatively actualized, species of good of its kind. If mechanical principles can be actualized without defect so can other norms and "generals," that is, goods of their kind.

The question then arises of whether creative actualization is always of a functioning kind, that is, an instance or bearer that functions according to a plan, norm or principle. Did the Wright Brothers actualize a single plane or "flight?" (I am presuming flight is a kind. Flight is a kind in distinction from what is not motion, for example, paintings, pleasures, and other kinds. Whether flight is a distinct kind of motion is a question for evaluation.) In a sense they actualized both and also more. The plane was singular and cannot be recreated as such; it can only be copied or duplicated. The flight in 1903 cannot be duplicated in 1903 only now. Flight was creatively actualized but also a single flight in time. Thus what can be duplicable is never absolutely repeatable, for its time frame differs and its individual content, that is, metal, wood, and so on. Good of its kind is a species of value as the most general term of differentiation. Norms in general can be creatively actualized as instances of a kind, the trend of the contemporary world. What is missed is the value of the difference between kinds, which creates a plurality of goods, and the unique or inherent value of the instance as a successful creative actualization. A further point is that quality often differentiates instances or individuals of a kind. Quality is a determinable difference between individuals of a kind, for example, two glasses, machines or art works. In terms of comparison, one is better, however we arrive at this critical evaluation (compare: the virtue of the ancients as excellence of a kind).

A bad plane is one that has not creatively actualized the principle or practice of flight. In a sense, a bad plane is not yet a plane, but only a defective attempt, since a plane flies. A functioning plane has creatively actualized flight, while a non-functioning one has not. However, improvements may turn it into a plane. Thus a bad plane is defective, while an incomplete one simply is missing parts. A defective one misses the mark, creating a gap between what has actually been accomplished and the goal. A bad plane falls short of its goal, end, principle, standard or norm (this view of "bad" is very close to that of Aristotle, although not tied to the metaphysics of substance and attribute).[11] Even in creative actualization of generics, principles, and other goods of a kind in mass production some may be defective. In this respect, creative actualization marks the difference between success and failure: goods and defective products. As I noted above this generic sense of bad covers good as disfunctional, that is, what does not work as it should or according to design. If the product does function as a type it is not up to standard, that is, the principle of the standard has not been creatively actualized. (This sense is primarily connected with instrumentality.) It then may be adjudged useless, except as scrap or a lesson for future endeavor. The standard may not be actualizable, either ever or at least for the present. The project may not be feasible, since it attempts to creatively actualize in impractical ways. It may also be evaluated as useless not because it cannot be actualized but because the project is not an improvement of the world or the type.

Bad as a choice can be choosing unwisely, in a choice that falls short but nevertheless aims at the good. A choice that is deliberately contrary to values is even more bad, that is, worse, since it goes against what has been established as valuable, whether social or individual. Of course, individual values and social values may be in conflict: value conflicts. Resolution of the latter is a matter for evaluation.

A value is a cause or ground for action oriented toward the future. The promise is that the action will creatively actualize good in some form. Since value is what is pursued through action, it is a cause or ground of action. However, value also ends action, as the consequence. This double aspect of teleology, both prior ground of action and result or consequence is peculiar to teleology and to value, since value regulates ends. Value as a ground of action is tentative, since the end is in the future and the outcome uncertain; it reflects evaluation since the end is chosen over alternatives. Value as creative actualization is the ground of superior outcomes, which justifies bringing a change into the world as improvement. The world of goods is created by this grounding of pro-jects in world through action, the valued world. Value is the point of grounding in the world of a construct and, subsequently, copies of a construct. Creative actualization constitutes the discovery of new and superior possibilities, and their actualization in the world through teleological activity.

Since value is distinctive in this sense, its grounding not analogous to cause in the mechanical sense, but more like a ground of action in the sense of ground and consequent. However, value is a "reason" for action only in a very qualified sense, since reason as logic is formal and has no content, including ends; and reason as ground and consequent can have anything for a ground, a desire, feeling, an object, an activity, and so on. Indeed, people generally behave for grounds other than rationality despite decision theory. Nor is value a reason in the sense of a principle as in Kant, since value differentiates what is unique in time and space as creative actualization of instances. With creative actualization, grounding is not in the "given" but in superior discoveries. These are evaluated superior both as actualizable goods that are not defective and which function better.

Is creative actualization identical with having been "grounded in world," or distinct? Creative actualization and grounding in world are two aspects of the same activity, since by grounding a project, idea, principle or norm, it has been creatively actualized in the world. The good itself has been creatively actualized in the world, in space-time and thus in relation to other actualities and goods. This relation is "grounding in the world," that is, creating actual grounds in the world for what had hitherto been merely possible, an actual basis in the world. The activity is grounded and the consequence is a superior world, with more good. This process of creative actualization is beyond "being" or "existence," since what has been grounded from the possible could have never "existed" without the novelty of creative actualization that goes beyond what "exists." Thus value does not refer back (*pros hen*) to being-as-substance, as the ground of the actual. Value is the ground of goods, not of substance. Value is not "grounded in metaphysics" or "ontology" as the ground of the actual. What "is" is grounded in value as the ground of what has been actualized. Value is the ground of genuine change in the world and thus changes what "is." Value is its own ground, as the source of a project, which creatively actualizes a good as consequence. As its own ground value is inherently valuable and does not refer beyond to an extrinsic source of value:

it is not an instrument. I will return to this point below in the sections on teleology and intrinsic value.

Is value the act of bringing about the good or an "attribute" of a good itself as a consequence? I noted above that value is only secondarily an "attribute," whether as a whole good or a quality. Value as creative actualization is primarily a ground, but it can secondarily become an attribute of what has been creatively actualized as an actual creation. Actual goods have been created. The ground of actualization of the novel becomes self-actualized as inherent when it is creatively actualized. The goal is creatively actualized as a good: thus creative actualization is both the condition of good and value itself. Does this imply some sort of substance as the bearer of good, as in Aristotle's account? Goods constitute a transformation from a mere possibility to an actual creation. The "substratum" may itself constitute a creative actualization, whether plastics or even metals. These are then the consequence of creative actualization, not the ground. Even if the dubious relation of substance and attributes is evaluated as valid, creative actualization reverses the relation of substances/objects with attributes/qualities. Required normative "attributes" or "qualities" are evaluated first: the end or goal of a pro-ject, for example, the capability to fly. The invention of means is next, in this case creative actualization of flight in a machine and actual creation follows. Evaluation of required attributes is prior to building the "object," since "objects" in the genetic sense are the objective, and require creative actualization. Practice is the condition of creative actualization of the norm, that is, the "objectification" of it as a good. I will discuss this point in more detail in the next chapter.

It was noted above that creative actualization is in general ameliorative. Novelties are produced with effort for the sake of some sort of improvement in the world. Value is distinguished from becoming by improvement: creative actualization is in one sense a holistic element of improvement, since a good plane flies as a whole and this is an improvement over one that does not. I noted that ostensive improvements might not prove to be actual ones, that projects may fail or only succeed in part or create more problems than solutions. Improvements may require critical and wise evaluation of if and whether they are improvements. An improvement must be just that, something that is new, distinctive, and a positive good, either a new kind or a relative improvement, superior to an older good or version.

A further test is that it is compossible with other goods, that creation does not destroy inherent value. A good must increase the amount of good in the world, or it has not improved it. Destruction of other goods is not such an increase. For example, since economic life is only one category of social activity, other categories are implied in which other sorts of goods form a distinct field of activity in their creative actualization. Different categories imply a plurality both of goods and of kinds of good; none can be reduced to the others. As I noted above, a prediction is good if successful, and this is a different kind or category of good than the goods of economic life. Goods are pluralistic and amelioration requires compossibility with previous actualizations, with different goods. Ostensive improvements may be so minimal as to not be worth the bother, for example, "new" nominally "improved" versions of some well-known computer programs. This is an issue for evaluation. Another is the waste created by riddance of the old goods.

One aspect of such improvement is whether a novelty is life supportive or enhancing, since life is a bearer of value in all theories of value. Another is that the world must be better in some respect with novel creative actualizations. New goods are added to the world's inventory, increasing the total. Comparative value is within kinds, but value is a superior actualization in the world, which thereby improves it as such. The world is not given but constructed by improvement: melioration.

Another issue we have raised is whether improvement or superiority is the same as creative actualization or distinct. Is creative actualization a value, but improvement of relative worth? The agent who chooses a project for creative actualization evaluates it as an improvement, evaluating the project as worth actualizing, a worthwhile goal. Thus creative actualization reflects a choice of superior alternatives, based on evaluations of worth. The evaluation is made that this project if creatively actualized would somehow improve the world. Improvement involves "value judgments," that is, evaluations of superior worth. Creative actualization also includes a critical evaluation. However, evaluation is prior to creative actualization or value, which in turn grounds such evaluations. The quality of the grounds of action is subject to critical evaluation in that a superior good, norm, quality or other value can be chosen. What is now actual includes past evaluations of quality and of what at that time would constitute improvement. These were actualized in the past and linger as durable goods. (What "exists" or is "real" does not take genesis into account. Thus these terms are axiologically inadequate. See above on the adoption of "actualization" instead of "realization.") Evaluation is the choke point of actualization since at least a seeming good has been evaluated worthwhile and created as actual. Value in this sense is the condition of actualization as the worthwhile, that is, as the ground. Imaginary projects cannot be actualized if they are no good. Value conditions actualization as ground, although it may be the same activity: ground as creative actualization.

The realm of evaluation includes ideals, desires, and other psychological factors, all of which must be evaluated for worth and potential. Ideals and desires that have not been creatively actualized have no value as such, since they are not grounded in the world, so they belong in the sphere of competing possibilities that should be evaluated.[12] An ideal can constitute a standard for comparative evaluation of actual and possible improvements or as a future goal. Desires, feelings, and other psychological urges must be critically evaluated for propriety, feasibility, amelioration, and other tests of value. Thus as such they are not even in the sphere of evaluation, much less of value, of creative actualization in world. A whim of steel will not make a plane fly.

Evaluation is the choice of priorities. Priorities are what we consider most important, and evaluation of alternatives uses axiological criteria, not psychological ones. Axiological criteria are the basis of importance as the measure of importance. They are evaluated as priorities as first in importance. I may not desire to accomplish some task, but it may nevertheless have priority on other grounds: required for work, for survival, to complete an important project, as a moral imperative, and so on. Desire and other psychological urges are often overruled in such evaluation of priorities. Only if they are evaluated worthwhile can they be transformed into ends, as Dewey argued.

Creative actualization is good, as a project would not be actualized unless it was evaluated worth doing, worthwhile. Value theory is in part the study of what is worthwhile in the world, what is worth creative actualization. What is evaluated worthwhile may become an end or a goal, which is creatively actualized if because it is rated as valuable, worth doing—or may even become a priority. Worth is neither fixed nor an attribute of some object: it is distinctive and independent. Not necessity but worth is the "mother of invention." We creatively actualize based not on "felt need," but because we imagine some superior kind or improvement on previously actualized kinds. These are evaluated as worthwhile working on, as meliorist, as improvements. Whatever is worthy has a compelling character, like beauty combined with duty: the attraction that Aristotle described regarding good ends.

Value as worth may also include esteem. Usually only what is worthy is esteemed: to be worthwhile is to be esteemed in some ways. Esteem may include personal evaluations however, while "worthy" is usually less personal, for example, education is a worthy goal regardless of personal evaluations. Usually the worth of something is argued on categorial grounds of inherent worthiness or usefulness. Esteem may well be relative to persons, although this reflects social and cultural values inculcated into persons.

Something is "worthy" of respect because it has worth or value. Worth differs from value as associated with the goal as an end, more of a standard or ideal that requires creative actualization, not value as such. Worth is a term of evaluation, which nevertheless can be creatively actualized in the world. Worthy projects can be brought about. Value is the creative actualization of this worthwhile *telos*. Worth is the ground of action, then, but not of consequences or results, which require success. However, revealing the worth in some discovery, invention, novel combination, art creation, or other good is itself novel. Revelations are made regarding true worthiness, even if the worth is conditional. New values are established that justify novel discoveries. Such discovery establishes worth or goodness, and is not purported or imagined but tested and proven reliable. We are justified in bringing such worthy goods about because their value has been established: we establish worth or validity of the value. This value determination is made each time we repeat this experiment. Goodness is now available, where less good or even nothing stood before. Values fill up our world and increase our possibilities of life. Again, value as the ground of change and goods cannot be accounted for by being, that is, the identity of duration. Nor is it "by nature." That a spice, too pungent by itself improves some food—that its "taste value" is increased—cannot be assigned to determination nor to nature.

In some cases worth can be quantified as a projection of what the value should be, for example, x is worth y in barter, trade or cost (compare Kant's notion of persons as "above all worth," and the phrase "freedom cannot be bought at any price"). A good may also be of demonstrable worth. Economic worth is a convention, but not an arbitrary one, as a facet of human culture. Quantification of worth involves critical evaluations of worth of goods, namely, costs, overhead, and so on, not subjective factors. Something cannot be priced at what I desire, since then all goods would be prohibitively expensive, as I would try to enrich myself by their sale. Worth must reflect a compromise between buyer and seller and an evaluation of genuine worth, a transaction that involves equality as a norm. The seller must determine the minimum price at which the worth of the object is represented.

The quality of what has been actualized is a fertile field for philosophy as love of wisdom: is what has been actualized truly worthwhile? Were the choices wise? Philosophy is in part a critical, hopefully wise evaluation of the quality and worth of the grounds of action.

In summary value may be characterized as creative actualization, making good on a possibility. Value is neither subjective nor ideal, since goods are created in the world for the purpose of improving the world in some way with more good. They are actually in the world and value marks the point at which good is achieved in the world. The value world is a project, a created world that is valued. Actualization brings a project to world as actual, not potential, while creativity brings novelty. Creative actualization is future oriented and the ground of genuine change, a challenge to philosophies of "being" and coming to be. The creative actualization of goods creates plural goods, differentiated from one another. They are distinct in time and space and thus value is the ultimate ground of differentiation as the ground of plural goods, including different kinds of goods. The differential elements are the unique situation, which includes the time and the place as universal for all activities. The test or criterion of creative actualization is success, functioning well, and practicality, from which feasibility is derivative. Value is the ground of action and regulates *telos* as result or consequence, which grounds ends in the world as the basis of preferred outcomes, evaluated superior.

Value is not an attribute, but can be the basis for qualitative enhancements as well as the holistic element of improvement in general: the inherent value of the good, its own excellence. Improvement must be evaluated, and one standard is whether a project is life supportive or enhancing. Value is the measure of priorities as first in importance, since value is the measure of what is important. Different possibilities are evaluated as part of creative actualization for whether amelioration will accrue. Evaluation is distinct from value in the relation of prior condition to ultimate grounding. Value is also distinct from worth although worth may be a ground of action, as what is pursued is evaluated worth pursuing. Value is creative actualization of a superior world, created through action, an alternative in which excellence is achieved. Value justifies bringing novelties into the world, the ultimate justification or ground of the changes that constitute a dynamic world.

2. Action, *Telos*, Meaning

I noted above that value is the ground of action and *telos*. Values are pursued in time as ends and manifested by acts. There is a close relation between action, ends, and value, which is worth exploring. The relation of action, *telos*, and value to meaning will gradually unfold also. The argument will be made that *telos* identifies the act as aiming at some end; values differentiate the act by finalizing it and thus giving the act finitude so that other activities can be taken up: new acts. The act is completed when worth is actualized or achieved. The endless stream of activities is thereby differentiated. The world is improved by the birth of new or additional goods. Good is achieved over time, at different times, through a historical accumulation. Thus values are not merely an analytic element: values can be singular as temporal, especially in personal valuation. However, the value problem is distinct from quantity. Indeed, one act can be distinguished from another by the worth achieved.

Action is not blind: in acting an agent must not only deal with the actual world, but with value. Acting is a complex of acting in the world, hopefully with skill, improving the situation and doing so while preserving oneself as an actor. Action, following Plato and Aristotle, can be the instrument of some further good or performed for its own sake, a means or an end. Action actualizes goods, that is, action is included in the root of actualization as the means, cause, and agency of creative actualization. Values may motivate our action but they are also the result or consequent of successful action. They are manifested, actualized by and through action. Goods acquire value through human action. They are pursued, made, improved or acquired. This results in actual changes in the world: action is the agency of change. Actions bring about different situations from those prior to when the action commenced.

Actions are not independent of value. Creative actualization as a theory of values is a theory in which values limit actions and thereby make actions finite. Creative actualization is the successful achievement of the action and thus finalizes the action. The achievement of the result ends the action and finishes it: actions are qualitatively ended when good is achieved. By finishing them, they constitute and mark them off as actions, differentiating them from mere coming to be, and also differentiating them from one another. Each act is limited and thus distinct from another action. Values limit actions by ending them, giving them a purpose. Values thereby make actions possible: they define action. By limiting action, value differentiates it, just as the action reciprocally is the agency of creative actualization of the value.[13]

Values limit value to finitude and thereby separate action from becoming.[14] Our activities are not tied to an endless coming-to-be but end as results or consequences that are improvements. A painter does not work on one canvas forever. When something of value is achieved, the painter generally begins the next painting. The painting or most other actualizations can be improved with minor adjustments, but there comes a point at which the goal has been achieved. The value is actualized and the action ends. Similarly, a building is completed, a project is completed, and an order to a factory is filled. In religion a ceremony is brought to an end. In science an experiment is completed. A medical treatment is successful; the disease is cured and does not linger. The actions and activities have accomplished something of value. We are then justified in ending the action by its having achieved value. The "end" is marked by the emergence of some good.

In the context of action, creative actualization constitutes an answer to George E. Moore's famous question about good. In answer to Moore's question, "but is it good?" or "is it valuable, does it have value?" creative actualization ends action by the achievement of good. Value is fully actualized as inherent. No further good can be actualized as *this* good, as this good is actualized and childlike repetition of the question is redundant. The end of a successful action is a practical end to a theoretical infinite. If the action falls short of the goal, good has not been achieved. However, good in relation to other goods can be evaluated. This question is one for evaluation, especially moral evaluation and environmental ethics.[15] Should we continue to make plastics? Use atomic energy? Is it worth it? However, evaluation is distinct from value, as I already argued.

Although value finalizes action, value as the cause of action does not bring all action to an end, only one action. By limiting action values make other actions feasible. Because actions are limited and come to an end, we can commence new actions. One comes to an end so another can begin. We can evaluate and choose what our next action will be undetermined by finished, completed projects. The world is thus improvable still in the continued possibility of more and better creative actualizations. The world is not in a finished or perfected state. Actions make our world possible: the world is not the flux of becoming or the endless cycle of biology and life—to speak nothing of the static world of being. Making a world means a better world is possible. Improvement is the enemy of perfection. Evaluation of what would constitute improvement is ever destructive of the "given." What is "given" to perception are past evaluations of improvement, of successful creative actualizations.

Values bring actions to an end so that they are not endless and thereby futile (Sisyphus). By achieving something of value, action is given its meaning, purpose, and justification, differentiating the worthwhile from the futile. The action is worthwhile, for it is not ceaseless and blind but definite and melioristic. Actual improvement results unless we later discover the action was a "waste of time," "useless," "no good," and so on. In turn, action gives actuality to values. If action is justified by values, values are actualized by activity. Thus action is tied to value: they are mutually connected together.

If actions manifest values with creative actualization, can values also be the ground of action? Don't actions as cause effect creative actualization? It was argued above that values constitute actions and separate them from mere becoming. Values also ground actions by bringing them to an end and so giving them purpose and meaning. Actions are thus grounded in world, and the world is made good. Not a lack but a valued goal stirs us to action, for a lack is only perceived as such in terms of betterment. The goal of improving the world and ourselves stirs action. Value as creative actualization is the source of qualitative changes in the world, of genuine novelties and differences. Change is not a coming to be for creative actualization means that novelties are created by action. The world is indeterminate, for it is modifiable.[16] Values are the ultimate grounds behind actions. They are its sufficient ground for they constitute action and give it meaning. Ironically, due to the double direction of teleology, they are embodied through action. A successful action entails its good; it completes or brings about the value it aims at: its ground. Thus the relation is not a *pros hen*, a referring back to a justifying first, since value is creatively actualized as ground and consequent. Action is constituted and a necessary part of this process. (Also, values may not be the ostensive "reason" or motive. A person may work to "earn a living." We may not judge that "earning a living" requires justification since in our culture its value is obvious and does not need stating: it is implicit in the "reason" or motive. Or earning a living is of value of itself or superior to alternatives, starving, welfare, living off of relatives, and so on.)

If value grounds action, how can instrumental values have value? Value as "that which is pursued" covers only end value, not means. Instruments have inherent value, as Clarence I. Lewis first pointed out, as instruments. An instrument such as a tool has proven worth, as it is useful for the creative actualization of something else in a reciprocal relation. The means are justified and grounded in the end but the ends require the means. Also, as Dewey pointed out, ends can later become means and vice versa. Making a tool is a goal, an end. Creative actualization of a tool, if successful, achieves value as a good. Means and ends are fluid and reciprocal. Moreover, the instrumental value is itself a creative actualization: the tool was creatively actualized at some point in the past but not ultimately for its own sake. Inherent value as a tool is not intrinsic value. The creation of tools and other instruments is novel, and a historical achievement grounded in world. The instrument has value but as an instrument in a relation.

Creative actualization is a general principle, since it can be successfully repeated: recreatively actualized. Thus norms, principles, and other "generals" can be manifested by action. A good of its kind is actual as an instance and creatively actualized by action. Action produces instances of the norm but recreation in world produces further instances. Action thus mediates norms and world by creatively actualizing norms in the world so that they "function as they should." "Oughts" become "is" or in our terms, actual, since their duration is finite and variable. The distinction of "ought" from the actual is merely theoretical and abstract, since practice is frequently the creative actualization of norms and principles. Action attenuates the distinction and creatively makes an "ought" actual.

As distinct from becoming, a relative improvement can be gauged. A painting may not be perfect or make life bearable, but is inherently valuable by itself, for it is better than no painting. The new painting is melioristic, making a small, or relative, not an absolute improvement by making the world a better place, even if only slightly. Values are pursued as grounds of acting since their actualization is superior to the present situation. Creative actualization must somehow produce a superior world, or the action would not be made into a project. Actions must demonstrate the value of evaluations by first grounding them in the world, creative actualization, and thus demonstrating their superiority over the past. Evaluations are judgments of potential value; their actual value is subject to subsequent critical evaluation of their success. Actions are guided by value but have their own mode of inherent value as the general means of bringing the potentially valuable to world as creatively actualized. The goals of action are to achieve something worthwhile and meaningful in the world, overcoming futile becoming (Sisyphus). A basis in the world is not in need of a further ground, as it is in the world as a value or a good, self-grounded, inherently valuable.

Values are often in the realm of time and space, since they are creatively actualized over time through actions that take place in a space. They are concerned more primarily with the actions that creatively actualize, not the goods that are the product of creative actualization. Thus values are often bound up with goals or principles regulating future acts, not descriptions of present goods, although they can refer to the latter also, since goods continue or remain as the act of creative actualization. Good is achieved over time, at different times in a historical, pragmatic accumulation. Value is revealed over time, and thus is not a priori or transcendent. Creative actualization is an immanent theory of values.

Good is frequently cumulative and therefore neither permanent nor an endless cycle. Value may come to an end (the beauty of the Parthenon, or the Colossus of Rhodes). Such goods must be replaced or are superceded. Time may also play a part in differences in value. What is adjudged scrap at one time may be a vital component in a broken machine at another. Further, we can have different virtues and exercise different talents because the same kind of value can be actualized at different times by different actions. Since action is completely bound up and constituted by values creative actualization is historical. However, such action is not tied to the endless flux of becoming, since time is utilized for improvement and change. Time is constituted as historical. The "ontological" problematic is superceded in the use of time for improvement. Values differentiate and give worth to moments as distinctive. The temporal flow is not similar to endless others—becoming—but have an end or limit created by actualization of the goal

The relation of action, *telos* and value has many different aspects: value is multi-dimensional. Firstly, there is the dimension of space and time: the locus and historical context of creative actualization. An end lies in the future and is actualized over time in a place through action. Another is quantity: that an act is a unity with its end and creatively actualizes a value as a good. Another is quality, which distinguishes the worth of an act. Quality is actualized as an end of the act. There is the causal dimension, the causing of a new good in the world as result. An act has a cause, which is its sufficient ground, thus the end as a cause or ground is presumed. However, the end is also a result or consequence. There is the dimension of the relation of the possible to the actual and the extent to which what is actualized manifests the possible. Further, does actualization of instances manifest general values, norms, and principles? Value comes in different modes. There is the worth of ends as justification of the pursuit of an end in action. For who would pursue worthless ends? The good that an act aims at is a distinct aspect from the end as result.

Values are pursued as ends. As I argued above, bringing an action or activity to an end, limiting action to a finite goal allows for other goods: other activities. Limiting action is the cause of more goods and different kinds of good. Further, teleology as an end is actualized in the world. The end as result is separated from mere subjectivity—from mere whim, wish, desire or feeling—by its creative actualization. The goal of action in the world is to creatively actualize this goal as actual: as a part of the world. The result is actual creation of a good, not a thought of a possible one. Achievement of the goal separates the actual from the possible. Technology is not the result of such psychological factors: not a matter of feeling, desire or will. The goods that are made are in the world and their good is part of the world. Norms as "oughts" are similarly creatively actualized. "Telos" is involved in aiming at bringing something into the world.[17]

Value, however, is the ground of *telos* and action. If value is the cause or ground of action, how can it at the same time be the result or effect of action? Aiming at the goal or purpose begins action but achieving the end is the consequence or result. This distinction is not always clearly drawn, but is significant because of the time relations it allows. Value grounds both at the beginning of an action and at the end, when a good or value is creatively actualized. The paradox of ends is that they are a first or beginning as the ground of action, since a ground precedes a consequent, a paradox.[18] Then the end is actualized creatively as result and the action comes to an end as completed. The paradox is that an end is both a cause or ground of action, and its consequence, when the goal is creatively actualized as a result, a good. (We might distinguish these two senses by saying the "end" and the "goal.") Thus value limits by ending the whole sequence as the ultimate ground. Value is the "ground" as the goal in the world, which becomes grounded through or by creative actualization. Value has its beginnings or first in its ends. Unless a goal is creatively actualized, the attempt is "no good." To creatively actualize is to ground in world as a discrete, autonomous instance. The goal is actualized in an inherent instance of good or value, which is successfully created in the world. The good differentiates the goal as a successful one that is actualized, thus grounded in the world.

The "state of affairs" that is created is the consequent of its creative actualization: value is the ground of action that creates new actualities as a consequence. Thus value can only be an "attribute" of a "state of affairs" as a consequence: value is the ground of the "state of affairs." Value constitutes a "state of affairs" in a historical sequence: creation of a new, better state of affairs. This presumes, of course, that we acknowledge and give credence to the norm "states of affairs." The latter is subject to critical evaluation, since "states of affairs" is a clumsy way of saying "actual" that, like "thing" and "object" ignores the genesis of the actual in creative actualization.

As the ground, evaluated worth pursuing, values are potential good; as creative actualizations, values. There is no value without creative actualization. As I noted above, the point and locus of such a creative actualization is the emergence of a new, distinct value, which is separable in the abstract from end, action, even good, although creative actualization may in some cases linger as a good. Value is not identical with enduring good, since value is sometimes achieved in different modes, for example, musical notes that are hit correctly but do not linger. However, creative actualization is also the consequence of acting for the end that is creatively actualized. Thus if the ground coincides with the result, the achievement of value is successful. In this sense, "ground" is actualized differentiation. By creatively actualizing we have created actual grounds for possibilities, that is, grounds for their actualization. They are not merely of potential value but of proven worth. They are actualized in the world: they are not a dream, a hallucination, a desire or wish. The value of a possibility is actualized in creative actualization. Value ends or concludes activity with actual value in the world, which remains in the world. Value continues grounding thus continues good or valuable. Value as creative actualization is reliable.

Values both justify and ground ends since we finish or complete an action when the worthwhile goal has been achieved. Creative actualization grounds by bringing action to an end when the goal is achieved. Our ground in the future, our aim has resulted in an end. We are justified in stopping an activity by the value that has resulted, the good that has been achieved. In sum, value as the ground includes the evaluation of worthwhile possible goals and actions as means to them. Value is the creative actualization of such possibilities whose result ends the action. The possible value is then grounded in world as actual: the possible is creatively actualized. Value also justifies the whole process as its ultimate ground, differentiating the worthwhile as actual. Distinct goods and values are creatively actualized, changing the world for the better. Values justify as the grounds of teleological action.

Values are the "invisible movers" (Friedrich Nietzsche) as they give action and its goals a focus and significance even before they "appear." They are both regulative over acts and made: they make history possible, but are in turn creations of history. An art becomes performed through the significance given to it by evaluation. A person writes a phrase of music, cooks a certain meal, marries a special person, and even picks up an attractive shell along the beach as acts of value. Each act focuses on determinate and particular end instead of other ends due to value. Just what ends up proving worthwhile differs in time, which allows different actions, and in quality and kind. Such differences are values in the sense of the most general term of differentiation.

Values differentiate goals as singular ends. Knowledge, wealth, art, and other, plural goods can be pursued, aiming at different goods. Not just any goal is pursued but a worthwhile one. Value is the determinate differentiation of goals in general: values differentiate goals. In turn, values are discovered through teleological action. Experiment, practice, work and other activities reveal the actuality of new goods. The possibility of such discoveries in turn generates new actions stimulated by new goals in historical accumulation. By differentiating goals, various different goals can be related as mutually feasible.

Values mediate means and ends, harmonize the possible and the actual, that is, imagination and action. (Compare "thirds" in Charles S. Peirce). However, as I argue below, values are ultimately a differentiation as instantiation, so are not mere "generals." Ends are instances of generals, which gives axiology its "logy" element. If a good end is evaluated as worthwhile pursuing, the action used as its means of creative actualization is well performed and the goal is achieved as a consequence. As singularly actualized, the good is distinctive, not merely an instance of a principle. This good is distinguished as distinctive: a good. Thus the end has a "real" basis in the world; it is grounded as creatively actualized. Values, as regulative of ends, have no actuality or "existence" prior to their actualization as an end. The goodness or worth of ends gives value to the world, but their separate "existence" apart from value would be worthless. Value, in the form of evaluation of potential value, regulates by differentiating which ends are creatively actualized in the world: they are put into practice. The world is thereby changed, if only slightly.

Evaluation, then, is in part normative. Norms are actualized as ends and this includes principled action. "Oughts" as a species of norm are also the ends of action: they involve acting for the future if they are evaluated worthwhile and are not impractical. Oughts are also creatively actualized as a goal over time and in a place as a pro-ject. Genuine oughts as a species of norm predict what will be but is not, unless an ought coincides with the actual already, for example, bravery as a trait of an individual's character. If we ought to do something we will and this creative actualization of the ought "confirms" it.

Values as creatively actualized ends, always involve temporality for achievement. Goals are achieved as a result over a period of time. As ends, potential values remain in the future not the present. They have no actuality—no "existence" prior to creative actualization, thus are in no sense "ontic" or "ontological." Thus they cannot be attributes of a substance. Substance presumes the being of the substance. However, goods must be creatively actualized with valued qualities for the actuality of a quality to inhere. Plants creatively actualized their own lives by creating a toehold among the rocks and with animals created the biosphere. A world of "being" would have been stable, thus life would not have arisen, a decisive argument against "ontology." Differences cannot arise from stability. Whereas "being" is oriented toward the present (as "presencing"), value is future oriented as creative. Duration is not the temporal mode of value, since value marks an improvement, amelioration, not a becoming. Differentiation in view of a *telos* is enacted, one that "was not." The "present" is what was enacted, or aimed at in the past. The present reflects norms and evaluations from the past. However, the other dimensions of time are not excluded by value. Instead, value as creative actualization involves a beginning, a middle and an end. The worthwhile goal regulates the whole dynamic creative activity and gives it a unity. The dimensions are united and given worth and meaning. Creative actualization involves evaluation of worthwhile goals as a start in time. Making or other actions and activities are the middle, while achievement at a point in time is the end. Creative actualization of the goal is the end of the action or activity, which limits it as a finishing. Thus there is a "narrative" to creative actualization, which ties it more to Aristotle's *Poetics* than to his *Metaphysics*, although the notion of potentiality and actuality is incorporated into the latter. However, creative actualization is not literary, since creation of literature is only one mode of creative actualization: the literary is one species of creative actualization. Narrative is not privileged, but merely representative.

I am presuming the differentiation of value, action and end. "End" is an equivocal (ambiguous) since "end" can mean both a goal, that which remains to be done; and finish, the achievement of the goal. The goal is separable from achievement as value. Action comes to an end when the value is achieved, so value can linger after action. Value is not identical with the end as a goal, however. Value inheres as inherent to what has been creatively actualized as a good. Value as either intrinsic or inherent constitutes an "in itself," and as such is not a goal, not teleological.[19] The goal is what is aimed at by the activity. Achieving the goal is the result. Value is the point at which the end is reached: no further effort is required, since the end point is achieved because it has been creatively actualized, that is, has value. Value regulates both goal and result as the test or point for actualization. However, value is not "present" as an objective or goal. Creative actualization here remains as a project: a projected end. Purposes and goals are pursued by action until they are actualized: until their accomplishment puts an end to the discrete activity. Values in relation to ends are the worthwhile element in any end that is pursued. They justify the worth of ends: whether some end is worth pursuing.

Creative actualization involves practicality as actualization, that is, feasibility. In relation to action and ends, practicality involves the discovery of values in practice, in action or activity. Novelties are discovered through experimentation, trial and error, and testing out hypotheses and other norms in the actual world. The actual feasibility of possibilities in the world is discovered by testing and creation, not by theory.[20] Practice is distinct from practicality as action is to feasibility. Of course an action can be feasible while its end is not: this further reveals the distinction. The discovery of value cannot be a priori; it must be tested. As with goals, practicality is temporal, the attempt to solve problems over a period of time. Value emerges over time in historical advances. Values are independent of conceptualization, ideas and other possibilities from the realm of evaluation. Conceiving a possibility as worthwhile is very different from achievement, which requires creative actualization, that is, the feasibility of the possibility. Imagining a project and achieving it are separated by a time in which the possible is made actual. Thus we cannot investigate the value of what is valuable starting from consciousness, since value in the mind or imagination is in the realm of possibilities, not value. Value as involving consciousness of any kind, including feeling, attitudes, and so on, is in the realm of evaluation, not value, that is, whether ideas or feelings are *worth* acting on. Value is distinct from evaluation, since evaluation governs the possible, which has not yet been actualized as feasible or practicable. Evaluation as a critical notion regarding the value of a value does not make reference to consciousness except in the personal mode of value.

The pragmatic aspect of value is staying true to practicality. One nail of the same lot bends when struck while another does not. The one that bends is "no good," although they are conceptually and perceptually identical as nails. Practice differentiates their worth. Even the difference between individuals may be revealed by practice. Pragmatics means that principles are normative but that circumstances are the context of creative actualization. Pragmatics are true to circumstances as much or more as to principles. The critical evaluation that "practice makes perfect" would be reformulated as "practice makes actual" and "practice makes better." Pragmatics means that actualization has its own sphere that requires as much attention as inspiration: the realm for critical evaluation. Without such attention to the practical sphere no value ensues. Actualization as melioristic is a pragmatic view of good: good as improvement. The slow accretion of such goods is a bettering, the only good available to us as finite creatures. Pragmatics simply recognizes that we are not God: we cope as best we can, but are limited in our powers. Our results are not perfect but limited by the constraints of feasibility, time, funds, circumstances, and so on. (A perfect agent, such as God described by the superlative mode, would evaluate perfectly and thus not require pragmatics.) However, a Utopian solution is nowhere forthcoming or practicable. By contrast, improvements, where practicable, are better and so should be adopted.

Since value as the creative actualization of a goal ends action, value ends a causal chain of action to consequence, means to end. The action comes to an end with creative actualization, a successful project. The act is completed when the end as goal is achieved and the worthwhile creatively actualized. Thus new acts can be undertaken; the act is differentiated through achievement of the end. By ending an activity, creative actualization gives the activity meaning. The action or activity is finite and thus bounded: neither infinite nor endless. The action is grounded in the world as completion, the activity finite and thus meaningful since it is not without purpose and significance. The value justifies as it ends the activity by its grounding in world as result. The world is improved and given new meaning by the birth of a new good.

Meaning is a species of value. Activities are meaningful because they are valuable. Value differentiates the endless stream of activity with a genuine change for the better. Even labor comes to an end with the completion of a finite task. Grounds for meaningfulness are ends of activity, which justify the activity, differentiate purposes, and thus end the futility of mere becoming. Agents create change and action moves from futile becoming to the meaningful, since the world is improved. Alienation from the world is ended with amelioration. The change also moves toward amelioration as the number and kind of goods are increased. Change is not for its own sake, but is purposeful and valuable. Thus contrary to Arendt (1958), labor also has a "fruit," in completion of the task. The "fruit of our labor" completes the harvest. Agriculture is not an endless becoming, since it improves the food supply and is more reliable and fecund than gathering. We evaluate agriculture as an improvement over a haphazard or unreliable food supply that achieves the goal of an adequate food supply.

Obviously, I am using meaning in the sense of value, not the analytic sense of the meaning of a concept. However, the latter is actually based on the former, since meaning is a norm and its creative actualization as a word in an utterance is a species of creative actualization (compare "speech acts"). Further, since analysis or division can go on ad infinitum (compare splitting the indivisible, the a-tom), an "end" to this activity is required for use of language at the point where meaning is reached. Value is the ultimate basis for analysis into "simple" or singular elements. Simple itself is defined normatively and this is the unit of analysis, the end of the process.

Only values can give significance to time by differentiating temporal moments as significant. Value differentiates moments and thereby gives worth to moments. Being or existence, as static, makes time irrelevant. Although becoming is temporal, time is not actually significant for becoming since there is no beginning or end to becoming only an endless flux of cycles. Becoming as a coming to be is an endless cycle unless it is altered, changed in accord with values, a goal that can alter action for the better, differentiating moments in terms of quality or some other mode of value. One starry night resembles another in most respects but a work of art is unique. The temporal flow is not similar to all others— endless becoming—but has an end, a limit or goal that makes it unique.

Just as activity is given meaning by connecting actors to their world, so do values connect the future and the present. The present, or what is actual reflects past evaluations and their successful creative actualization. Critical evaluation of the results may stimulate further action with the goal of future improvement. The future becomes clearer with valued goals. However, the value of the present is not overlooked, since it holds the record of successful creative actualizations. Those who value are the judges of time. One activity ceases and they must judge or evaluate how the next time frame will be used. The time can be spent well or futilely: it can be wasted by mere becoming or acted upon wisely in a valuable manner. We are freed from futile becoming, from a Sisyphean fate, by values.

In summary, the passivity of previous value theory as affective is ended by creative actualization. The world is improved by action, not by contemplation. Meaning is not imposed upon us by a transcendent or historical agency: meaning is not determined. Meaning as a species of value is discovered in the world: investigated, revealed, uncovered, and created. Values give action and time meaning by bringing activities to an end. This end is a change and improvement. By distinguishing a time period as important, time is given meaning, which it lacks in mere becoming. Improvement is the end of a cycle of coming to be.

Values, then, are not transcendent ideals in the Platonic sense, since they make a practical difference in action. Among other grounds, values are not similar to Platonic ideals since they do not have "more real being."[21] They are not apart in some objective, intelligible realm (Scheler, the Neo-Kantians), not purely normative, since they are actualized in and through worldly successful action. They are not objective, as they are not "beings," since value is independent of goods, and more general. They do not persist as "beings," but mark change and improvement. Again, they are not subjective ideals, since they are actualized in the world by efficacious action. Value is indifferent with respect to subjectivity and objectivity, the problematic of Cartesian consciousness. This problematic is only a small part, not the whole: one difference among others of the larger whole of values. Values are not "concepts" since they are manifested in world through action as differentiations. They are not imaginary although they are not present until creatively actualized. Because they must be enacted, they may be imperfectly actualized in comparison with the imagined goal.[22] They are ambiguous in some respects since what is truly good may be problematic.

3. Value as the Ground of Change: Improvements in the World

Values are not only the ground of change but of improvement, since the changes are marked by improvement.[23] Values are the ground of change for the better and amelioration as improvement marks the difference between becoming and change. Differentiation over time is based on the evaluation that such differentiation is worthwhile, a projected improvement. The improvement is temporal, a change from a situation to betterment. A static world of being is escaped through changes for the better. Improvement is open-ended, since any improvement disproves a static world. The world has proven flexible, fertile, and capable of change. Further, goods can be created in addition to other goods, that is, differentiation of new goods is possible. Most values are cumulative and compossible, not mutually exclusive. Creative actualization of one good is not at the expense of another, if the novel good is genuinely valuable. Thus a world of values is a world of improvement, of meliorism. This is the good of the world: the world is opened and revealed by improvement even as it is improved. Grounding in the world results in the creation of a superior world.

Value in one respect is the holistic element of improvement. A better car is better in several respects, but specifically, improvement of a part is of the whole part and also improves the whole. This may take several forms, for example, more power with greater fuel efficiency, more options such as delayed wipers to cope with different situations, and so on. Improvement is historically situated. What is genuinely better may be ambiguous since more power may mean more pollution and less fuel efficiency. Improvement requires critical evaluation. However, improvement in a specific respect may be demonstrably better, that is, that a car actually gets better mileage than a previous model. The improvement is not imaginary, but can be measured, tested, and utilized in the world.

It was noted above that value is the source of genuine novelties and these can constitute new goods. Value is also the basis of differences in qualities, which are another mode of improvement. Principles may also be creatively actualized as copies of demonstrable improvements. Since goods compete for attention in a world of improvement, each new good must actually improve, especially to replace its forerunner. Values are self-corrective, since a mere novelty will not take its place among the goods unless and until it has proven itself in some respect. Different tests for value may have arisen in different fields: a good painting may be unique rather than an improvement in quality. However, the painting has received critical evaluation as an advance or success, that is, a different sort of improvement, a genuine novelty. Meliorism is the constant surpassing of old projects through action creating novelty, the goal of new values actualized. This axiological view of improvement provides tests for evaluation of projects.

Improvement cannot be separated from evaluation. What has potential value consists only in an evaluation of potential value, namely, that some future outcome would be *better*, based on some standard of comparison, possible improvement, novelty or difference. Possible values are not yet creatively actualized as values, not yet grounded in a world through actions or activities as means. The potentially valuable can be evaluated as possibly valuable, projected as an improvement, but not yet of actual value. Value itself is the measure of historical improvement, of actual changes in the world over time. Since different possibilities may be actualized and thus different goods and changes to the world created, evaluation is a critical activity. We evaluate both which actual world we should create and which improvements in present actualizations are called for. Thus qualitative and novel improvements of both the present and of the future involve critical evaluations. As I pointed out above, improvements are measured by the distinctiveness of a new kind, which is an improvement to the world in general or some partial improvement in the quality of previous actualizations. These must be compossible with previous goods, so an increase in the quantity of good can accrue.

Action is undertaken with the goal of relative improvement, a bettering of a situation or a previous good. Value as the ground of change is dynamic, altering the present as the future through projects. The future is creatively actualized and thus differentiated from the present. Creative actualization incorporates change and thus argues against a static universe and the "values" associated with such an imaginary universe: "duration" of values as actualizing their "being" and catatonic contemplation of such stasis. Values can be revalued in a world of improvement. Values are not permanent, which would represent the interest of "being," since revaluation and improvement undercut the value of the actual. Values are the ground of change as improvements, the measure of the former. The expansion of good involved in amelioration argues against a fixed sum of good in the world: the "being" of good. Improvement is anti-ontological. A better world is not a static, fixed, enduring world. To be sure: that a good can sometimes endure as a value is also to the good. The creative actualization of a good that endures throughout many adverse circumstances may indeed be valuable. A car that lasts may be an improvement over "dynamic obsolescence." A tool that does not break, a painting that does not fade quickly, a diet that preserves our health are all melioristic. However, endurance or duration is no mark of value as such: the Ku Klux Klan persists despite its many axiological defects and those of its members.

Duration may require critical evaluation. Orgasms do not have duration: duration does not define the value of pleasures. Nor do musical notes, performances, and many other goods. Creative actualization of pleasures, performances, and other temporally relative values may require the opposite of duration, namely limitation of one for the sake of another. (Compare the *Dao de Jing* in which the "non-being" in a space or void creates the being of the cup). One note must end for another to begin. Duration of one note may prevent another note that adds to the beauty or harmony of the whole. Duration may be appropriate or feasible to some actualizations but not to all. Endurance may also require maintenance, which is a mode of creative actualization, that is, recreation of a good as functional or the like. Maintenance argues against the duration of goods, since they lose their value with time. Maintaining value requires replacement or repair, an argument against duration of value.

Another factor in evaluating amelioration is ultimate values versus immediate ones, long-term effects as compared with temporary ones. The environment is a long-term good, which nevertheless undergoes changes, for example, from "pioneer" to "climax" species in a habitat. Such an ultimate value may be evaluated of greater importance than immediate gains from killing all the trees in an area for trifling profits or other trivial reasons. The evaluation is made that preservation of the environment is better than destruction, since an important good is preserved. Destruction of the environment does not add to the good in the world but instead reduces it; quality is affected as much as quantity including quality of life. Here preservation may be the greater good, the better.

If Nietzsche's Zarathustra were to look down from the mountain today, he would not see a flux of forces, or "beings," but land evaluations: land that has been evaluated in terms of some valued end it might serve, whether farm, residence, business, airport, wilderness preserve, garden or what not. The evaluation is manifest in the intelligible order of the land and how it has been "regulated." The value it has been put to is the change from wilderness into "productive" land use. Past evaluations, wise or not, "appear" as divisions of land into properties, marked by borders, changes in vegetation, buildings, and the like, which serve one or another purpose. Evaluations have been creatively actualized in the world, transforming the world. Of course looking down from mountains in the wilderness will give one a different view than from mountains close to settled areas. But this division is itself a creative actualization of an evaluation. Clearly, a better policy would have been less destruction of the wild. However, this is a topic for a later chapter.

Transforming the world through actions that creatively actualize is creating a new world: creation of world. The world is created through the conjunction of action, ends, and norms actualized in a world. Value is the historical point at which the novel becomes actualized but also involves a place in the world. The world is involved in all creative actualization as the place for bringing about new values, for bringing them into the world. Thus the world is the place of value or where creative actualization takes place and is thereby transformed into a new world by value.

The actual world is the world of goods. Tools, art, medical instruments, scientific results, and economic goods have all been creatively actualized as actualities in the world. Such goods are perceivable but only as the end product of action. The world is the collection of all such goods, all actualizations. Thus the world is the value world, and value is not a realm, a distinct part. A "realm of values" reflects a part, in which there is a world that has not undergone transformations. However, such a world, a static world, is a fiction. The "present" world reflects past evaluations of what ought to have been actualized or what was worth actualizing.[24] As I noted above, we have tools because our ancestors evaluated them worthwhile and creatively actualized them as goods. The created world is valued and also a project that has come about.

The world can be improved and this implies that the world as some sort of (final) state or standard is a fallacy or myth, since the world can be transformed in different ways, creatively actualizing quite different "realities." (Compare the popular expression from the 1960's, "reality is a crutch." Reality is an incoherent notion in this view, since transformation of "essential attributes" makes what is "*in rem*" an accident. If there is no final state of reality, there is no reality since there is no standard of reality. The world, then, has no final state since it can be changed into a very different actuality. Yesterday's "reality" of horses and buggies has been transformed into today's actuality of traffic jams and pollution. Goods are differentiated from world for a purpose, whether for the sake of knowledge, "making a living," and so on. Such differentiations are values, for value in one sense is the most general norm of differentiation. The possibilities transform the world through creative actualization: value. The world is not fixed in ontological stupor but can change, an inferior "reality" left behind and superceded by new more accurate descriptions, technologies, and other modes of value. Even goods that worked well in the past are replaced with better goods. While the duration of some good after its creative actualization may have value, its decay may lead to superior values, to improved makings. Change, the bringing together of imagination, practice, and the world to separate goods, means that improvement is possible.

The transformation of the world as an improvement means that we are not locked in ourselves, in Stoic denial of world, wishful but impotent to change the world. On the contrary, value as creative actualization is transformation of the world for the better. We are part of the world that we transform and belong within it, even as we transform it: such creative actualizations involve ourselves as agents acting within the world. The world is our home in which we are at home as agents of its direction and destiny. The world as an alien place is transformed into a better, more intimate place of which we are a part. The world is transformed by our values and thus is no longer foreign: we improve it to reflect ourselves in part as our home. The remainder is the condition of what is our own, reflecting a division of the indeterminate into own and alien, what someone values as good and what she merely tolerates, ignores or devalues as bad. Transformation of the world for the better ends alienation from it.

Art provides a paradigm example. Art is the creative actualization in the world of beauty or of some work valuable or good for its own sake (Pollack). Our imagination is grounded into the world in the work of art, in which our fancies are brought into actuality. The imagination is brought to earth out of the self, but a self that has been creatively actualized out of the earth. Creativity unites imagination with world in actualizing the work of art. However, beauty also involves a self in the world who can observe actual beauty in the world. Beauty thus involves a content, action, imagination,[25] and perhaps evaluations of beauty. Thus a bridge is creatively actualized between the private self and the world in which the isolation of the self from the world is overcome. The self is opened up to the world that he inhabits, and self mingles with world. Beauty occurs to a self who observes it and differentiates the beauty out of a world while in the midst of the beauty in their world. The reverse of this process is creative actualization of art. While beauty is creatively actualized in differentiation of a beautiful content and action to observe it, art is the creative actualization of a mode of value in the world.

Aesthetic values are not a mere projection of the imagination. They are grounded in the world, the actual source of such beauty. The beauty of the Grand Canyon is not a mirage or illusion, even though this beauty is perceived. The Grand Canyon is actually beautiful, since it can be perceived as such by contrast with the surrounding landscape, which is less scenic. To speak nothing of not perceiving it and thus not perceiving its beauty. The ground of the beauty is in the world, the actual Grand Canyon. Perceiving the beauty of the Grand Canyon involves the action of going to the Grand Canyon as a condition of viewing its beauty and the activity of observing it, creative actualization of viewing the scenery. Thus beauty is not a "subjective" illusion. Beauty is grounded in the world, for it is not a representation of nothing.

Value is the predicate of a kind only as a *consequence* of creative actualization. Value is primarily the ground of what possibilities are actualized. Prior to actuality as a good, the instance must be creatively actualized. Only as a *result* of creative actualization can goods can be judged with value as predicates. Such goods and actions valuable in themselves have been evaluated as worth actualizing in world over competing possibilities, including competing objects of desire, feelings, attitudes, and other psychological factors. Desire and feeling are overcome in many such evaluations, since they are not evaluated worthwhile. Value is the ground of preferred outcomes, that is, the cause of actions and activity that aim at an improved world. A world in which a good result has been brought about, accomplished or achieved is judged superior to a previous situation. Thus value justifies bringing a novelty into the world. (Inaction is the evaluation that the world is sufficiently good as it stands, that available potentialities are not worth the bother. In other words, the world is as it ought to be.)

Value is the grounding of possible ends in a world of superior actualizations. Value as ground is the convergence of act and a world: the plan is successful, the novel is realized. In this sense value is the cause of goods to which value is attributed. Value is the ground of their actuality, which has been creatively actualized from a past in which they are not actual: the purpose of bringing them about. But thereby the possible, including norms, principles, possible ends, novelties, and others are grounded. They are given an actuality that separates or differentiates them from the possible and potential. Those that are creatively actualized have been evaluated worthwhile: superior to competing possibilities and worth bringing to a world. Thus they are justified as superior actualizations.

Value as ground, then is the "cause" of creatively actualizing the possible, the purpose; actual grounding of the possible in the world, that is, successful grounding; the result as a good or valuable action, and the justification of such grounding. Value is the "cause" as the goal of action, the ground as the consequence or success, and the justification as improvement of world. Value is both "having a cause" or a ground, and also grounding the possible in the world as actual: the goal is actually part of the world when creatively actualized. As I have indicated elsewhere, cause in this sense does not match the modern sense of cause as power to effect, except metaphorically. Ground includes cause but is not reducible to it. I have used "cause" in this paragraph to distinguish it more sharply from grounding in the sense of bringing something new into the world and thus grounding the possible. Value as a standard of comparison, involving evaluation of competing possibilities, is a ground in a sense similar to value as the end of action. The actual is created through novel value events, the discovery of the novel as a creative actualization. Values ground the possible with the actual but create the actual in the novel. The creative must be actualized but the actual is created. The creative is inseparable from the actual. The world is changed, hopefully for the better. Achieving a value is to ground the goal while at the same time improving the world with more good. The goal is no longer a project but part of the world created through action. The value world is the result of superior alternatives achieved. Since the world is transformed by such novelties, however subtly, value is the ground or "cause" of a new world, the transformed world. Enough of such transformations remake the world.

Since improvements are evaluated before they are manifested through action, the world is transformed by choice. The valuetive is contingent: it "could be" or "could not be." As different possibilities can be actualized, the result is not determined, since what has been actualized reflects past evaluations of what ought to have been but was not at the time the project was undertaken. Since it was not, and is novel, it cannot have been determined by what was.[26] Further, values are oriented toward the future, not the past. Creative actualization is the bringing forth in the future what was not actual in the past, and so cannot have been determined by past causes. Creative actualizations are novelties, for example, flying machines, so they cannot be determined by past conditions. There were no such machines in the past to condition their creative actualization.[27]

Creative actualization alters the world slightly from what it was before. The world is different with the creation of new goods. This making results in the multiple goods of the world, since each is finite and different from the rest. Creative actualization is a finite activity, which can nevertheless be repeated. Values bring actions to an end and differentiate the actions by ending them: they give shape to the world by limiting actions. One good is limited in time, space, and effort, leaving space for others. Through differentiation, multiple values are compossible, including different modes and kinds of value. Each instance actualizes a kind in the world, a principle, since it can be copied. Thus a genus and species can be normatively ascribed to an instance that has been creatively actualized, for example, a species of flying machine. This actualization is a further kind of compossibility, since multiple norms are combined in an instance. A norm is actualized as an instance compossible with world. An identity is combined with a difference, a distinct individual, in an instantiated kind. Norms are creatively actualized in world: grounded in the world.

Values do not, then, exist in some "ideal realm" since if they did, they would never be actual and there would be two worlds, the ideal realm of the ought and the actual realm of the is. Max Scheler argues that values exist in an ideal realm; he is probably thinking of standards. However, contrary to Max Scheler, standards that cannot ever be creatively actualized are Utopian. Standards are goals of action, not ideal entities (Scheler, 1916). But creative actualization attenuates this distinction by bringing what ought to be about. The goal or meaning is achieved in the world. Creative actualization brings norms and principles to a world, thereby improving the world. There are not two realms but one world, which has gradually been improved by creative actualization of change. Values as the ground of such changes are neither "real" nor "ideal," since value regulates the "real" as novel actualities and the ideal as feasible or practicable. Which ideals are actualized in the world is grounded in their actual worth. Similar considerations apply to subjectivity and objectivity. The subjective valuation is a moment in a larger project in which values are brought about in the world as novel creations.

The "real" is a historically and culturally relative term, which gives it limited worth. Spirits are "real" to an animist but not to modern science. Airplanes are real to the twentieth century but not the tenth. Everything imaginable that could conceivably be put into practice may ultimately be "real," but may not "be" now. Thus we can evaluate the term "actualities" as axiologically superior to "realities." The real is not "given," since reality is creatively actualized over time and is the consequence of changes brought about by action. This view is not "anti-realist," nor is it realist, since creative actualization of novelties transforms the actual world, sometimes in unforeseen ways. Ideas envisioned are "realized" but thus transform prior reality. But as teleological pro-jects or imagined worlds, prior to creative actualization, different values can give us alternate worlds. Reality is *not basic*. There is no "reality," only whatever we have actualized. Thus creative actualization supercedes the "realism-anti-realism" debate. Moral progress is more basic as a transformable world is an improvable world. "Reality" is a value, as F.C.S. Schiller argued, since reality is creatively actualized.

4. Evidence for Value

What justifies the evaluation that value is inherent not "subjective?" What evidence do we have for such a wide scope for value? For its various modes? I will argue first for direct evidence of value and later for normative and logical evidence. Value is revealed in actions even more than perception. I will cover perception below; at this point actions as creative actualizations as the main focus of value will be examined.

Philosophers have misconceived the problem of knowledge of value, the so-called issue of "cognitivism." Knowledge is itself a value. Achieving knowledge is an end, done more or less well. A *good* prediction is a prediction that works well—but it is not "a" good in the same sense as economic goods. Knowledge is a distinct kind, but no less good. Value grounds knowledge in the sense that knowledge must be creatively actualized from mere opinion, belief and claims by means of gathering accurate evidence and experiment(ation). These means are tools, having their own value, as revealed by past successes. Knowledge cannot be value free if is a kind of creative actualization, reflecting past evaluations of the worth of bringing new facts to light. Facts too reflect past evaluations of worth and may themselves be improvable in the future. Thus if we can have knowledge, we can have knowledge of value. We can know past creative actualizations that as such were successful and worked out, e.g. knowledge of the exact parts, functions and the like of the original Wright Bros. Aircraft.

What of other kinds of value than knowledge? Are these a form of knowledge? Or do other kinds invalidate this connection? All the categories, or kinds of value have a practical knowledge that distinguishes them. Artists know how to paint, to play an instrument, to dance. Businesses have the knowledge of how to manufacture or other trade secrets and "intellectual property." They use this knowledge to creatively actualize products—or art works in the case of the artists. Again, we have knowledge of value as knowledge of creative actualizations that worked.

In another sense, the sciences of sociology and anthropology have answered the question of cognitivism in value theory. Insofar as these sciences have described the morals of our society or of different societies, they have described moral facts, patterns, and principles. Moreover, the arguments of relativists like Ayer, Stevenson, and Sartre themselves depend on the knowledge of such facts, patterns, and principles, in their use of hard cases to argue for relativism. As C.I. Lewis argued, we can predict value states, which for him are what will be the source of continuing satisfaction. The sources of such value states have inherent value to cause such states and since we can predict them, they are thereby cognitive: known.[28] Lewis states:

> It is also important, for the interests of ethical critique, to observe that any intention, being a prediction, is subject to criticism not only of its moral worth but also of its worth as cognitive. It may be morally right or wrong, but in any case it is cognitively correct or incorrect, valid or invalid, and true or false. (Lewis, 1969, 163)

Actions are bound together with value, since values make actions possible. To value something is to act on it: pure entailment, not simply ground and consequent. If an action does not follow, the good or goal is not actually valued.[29] Unless value has consequences in action, it is meaningless. Values are revealed through action since they are creative actualizations utilizing action. Value may even be creatively actualized in the action itself, as I noted above. Value can be a quality of practice. Creative actualization may also be verbal, that is, valuing in the sense of esteeming. Esteeming or valuing is creative actualization over time. Value in this sense is evident in action, since the action aims at a valued end. The action receives its type and direction from the value, which is exhibited in choices of action as well as how time is spent. Value also differentiates the action as distinct, as having its own goal. Still more evidence of value or "how do we know if it's good" can be derived from considering whether we would be able to give up a good. If we would not then the value of the good is not whimsical but inherent, and the evidence is that we do value it enough that we do not wish to give it up. (This point is a variation on the pragmaticist criterion of value discussed above, since values are revealed by consequences for action).

A person cannot be said to value any good unless it shows in their actions. If someone personally values any good, they will pursue it. I have borrowed—kidnapped?—Peirce's notion of pragmaticism in this context to describe values in terms of "conceivable practical consequences." I call this the pragmaticist criterion of value. If a value does not result in action the goal is not actually valued, since the ostensive value does not issue in practical consequences. This in turn has consequences for evidence, providing evidence of what a person actually values: "actions speak louder than words." (It may also be therapeutic, since a person may not realize that their pragmaticist choices constitute their actual values. A person who "does not know what they want" reveals their personal values in their activities.) Values, then, are not just beliefs, but involve creative actualization of goals in actions and activities. The point here is that values are revealed by action, not by will or deliberation. The person acts to bring about the good as a goal. The same person could have pursued other goals: they chose this goal and this reveals their values. Thus choice may be involved between alternative goals, a point I will return to below. Again, a person may acknowledge the value of alternatives to other persons without pursuing such goals by themselves. The value of alternatives, as revealed in the actions of others, is evident, that is, socially shared values and cultural values. Since that that is pursued has value, human activities prove the validity of value as creative actualization, either as personal or as a type or category of value.

Value as a quality of a whole of a good or of a part is also revealed through actions. Using a hammer to drive in a nail gives a practical test of the inherent value of both the tool and the nail in action. One nail bends and another doesn't, although they may even be identical in appearance. If the nail breaks it is defective. This is not a matter of subjective perception: the nail itself actually bends or breaks. The worth of the nails becomes differentiated in and by practice. The nail has been creatively actualized through actions and its inherent worth is proven in action. Thus the discovery of value cannot be a priori, since it usually must be tested or at least critically evaluated. Further, creative actualization as involving action and the result as a good are interrelated. An action is "no good" unless a good result is achieved. Achieving the end is a standard for creative actualization, since no creative actualization has actually occurred if the end is not achieved. Practice and action, then provide evidence for value, including personal values.

Our attitude, will, desire, feelings, emotions, and other subjective states have nothing to do with how well the nail functions. Value results in goods that function well or properly, or it has not been achieved. The value of the good is revealed in practice in this other sense that if the nail functions well its functioning is independent of subjectivity of any sort. A nail does not work well because we "will" it or because we have pleasant feelings about it working well. Its functioning is inherent to the nail as a good. Otherwise our feelings alone could fix a defective nail and bend it back into shape: the stuff of science fiction. The inherent functioning of a good constitutes more evidence for values, since the creative actualization of functioning goods in the world is independent of the subjective states or psychological powers of agents.

Value is proven also in *often* functioning well. A good engine may operate for extended periods at thousands of revolutions per minute without failing. The engine may be turned off and on without problems, proving its reliability over time. The reliability has nothing to do with subjective states, any more than breakdowns have to do with subjective states. The engine functions well on its own, regardless of our feelings or other subjective states. Moreover, this functioning can be observed, tested, and improved upon. Knowledge of techniques is also a form of knowledge, tied to practical activity but also to the most arcane theoretical sciences. Technical knowledge is a successful means to a valued end. That we can copy this knowledge in different times and places argues against its mere subjectivity. Indeed, as C. I. Lewis has argued, knowledge is a component of many actions, since they involve known techniques. Once a technique is mastered it can be repeated and creative actualization duplicated.

Not all qualities are perceptive qualities. A plane that is capable of flying but is "sitting, waiting" on the runway may not exhibit the quality of being a good plane to a perceiver. Staring at a plane on the runway to analyze the "nature" of a plane will not determine if it will fly, or even if the plane has the internal parts, in good working order, necessary for it to fly. This quality is revealed in the flying, the creative actualization of flight. The quality of its engines may not be apparent to an observer, but only to a pilot. Both holistic qualities and particular ones may be revealed in creative actualization of a function, then, but are no less valuable than "perceived" qualities. They may be inherent without being perceived, although the flying as such is perceived. Applying a narrow model of perception as some sort of epistemological test of value is completely misconceived.

Other modes of value may not lend themselves easily to perception as well, namely, standards and categories. Standards may be achieved without being perceived, for example, in tools that come up to standard. As in functioning, the good of the tool is revealed in action and thus whether it meets a standard requires the actions of testing. Such value may be relative to a standard that is not apparent to a "subject" by merely staring at the "object." Other norms are revealed in the identity of kinds creatively actualized, for example, identical nails, autos, and so on. Any individual good is of a kind with only few exceptions. That a good is a nail and not another kind such as a screw, brad, tack, and so on is evident from perception, however, and thus its normative identity may be a perceptive quality. As for categories, a diamond is a good both for economics and for several other kinds: esthetics, scientific uses, even as an expression of social intimacy. None of these but the esthetic is apparent from mere perception. Thus approaching value from "empiricism" or other philosophies in which perception is evaluated as ultimate can never do justice to value.

Action is also involved in creative actualization and action involves actualizing projected value in the world. This sort of change cannot proceed from theoretical perception: it is practical. Prior to creative actualization, no value may be perceived. The value can be hidden from perception, but as creative actualization is grounded or made part of the world with the justification that the good will improve the world in some sense. This good is inherent to the instance and thus of inherent value if the projected worth is actualized, whether this good can be perceived at any one moment or not. A plane may be perfectly good and not in flight. Perception alone could not account for or ground such goods, since they are not perceived but imagined, evaluated and acted upon. Quality is actualized through creativity. Value may not be a "property" in the sense of a color that "anyone" can see. Similarly, comparative value may not be evident to perception alone, in the absence of standards. An empiricist untrained in engineering might be totally unable to evaluate two different planes for superiority. Nevertheless, one plane might be superior to another in the inherent sense, that is, based on the features and capacities of the plane, not the whims and feelings of the "observer."

Values may be the source of facts in the sense of facts that means knowledge claims, since "knowledge is power." After trial and error flight is achieved and it is then known that flight can be creatively actualized. This may create new goals, namely, repeating the success. However, value is independent from knowledge as its origin. Similarly, creative actualization creates the goods which knowledge as "facts" is based on. Indeed, as I stated above, discovery of such novelties reveals a whole set of aspects about the world unknown to that point: value is revelatory. New goods as feasibilities reveal something of world previously undisclosed. In other words, practice/action as value driven reveals far more about the world than perception. The practical knowledge gained from action, trial and error and other creativity is the basis of all knowledge.

However, many goods are perceivable, as are the actions that creatively actualize them, if the empiricist could get over the catatonic model of perception. If a pot or other container is not broken, then presumably it is good, in the sense of not defective. House paint may reflect the personal values of the owner, but is nevertheless creatively actualized in the world for all to see. Colors are both perceived and value-laden. Again, the perceived colors of the Grand Canyon are beautiful. Above all, what is perceivable are the novelties and changes that creative actualization has made in the world. Experience, perception over time, is historical, and historical changes caused by creative actualization can easily be experienced over a lifetime. Personal computers "were not," did not "exist," but now "are." Before they could not be perceived, now they can. Since perception as such leaves out this historical element of experience, or some models of perception do, they are seriously deficient as models.

Values as value regulated goals may also be perceivable. Both action and actualization can be achieved: the evidentiary test for goals that have become actualized. The value or deficiency of such goals immediately becomes evident in terms of success, functioning, and the like. Such a consequence marks the achievement of the objective, the end of the action since the pursued value is creatively actualized. Values are "objectives" as the end of action but not an object of perception. Objectives are goals regulated by worthwhile possible values. Thus value as the objective of action is not the object of a subject. As I argued above, achieving an objective is entirely independent of desire, will, interest, attitudes, feeling, emotions and other psychological-subjective factors. A plane does not fly because we will it to fly. We can "perceive" this objective externally: the house is built, the break is repaired, the fat is shed, the meal prepared. Perception *follows* creative actualization: value is perceived as a *result*, which is equally evident as soon as we consider the actual origin of most of the goods of the world. These goods were creatively actualized in the past, out of an even more distant past in which they had not been actualized and thus could not be perceived. The houses, planes, roads, and so on of our world reflect past values in that agents considered them worth actualizing. The world of goods reflects past values that, as present, can be perceived. Past choices are now actual as goods. Even some normative characters can be perceived. I mentioned above that we can identify an instance as of a kind, for example, a nail, not a screw, brad, tack, and so on. This normative identity is perceived, as the result of creative actualization. Value "qualities" may be normative characters. They are normative if they actualize normative standards as well, for example, some standard for nails in terms of size, durability, resistance to rust, and so on. Such normative standards may be manifest in practice, that is, the use of a nail and its functioning well: a good nail. Normative characters are characters by analogy with trust: as reliable. They actualize standards that prove reliable, that is, a screw shape that can be used reliably as a screw: a good screw. Thus such characters are not an "attribute" as there is no "universal" entity or substance. Norms are prescribed of kinds in definitions.

Beauty is still another good that can be perceived. Not only is the distinction of beauty from other goods evidence of kinds, but beauty itself is manifest. Beauty indeed involves perception, but it does not follow that the experience of beauty, or judgments of beauty are "subjective." Anyone can observe the beauty of the Grand Canyon; thus its splendor is not private. But more, we see beauty of the Grand Canyon, not of the perception. This beauty is grounded in this canyon: it is not illusory and thus not dependent upon a subject. If value depended upon a subject then the evaluation that the Canyon is beautiful could not be made, since it would depend upon the subject, not this particular canyon. But the beauty is tied to this individual canyon as outstanding, not another. It is thus inherent in this canyon. Nor could it be tied to the subject, since then the feeling, emotion, attitude or will would be enough and no one need journey to the Grand Canyon to perceive it. People could simply cultivate subjective states. But the beauty of the Grand Canyon is actual, not a subjective mirage or a reflection of "proper attitudes." Perception of its beauty is grounded, while judgments about its beauty are evaluations. To confuse an actual good with a judgment is to confuse value and evaluation.

Thus the world of value is larger than the world of perception and reveals much more about the world. Perception that does not incorporate changes in the world, that uses an ahistorical model of perception, is blind. The practical changes in the world that are "bracketed" by attempts to abstract a "pure" or "contemplative" world of pure "essences," or "pure experience," are blind to the historical changes that have creatively actualized the world that is the condition of the subject. Such abstractions from practice and history leave out the changes in the world that reveal values. They are blind to their own conditions, especially the value of such abstractions that is their motive and creative actualization of a world that is the condition of such activity.

I noted above that specific value commitments reveal choices, whether a personal choice, the choice made by a society or culture as a whole, or the choice of one good over another as more suitable for a particular task, for example, a Phillips head screwdriver instead of an Allen head. Such choices imply alternatives, that is, that other values could have been creatively actualized. The values that have not been actualized are in the evaluative, normative and logical realm: possibilities. Such possibilities may remain as possibilities but may also be impossible after a time, for example, money spent on one and not another item. The normative realm has a whole other set of value possibilities for which there is logical evidence.

The first is "polarity."[30] Polarity works in terms of simple negation, but in more practical terms. That is, for every value there is the logical negation of the value, either what is not that value, or in more practical terms, what is bad or what is defective. Polarity is a logical character of norms, and distinguishes norms in the abstract from actuality. Since polarity is a logical character, it can be applied to every evaluation of good. A good pot for boiling water is not a bad pot for boiling water, since these are practical negations. Or, what is much the same, a pot cannot be both good and bad for boiling water. The evidence for polarity is, again, in practice, that is, in testing. A pot made of metal works for boiling water; one made of wood catches fire while a paper one collapses when water is placed in it. Since this quality is not apparent from perception of such pots alone, without actual tests, it may not seem like there could be evidence for it. But as with a plane, the evidence is in practice, that is, involving action over time. Metal pots work better than paper ones even if this superior functioning is not apparent from staring at a metal pot to determine its "attributes." It would become clear even to empiricists if they attempted to boil water in paper kettles when they "felt" like it.

Rank[31] or comparative value is another normative character for which there is evidence, as the above example of the kettles indicates. Not only are metal kettles good and paper ones bad for boiling water. Metal kettles are superior to wooden ones; some types of metal conduct heat better (silver, copper) and thus are superior where conduction is valuable. Similarly, we use a shovel to dig not a stick, because the shovel has generally proven superior, and this can be demonstrated in practice. Most people will use a vehicle instead of walking. The evidence lies in practice, testing to see which alternative is superior. However, rank is evident, not merely subjective. I challenge those who regard values as merely subjective to use a paper clip to hammer in a nail instead of a hammer. Or a paper hammer. The metal hammer is inherently better. Rank emerges out of competing claims, theories, conceptual schemes, opinions, and beliefs. These must be evaluated. The question of rank arises in both practical and theoretical situations where relativism is not satisfactory, for example, deciding how to build a bridge as opposed to what flavor of ice cream to buy. The question is "will this bridge collapse if I cross it?" We could ask everyone how they feel about it, or follow a more practical procedure by testing materials and designs for bridges.[32] Rank is emergent from comparison, through evaluation. However, evidence for rank is not subjective except in the case of personal values. In justice, some things should be left to personal taste, since a world with many flavors is better than one with fewer, although such flavors must also be creatively actualized. However, the value of survival may dictate that bridges not be a matter of personal taste, that is, concessions to tolerance for different opinions. Even in personal taste, however, there is a ranking.

Kinds or categories of values are still another mode for which there is evidence. Categories are general or normative abstractions from instances of value. That the norm is actual is evident from consideration of the above example. A screw is not a hammer; we can distinguish one from the other easily. Similarly, we can distinguish beautiful landscapes and the beauty impresses us apart from the economic value of the landscape or whether it is "sacred ground." Such categories or kinds are actual in the world, since land may actually have economic value that is independent of what any one person is willing to pay for it ("real value"). We can "verify" instrumental values by testing them in practice and categorial values in a vain attempt to do without them, for example, truth as a value in knowledge or prosperity as a value in economic life. Similarly, rules, another type of norm, can be "verified." Children can "verify" parental commands by disobeying them and receiving punishment. Criminals can "verify" legal punishments by being "caught in the act." Speeders can "verify" traffic laws by speeding in front of a policeman. Practical rules for making can be verified by results.

Nevertheless, values in some respects remain hidden. Values are hidden as future possibilities that have not yet been actualized. Value emerges over historical time, and the creative actualizations of the future, and thus the full extent of value, remain unknown to us. The grounds for success in some cases and failure in others may elude us (cures for AIDS). In the case of novelties, nothing can be known to be valuable a priori. The goal is projected as valuable, but the ground is not known as such. Inherent values are not known as a ground, but only as grounded, that is, as a consequent or that which has been creatively actualized. Even the transferal of value from goal to successful action to good hides value as such in the synthesis of norms and differences that is the instance of good. Value remains in some respects a mystery; value is not present as an entity and thus may be less perspicuous than the smaller field of the actual, even though value is the hidden ground of the actual and manifest as consequence, that is, in successful actions and functional or superior goods. Value as the ground of action is a goal, not an attribute.

Nor is creative actualization a form of knowledge, although it leads to knowledge as its basis. What "is not" cannot be known, only imagined. After a successful creative actualization, the process that has proven successful may be repeated, and learned. But the original is abductive, to kidnap still another phrase from Peirce, a hypothesis or pragmatic attempt to creatively actualize a novelty. A valued goal is projected and becomes a project of action, that is, bringing about the good through successful action. Thus the pursuit of knowledge as a goal is not known except as result. The ground is not known. This analysis applies *a fortiori* to knowledge itself as a valued goal, in the form of research. If the results of the research were known, they would not be pursued as a project.

Nevertheless, since the actual world is creatively actualized with knowledge—fluorine as an element; Neptune—"reality" is a value as creatively actualized, as Schiller argued. Making "reality," that is, creatively actualizing is making knowledge of reality. Knowledge is also creatively actualized, providing decisive evidence of value cognitivism. If we know anything at all, we have evidence of value, since the knowledge was creatively actualized and creative actualization is value. The value of experiments in science, and value in general as creative actualization are the same in procedure and form: a bringing about decisively. Results of experiments succeed in determining knowledge; technical creative actualization succeeds in actualizing a method. Thus the movement from potentiality of knowledge to its actuality argues for "actuality" as an axiologically superior term to "reality."

Values are not "present" but hidden and obscure until they are creatively actualized. Yet because they are pursued they are relatively superior even if we sometimes fail to realize or attain our ends, discover new problems that emerge as a result of our creative actualizations, such as pollution and traffic jams from automobiles, and attain results that are a mixed blessing, that is, ambiguity in value.[33] Wealth and knowledge are not unmixed blessings since wealth is defined in terms of inequities in income or property; knowledge creates elites and problematic creative actualizations such as atomic weapons. Improvement may be incremental; it is usually finite as well. Values are not an escape from the "human condition." While we create our world with creative actualization, the ground that enables us to continue to creatively actualize eludes us.

Nevertheless, creative actualization meets Moore's test of "is it good," since if the good is defective, then value is not inherent in the bearer. Creative actualization is good as inherently good, as instantiating a kind, as non-defective, as having been brought about, and so on.

In this chapter I have characterized value as creative actualization. Creative actualization is bringing about novel goods in the world through creativity and activity that successfully achieve the goal. Such goods alter the world, and value is the ground of actual changes in the world. The world is thereby improved and given meaning. Value was distinguished from evaluation; evaluation regulates which possibilities will be creatively actualized, since they are judged worthwhile. Successful achievement marks their value, while failures are no good.

Two

MODES OF VALUE

What is the relation of creative actualization to inherent value? I will treat this issue in this chapter as well as related issues, such as whether value is connected with the individual act or a novel type. In part this topic has been touched upon. However, inherent value is a rubric for a set of problems that will be covered in this chapter, including the bearer of value, its locus and scope, and different modes in which value is manifest, including normative identities. The analysis will cover modes of value, the different ways goods are actualized. I will also cover some terms from the tradition of value theory, such as evaluation, intrinsic value, inherent value, and instrumental value.[1] All of these issues center on inherent value, as a bearer is an inherent bearer of value in which value is located. The question is what is the scope of such bearers and loci of inherent value? Does it include an inherent good of animal and other life? How can one genus, creative actualization, be inherent in different modes, for example, norms and also predictions, copies and the unique? I will cover this topic as well.

1. Inherent Value

A distinct creative actualization is novel in the respects that were distinguished in the last chapter, either entirely new or a new copy or close facsimile. As novel, the value is different from what was previously actualized and of distinct value, that is, its value is distinct from any other individual or kind. An instance of the value has the value inherently, or it could not be a distinct instance.[2] For the value is the ground of actualizing the good. A good is the achievement of value in one mode of creative actualization, making a good. Achievement of value of any kind is achievement of inherent good or value, since achieving the end creatively actualizes successfully, and thus achieves value as a goal in the world. This achievement is inherent in the consequence or result, or the project is a failure. If it does not reach the goal, little or no value is inherent in the result. However, creative actualization meets George E. Moore's test of "is it good," since if the good is defective, it is not inherent in the bearer. The plane flies or it does not. Those that fail have not completed a project, the goal, since the norm of flight has not been actualized through creativity.

"Inherent" value is the value or good that any instance of value bears. This value or good is inherent in the bearer as a whole, although it may focus on a part. Inherent here is distinct from "extrinsic" value, since if an instance is inherently valuable, this value does not derive from an extrinsic source, but belongs to the bearer or locus itself. Inherent value and good in a bearer differs in quantity, content, and quality from other inherent values and goods. This distinction is more basic than instrumental and intrinsic value, since both of these are types of inherent value. A tool that functions well or properly is an inherently good one, just as intrinsic value is inherent in living bearers. Inherent value and good is value that is in bearers "by themselves," that is, apart from other bearers and as distinct, novel goods.

"Inherent" does not mean "within," in the sense of inside, as it may involve the whole, a part, a function, and also involve operations in the world. Thus inherent value is not hidden from the world. Inherent value may attach to some instance, however, and thus be within in that sense. Just as science has been described as "unlocking the mysteries within nature," so does creative actualization bring new possibilities to the world that lie within the potentialities of the world but have not been previously actualized. Thus the value is inherent to the novelty as it was hidden within the world as a possibility, but now inheres as an actuality in the good.

A bearer of inherent value has proven its worth and thus is in the world. Tools prove reliable by themselves; the judgment that they are equal to the task is a prediction, and if this proves itself, creatively actualizes the hypothesis in the world. Thus inherent value cannot be subjective, even granting the validity of the norm "subjectivity." The value of its functioning properly belongs to the tool as its character, and the tool would not otherwise be used. It would be absurd to argue then that inherent value is subjective since if this were taken literally, marshmallows would be as good as using hammers. Use is evidence; using marshmallows to drive nails would be "no good" or even folly. If the hammer works, it is of value for driving nails and this value is inherent to a hammer by itself. Inherent value is of proven worth by itself, not derivatively from any external relation, whether to a subject or a substance. The hammer as a whole is inherently valuable, not a substance with hammer-like or hard attributes, although any hard body might do in a pinch. Theoretical, speculative or hypothetical value that has yet to be pragmatically tested has no inherent value; this sort of value is in the realm of possibility, of imagination and evaluation. Similarly, subjectivity is subject to evaluation, since subjective whims must be evaluated for possible worth, to speak nothing of feasibility. Inherent value cannot be value only for us or to us. Inherent value or good must be in the bearer by itself. If value is inherent to some bearer then the value is not "subjective," but inherent in the differentiation of the instance in the world apart from us, which is the condition of its use value for us. If the bearer did not function in the world properly according to its kind, inherently, it would not be used for that function. A plane must fly before we can make personal decisions on whether to go for a flight in it. The plane must be inherently worthy. Such personal values are distinct from other kinds of inherent value, although they have an inherent value of their own.

Inherent value and good are self-justifying since the value is inherent in the bearer: it does not refer beyond itself for justification. Either the bearer is inherently valuable or it is defective. Since in this view all values and goods are inherent, their creation the ground of their actualization, inherent worth differentiates each instance, for its value does not refer beyond itself. The value is inherent, differentiating the bearer of value from the inherent value of another bearer. Further, the bearer is valuable for its own sake, not for the sake of another, that is, on its own, not valuable as a consequence of evaluation, but has proven valuable in itself. The ground of its value is itself, and provides justification for evaluations of its value. Inherent value is self-grounded, since justification of discovery must ultimately have a value of its own.

Inherent value is distinguished by finitude in the sense of limits, since its value as such is valuable in contrast to other values. Inherent concentrates value, excluding extraneous extrinsic elements and limiting value to what has been creatively actualized. Distinct, different value is limited value. Its value is not infinite, but within a limited sphere or realm which is determinate. Since multiple values exclude each other as instances, finitude is an element of the character of inherent value (but compare a Swiss Army knife!). This finitude makes distinctive activities actualizable, each with its own proper time. The time of each is limited to its actualization. The value can be historically superceded if its limitations are overcome.

Is a good of a kind not inherent but relational, that is, does it lie in the relation of the instance to other like instances? Or in the relation of an instance to a kind and thus not inherently valuable, good in itself, but extrinsic? How then can value be inherent in a type of good, if it is inherent in instances? The value inherent to a distinct kind is that value which is peculiar to that kind, a distinct creative actualization. As a unique kind of good or of value, value can attach to the kind qua value. For this is to consider the kind in the abstract as one instance of different kinds, that is, in the realm of evaluation or norms. And such a kind as an instance can constitute a distinct good. This good is inherent to the kind as distinctive from other kinds. However, such kinds are creatively actualized as instances in the world. The norm is manifest as an instance of the norm, for example, in mass production of units of a good. To increase kinds of inherent value is melioristic. To lose that value would be to reduce the kind and amounts of good in the world, which is anti-meliorist, since the world is not bettered but worsened. If a bearer is a unique instance, for example, individual personalities and most art-works, or even individuals of rare species, the individual is irreplaceable; in mass production of simple tools, such as nails, the value of the individual is much reduced, since it is easily replaced. Thus the value of the unique kind in the world is actual, not merely hypothetical. "Inherent" value is then revalued in creative actualization as including the value distinctive to a way of creatively actualizing and inherent to such a kind not incidental or extrinsic to it. However, inherent value is manifest as instances, common or rare.

Inherent value is distinctive from other values, since the value of inherent value belongs to the instance. It is not an instance of an identity in terms of its distinctive value, which is different from other goods. The value of a bearer of inherent value is not extrinsic to it in relation to a kind. The value of the individual bearer is unique in some respect if it is inherent, ultimately different. The problem of individuation can be treated as the problem of the value of individuals, including especially individual actions and functions, that is, the degree to which they are not specific instances, but individually distinct.[3] "Unique" is more a principle of difference than identity. The "species nature" or kind does not capture the individuality of the individual and is only a normative identity. Further, the kind requires actualization through and as an individual instance. However, difference in value does not of itself create a ground for hierarchies of value, as I argued above, since such hierarchies subordinate difference to a supreme identity. Difference alone creates plurality: values.

The normative is actualized in a unique instance, that is, an inherent good or value, actions, and functions good in themselves. If it were not inherent in the good, it would not have been creatively actualized, that is, the action of actualizing the norm would be unsuccessful. Thus grounding a norm is grounded on or based in inherent value. Creative actualization is good in part as bringing norms, ideals, standards, obligations, principles, and the like to actuality each time, that is, as a new instance which is not defective. The instance of the norm is separate from other instances, from other inherently valuable attempts. Thus norms are grounded in and by inherent values, that is, creative actualizations of their bearers in the world. They are inherently valuable as a temporally, spatially, and unique instance. Value does not derive from an extrinsic relation, but in the relation of the goal to its achievement as an instance which is successfully attained. This particular case of the norm is not defective, but has accomplished the goal of reduplication of the norm in the world. Thus value captures the individuality of the individual: what is different and unique about it.[4] This can never be adequately conceptualized as a normative identity, for example, in terms of "existence." Value is what is especial to the difference of the individual, what is most distinctive about it: the unique.

If value is inherent and not relational, what unites the different values as, identically, values? What is the "one-in-many?" The answer is again, inherent value as creatively actualized, since each value has inherent value as a successful creative actualization. But inherent worth differentiates, for such value does not refer beyond itself to an extrinsic source of value. The value is inherent, differentiating the value from the inherent worth of another value. As I argued in the last chapter, pluralistic values do not form the relation of values to a first, either a *pros hen* or a *summum bonum*. A hierarchy of values, of which participation in a *summum bonum* or *pros hen* to a first are species, implies that all value is derived from the single value at the apex, thus constitutes a monism of value, that is, only one value. All values receive value from the one value: value not values. This would constitute the reduction of values to value and thus would be anti-pluralist. Inherent worthiness cannot refer beyond itself to a first (*pros hen*) in value, nor derive its value from a higher value (*summum bonum*) since its value would then not be inherent, but extrinsic. These models contradict the notion of inherent value, since no value can be inherent which receives value from an extrinsic first. Inherent value is the value peculiar to a distinct instance, and thus cannot be relational in terms of its value. Otherwise, the instance is defective, for good or value has not been creatively actualized as the instance. Thus the instance has no value. Again, a man does not need a higher value to pursue a beautiful woman, but mere acquaintance with her manifest worthiness. The attractiveness of value is inherent, not relational. The value of fine music, a sunset, health, and other goods is similar. The value is inherent in the bearer, not in the relation of the bearer to some higher good. Hence the attractiveness of the worthy as a project or *telos*: the value is good by itself, its creative actualization inherently valuable.

To differentiate a distinct good sets up a relation of what is different from other goods, from which it is differentiated. This difference can be justified by inherent grounding for which inherent good is required. (As I noted above, inherent value should not be confused with intrinsic value). That is, the differentiation is worth differentiating for its own sake. Thus grounds for differentiation are provided by values, if the project is worth differentiating. The worth inherent to what is creatively actualized offsets and enhances its distinctiveness sufficiently so that no further grounding is needed. The value is inherent in creative actualization, the functioning of what has been creatively actualized, or the good. Difference is grounded by inherent value.

William James tied pluralism to meliorism and both the pattern and the relations between members can shift over historical time due to amelioration. What one age deemed important may be devalued in another. New values can replace old. The values are not determinate as novelties. Nor do they form or emerge in a precise pattern, or we could predict social, political, and technical developments more precisely. Pluralism does not form a "constellation," since the latter implies a determinate pattern, which substitutes a "center" for a "first." True differences in value are inherent in distinct kinds and unique instances; they have no common ground but only relations established according to worth. True differences are multi-centered, having different relations to various kinds. Such differences are also distinctive, as they bear no relation in value to other values, but are inherent to the novelty. While such values are inherently superior to those which they have replaced, they still have increased the plurality of values in the world. Evaluation is reflected in such a world, which has been evaluated superior to alternatives. However, the plurality of inherent values that has been formed is an independent variable, since whether a project evaluated worthwhile can be creatively actualized is as much subject to feasibility and success, as evaluation.

The world is rich in the number and kinds of value which have been creatively actualized. However, richness is a consequence of creative actualization, of value as inherent in an instance, instead of a character of it.[5] True, creative actualization is potentially rich, since so many projects have proven feasible and have resulted in plurality of value. But the richness of value that which has been brought forth into the world does not attach to kinds or individuals but the whole. It does not mark the distinct value of a kind or an individual as unique except tangentially. Richness is not inherent as a character to values, although the world is indeed richer for values.

Value has been presented in this chapter as having a bearer of some kind and as inherent in such a bearer. What constitutes the "bearer" for each philosopher is based on the metaphysics of that philosopher, whether the bearer is a being, an ideal realm, complex wholes, feelings or other subjective states or in a relation. That philosophers have posted value in different bearers may constitute a problem, but this is a metaphysical problem, not an axiological one, since some sort of bearer or instance of value is upheld (McDonald, 2004). The stalemate is not in the value relation, but in metaphysics between those who require value in an object and subjectivists who insist it is not an objective property. However, as the persistence of the value relation may indicate, problems of value may be an independent variable, since the instrumental-intrinsic relation may not be affected by ontological problems of being or the metaphysics of the subject, which it predated.

Does inherent value refer to the quality of some referent and thus constitute a *pros hen* to "being," the "substance" which bears "attributes?" I will argue below that Aristotle's conflation or confusion of the problem of value with that of being has been deleterious to value. This section will also answer the question of whether the "good in itself," or "end in itself" are "ontological" constructs, that is, whether they presume the "being" of the "in itself." If pleasure is subjective but good in itself, as in hedonism, then an in itself can be isolated which is not a "being," that is, an attribute of a subject.

Value "per se" is not "ontological," since we can never grasp value as such, as a "thing." Firstly, value in the primary sense as creative actualization is regulative of actions, not things. Good things are the consequence of such actions. Another significant sense is the function of goods, for example, the use of a tool in action. With its origins, value is temporal and creative actualization is the mark of change, not duration—the mark of "being." Value is historical instead of ontological: comparative improvement through past and future choices that are manifest. The "nominal" understanding of good would be as actualization in an instance. Moreover, value is not an entity in any sense; we can only abstract the value that becomes actualized through activity with the imagination, as superior to other likely outcomes. Value is not grasped as an entity but as the creative actualization of goods and other modes and bearers, such as actions valuable in themselves. Value is what all such plural modes and bearers have in common as distinct creative actualizations.

Many goods are not "ontological" in the sense of tangible "things."[6] Knowledge is not a thing in this sense either in the physical or the historical sense. The value of knowledge as a standard is an ongoing project whose historical end may never be achieved. Knowledge as a project motivates us in the present such that we aim at a future in which some question will be answered. But we never hold knowledge in our hand, although knowledge as generic may have instances and bearers, that is, individual cases of knowledge. "Knowledge" is a regulative standard, not a "thing," which directs our activity but remains hidden from us while we pursue it. Such standards "are not" as a future project and "are not" as intangibles which are never truly attained as thinglike. Raising the question of their "ontological status" is the wrong question: some actualities are not tangible; those that are tangible were creatively actualized as goods not "beings."

As I noted above, such goods do not have an ontological condition. The "things" of the world are previous creative actualizations, that is, goods. There is no "existence" above and beyond these goods: to add such a "concept" would constitute mere metaphysical meandering, which we can dispense with using Occam's razor. "Existence" is itself a multiplication when "good" is sufficient. Good is not a property of being, since the ground of goods is their value, actual or potential. "Being" is superfluous since the actuality of the good is the result of values not the cause of it or basis for it. As I also argued, substance as being implies a hypothetical static world of beings. The actual world of change is grounded in value, as the cause of change, and thus supercedes what "is" with what "is not" until such time as the project is creatively actualized. Value cannot be a predicate of a "substance" as value is the ground of the actual. What is now actual reflects past norms, ends, choices, and other evaluations, thus "onto-logical" questions can be reduced to normative ones.

Philosophers and others refer to rational choices in logic, "decision theory," and other normative spheres without ontologizing the corresponding capacities, without tracing these back to a necessary condition in the mind which becomes reified as desire, the "rational faculty" or "free will." Similarly, values need not be fixed with ontological conditions, despite the literature on the "ontological status" of values. The distinction of substance from attributes is problematic, for it is not an ontic distinction, but a logical one. If the distinction of "substance" from its attributes played any role, marshmallows might be as hard as hammers, for then the marshmallow could take on a contrary attribute. This debate reveals the metaphysical presuppositions of their authors, but little about value (McDonald, 2004). This is to speak nothing of the inherent problems of "ontology," such as whether what has been creatively actualized as a novelty can have being? But a creation is a change, thus came from "not-being." Thus it "was not" etc.

The goodness of a good cannot be separated from the "substance" in actuality without destroying the goodness. A hammer that is destroyed to show that some "stuff" remains after its value is destroyed is not a good hammer. Thus goodness and its bearer are actually inseparable. The thesis of an underlying "stuff" is superfluous; the stuff has itself been creatively actualized as a good, either the metal or other material of the hammer. Even broken goods consist in goods. Thus creative actualization of a hammer from metal is not a new attribute of a substance, but "value added" to a previous good, refined metals, and grounded in value. Nor are all values attributes in the sense of qualities of a good. Holistic characteristics are not such qualities. Another case is a good that meets or comes up to standards, ideals, and other transcendent norms.

Further, a good is not an attribute or relation to a substance, its perfection, but inherently valuable. Value involves the actualization of a possibility not the perfection of a substance.[7] Creative actualization is not a perfectionist theory. Perfection of a substance as an instance belongs to hierarchical theories. The instance "exists" prior to perfection of being and is subsequently perfected. Planes that won't fly are neither valuable as a plane nor feasible as a flying machine in the world. Nor are goods as valuable instances "perfect"; they may lose value over time or new and more valuable kinds or qualities may supercede them (computers). An original success or creative actualization may not be perfect, but only adequate. A plane may function but need improvement (rough ride, pollution, comfort, etc.). The issue is not perfection of a function or kind but actualization of it in an instance. A good may be far from perfect but "good enough." Meliorism is not perfectionism. Perfection has a relation to a standard, not of inherent to an instance. Meliorism is comparative or relative improvement of an inherently good instance as compared with another inherently good instance. Perfection according to a standard is only one mode of creative actualization, which requires creative actualization and also may itself be superceded by new standards: amelioration. Perfection is the enemy of change.

Actuality always has its locus in goods or in actions valuable in themselves. Nor are goods entities as novelties, as I noted above. Potentiality may have no "existence" or "substantiation" at all: it may never be creatively actualized. A heap of junk can "exist" but not as a plane. "Existence" as a flying machine is distinct from "existence" as a failure. Thus a good is not a perfection of existence as the worthless "exists." To "exist" is not a mark of *value*. Even a heap of junk is creatively actualized; its actuality reflects its genesis. Actualization as creative actualization of a function is only secondarily inherent to goods, since the grounding act of creative actualization is primary. However, failure may precede successes; these failures "exist" or are actual but are "no good" (flying machines prior to the Wright Brothers). However, even this would be too great a concession to ontological metaphysics, the prejudice of our age. "Existence" is tied to perception, that is, empiricism, which can ignore genesis as if in a catatonic trance. The question becomes which is a superior term to use for the world's inventory, existence or actuality. Since the question is evaluative, the answer is obvious. Even a heap of junk has been creatively actualized.

Once actualized a good may involve a whole, as in a plane, which is no good without all its parts. Thus good is not a part or attribute of a substance. A substance is defined as capable of having contrary attributes.[8] However, even if this were granted, good and bad are not contraries, since bad, as Aristotle also would argue, is a defect or lack of good. Thus "bad" is not the opposite quality of good in any precise sense of an opposed but positive quality. Good and bad are not exact "polarities" but instead, one is the lack of the other, bad as the defect or lack of good. Thus their relation is not a matter of contraries of substance. A "bad" plane has not been creatively actualized nor ended a project, since the norm of flight has not been actualized through creativity.

Nor are inherent values perfectionistic, that is, perfections of being. To be valuable in itself, a value need not be perfected, but only revealed as good in itself, that is, distinctively valuable. Creative actualization involves bringing about a value as a goal, and thus achieving the end. The end may be but imperfectly achieved however, while achieving the good. Later improvements and developments are ameliorations; what seem like perfections may later be superceded either in part or in kind. Further, values are inherent as they differentiate instances which have been evaluated as worth differentiating. The distinct value of the instance is not a general predicate, that is, a universal in the form of a species or genus under which an instance is subsumed in the categorial proposition.

This is to speak nothing of the problems of an ontology as metaphysics. For there is no permanent substratum: the substratum is at best an abstraction, the "logos" but not "ontic." "Substance" is not an entity, since it cannot be separated "ontically" from attributes. The distinction is not ontic but logical, that is, normative. In the cosmic sense, all substances evolved from more primitive energies, and thus came to be. They "were not," and thus did not have being. In the axiological sense, the "substance" was creatively actualized, for example, the pure metal from the ore, as a good. Finally, "ontology" cannot account for principles, norms, and laws of science. These generals are not *per se* beings, that is, substances, since they are instantiated along with what they "govern." But they are as "real" as such "entities," even if they are "invisible." Thus "ontology" is inadequate as an account of the whole, as "metaphysics."

Secondly, inherent value does not constitute a relation to being, a substance, a subject or any other metaphysical bearer. Inherent value is good in itself, thus not in a metaphysical relation. For a relation implies that something is not good in itself, but derives its value from the relation. Aristotle correctly saw that good is inherent or in itself, but then he contradicted this notion of ends with the notion of subordinate ends. If a good is good in itself, then its value cannot refer beyond itself, to a first in value, for then it would not be inherent but relational. Inherent value is something valuable in itself, good in itself, not in relation to another good. This inherent relation of a good to itself applies to instrumental values as well. Even instruments have inherent value if they are not defective. Their good remains with themselves, since they are up to the standard of their kind. A hammer is not good because of our attitudes toward it or our emotions, but because it is capable of driving a nail.

Creative actualization is not a *pros hen* in the metaphysical or modal sense either, a "referring back" of a quality to a substance which "has" inherent value. Qualities are "equivocal," to borrow Aristotle's term, that is, of different kinds and modes: they are multiple. Qualities can be multiplied in the actual world, the creative actualization of the worthwhile at the expense of the inferior, defective or valueless. The actual goods of the world are an aggregation of qualities, that is, of normative identities. This can be achieved in several modes, without some common essence. Indeed, I would argue that attributing a character to an instance cannot be justified in terms of "nature" that is, an essence or fixed idea, since this is incompatible with scientific progress. Taxa are often reclassified as new information becomes available or older work is re-analyzed. Analysis cannot appeal to the "nature of the case," then, as just what this nature is may be problematic and "falsifiable" (McDonald, 2004). Knowledge and beautiful art can be creatively actualized, but are not "substances." The value of types of value reveals the multiple modes in which good is creatively actualized. Such kinds stand in a reciprocal relation as equally goods. Values may also be "co-conditional" in the sense that reciprocity may mean each good is the condition of the other, not only autonomous from the other. This is the relation of each element of a habitat to the whole.

I would also argue that intrinsic value is also independent of the problem of being or metaphysics (McDonald, 2004, ch. 3). The relation of extrinsic value is to intrinsic value, instead of an attribute or some other relation to a bearer, whether a being, a subject or an object. This relation of instrumental to intrinsic value has been consistently upheld, regardless of the metaphysics in which it is placed, even contrary ones such as subjective and objective. "For its own sake," a phrase synonymous with intrinsic value, is contrasted with actions performed for the sake of something else and thus arises by contrast with extrinsic values. That is, extrinsic involves the relation of different, distinct values. Since the value relation is independent of either metaphysics of being or of the subject, we can treat value independently from metaphysics.

Some might argue that a bearer of value has a separate "existence" than value since something which has been creatively actualized but is defective continues in actuality after the failure to produce a novel value. Thus value is a mere attribute or quality of what "exists" as a "thing." Bearers of value, then, exist as things to which value can be attributed. However, such "things" have themselves been creatively actualized: the metal of the defective machine did not appear out of nowhere by magic. Such metal is a previous good which was turned or can be turned to other uses based on evaluations of potential worth. Similarly, the other components are goods; new applications are modifications or functions of a good or of several goods combined. Thus any goodness that is added to a previous good is a new creative actualization, not a quality of an "existing thing." "Things" are previous creative actualizations. A metaphysics of "things" that ignores their genesis devalues their creation as novelties, to speak nothing of its inadequacies as an account of "things." Further, as I noted above, a bearer of value may have value as a whole, not in one particular quality. Planes have value as a whole, not just for their wings. What of the argument that flight is the condition of the value of a plane and that this is one quality, although not a part? But the whole plane flies, thus this characteristic is holistic.

Nor does anything "stand forth" (ex-sistere) unless we evaluate it as worth paying attention to. The rest is "background" and does not receive attention until worthy. Most of what "stands forth" does so because it has been creatively actualized and thus becomes part of our world. Thus there is no more value to be attached to nominalism than to its contraries. All we may know of an individual are its common features, as John Duns Scotus and others have argued. The attention we give it for a moment may be as a representative of a type, although the instance is no less individual. Its significance per se is a differentiation, that is, grounded in its unique value.

Inherent value is distinct from the means-end relation, since either means or end, as kinds of value, have inherent value as good in itself. A good tool has inherent value as a tool whether an end or means. Indeed, most instances of value can be used as means, but the value of the bearer remains whether it is in use or not. Their value is inherent as finished individuals. Thus both intrinsic and instrumental values have inherent value.

Intrinsic value is the value of a self, that is, a self as an end in itself, and includes activity that is self-actualizing. Self-actualization is creative actualization of the self as a value, and thus for its own sake, never for the sake of another value. The purpose of the activity is recreation of the self as such, not as an instance of a kind. Since it involves a distinct self, it cannot be the good of a kind. The creativity in intrinsic value is an end in itself which expresses the self but also contributes value to the world. The self is open to a future ordering in which its own creativity is bound to itself: a free acting toward the future. The self-actualization is intrinsically valuable and grounds self in the world. The works of a self unite the self to world as its own world, creation of itself in a world of its own making.

Insofar as an individual is unique, it is even more rare, thus precious. Just as a species constitutes a unique good, so do individual selves as self-actualizing. An intrinsically valuable individual is not only an instance of a species. An individual stone may be generic; an individual life can never be, especially in higher organisms with individually differentiated responses and internalization. Self-actualization means that purposes can be formulated, evaluated, and acted upon that reflect only a self: autonomy. Thus value is inherent in a self as creative actualization of itself for its own sake.

Selves contribute to amelioration as unique goods. One unique aspect of a self-actualizing individual is its location in a unique place and period. That a plant can actualize itself in this generation with its unique conditions differentiates it from the previous generations with different conditions.[9] Again, an animal must successfully avoid a predator to actualize itself: it must choose a successful strategy and not stumble or break a limb with each step, for this would prove fatal. The animal must find food in winter, drought or other conditions of scarcity, protect its young, and teach the next generation skills for self-actualization. Creative actualization of the self involves continual new acts in time and space, which are successful means to the end of self-actualization. Indeed, the idea of self-actualization means that each organism can care for itself. Not "nature"but each thing is itself responsible for its own development. Thus some seeds develop and some don't: the cause cannot lie in the species (=form=nature), that is, external causes. For then all seeds would develop equally under equal conditions. (Compare two siblings, one of whom does well while the other is a ne'er do well).

What is the relation of creative actualization and instrumental value? Instrumental values are creatively actualized out of the potentialities of the world. These are tools used by self-actualizing, intrinsically valuable agents to aid in their survival, which are forged out of the world.[10] They are differentiated or singled out of the world for use in the world. Other goods are instrumental by analogy to tools, since they are used as means, for example, "tools" of business management involving accounting, sales, reorganization, and the like. Instrumental value originates in what is good for intrinsically valuable agents, that which aids survival or sustains life. However, at a more advanced stage, instruments enhance life: they are instruments of luxury not necessity. Most consumer goods fall into this category, since they are not required for survival but may enhance life, for example, automobiles and washing machines. While reliability is a standard for instrumental values, it is not necessary, since a temporary expedient which accomplishes the job is immediately good.

As I argued above, instrumental values are inherently valuable. In this sense, everything valuable is valuable in itself, even purely utilitarian values.

For a tool or instrument is distinctly useful, that is, is useful apart from other tools, by itself. Thus instrumental value is inherently valuable as a differentiation of a unique good. A good hammer retains its value as a tool from use to use, whether or not we wish it, feel it, desire it, take an interest in it, have "proper attitudes" toward it, emote at it or have any other psychological intent toward it. We can stare at and have such psychological states toward the display of hammers at a hardware store all we like without changing the shape, the function, the reliability or the diverse uses of a hammer. We could even follow the New Agers and emit psychological or "rays" or an "aura," consisting of our attitudes, desires etc. But the hammer would remain the same in form and function. Nor will crystals that amplify such rays affect the value of the hammer. Carpenters will continue to come in to the store, buy and use such tools; they will ignore such psychological states as irrelevant to the value of the hammer.

2. Bearer, Locus, and Scope

The distinction of bearer and locus of value was made in the Preface to this work. The bearer of value is the instance, token or individual unit of value.[11] Why is the distinction of bearer from instance warranted? An instance could be an example which is not actualized or not yet actualized. A drawing or design of a plane which is going to be put into production but is not yet built would be an instance of the model but would not bear value, for it has not been creatively actualized, and does not function. An instance may be a future bearer, but can also be an instance of a norm which has not been actualized. All bearers are instances, but not all instances are bearers, since bearers must be actual. A bearer is an actual value, function or good in which value as ground and actuality have been united. Defective instances do not bear this value. A bearer differs from an individual by bearing a specified, determinate value. This good bears this value.

Locus is location of value in a type, that is, the sort of actuality, unit or qualitative level in which value is located: whether in individuals, actions, standards, or whatever. Locus is closely connected with the "metaphysics" of value, or rather the mode, as locus in individual subjects presumes some sort of metaphysics which includes subjects. However, locus and bearer may be distinct for other reasons, for example, the composite location of species in individual bearers in specific metaphysical systems, but the locus of value in the species. In this case individuals would bear value but its locus would be in species. Aristotle's theory may be of this sort; some radical environmental theories hinge on it.[12]

The bearer of value may take several forms, including goods, actions, standards, and ideas which have been actualized. These constitute different modes or ways in which good can be brought about, creatively actualized. What has value may be individual, particular, or an instance of something more general, normative. A particular work of art may be a unique bearer of value, while food may be more generic. The value of a unique bearer may be inexpressible as unique, since concepts are a species of norms. But the value a bearer has is its uniqueness, and since every bearer which is a good is unique in at least some sense, value is the ground of such uniqueness that is inherent in the bearer.[13] For example, quantity distinguishes individual bearers of a kind of value. Two of a kind of good may be alike in most respects but not in quantity. Each good is a good instance of a bearer of a kind of value: an individual at some level (compare "value of a variable" in logic).

A plane is creatively actualized as an individual bearer: the first Wright Brothers model. However, this can be normatively generalized in the imagination and then evaluated as worth duplicating. As this proved feasible in actuality, "aircraft" proved to be a generic principle, which can be creatively actualized successfully in multiple bearers, that is, copied. Many goods of the world have proven to be reduplicable, and thus creative actualization of bearers can be of kinds. Copying successful prototypes is generally feasible. Creative actualization of bearers as goods is a singular actuality of a generic principle.

Value does not necessarily attach to an individual bearer qua individual. Differentiation of novel creative actualizations is a commitment to plurality and thus to the possibility of different kinds of goods. Different kinds of goods may involve a good of its kind since a good which is an instance of a good of its kind actualizes the generic in a bearer of value. This possibility of the world is one that discovery has revealed: that kinds can be creatively actualized as copies (although they are not "natural" kinds as created. The question arises whether a good of a kind is a distinct type or level of value, or instead a necessary quality of value. Since creative actualization is generally reduplicable, is good of a kind a necessary character of value: creative actualization of a functioning sort? Is "principle" necessary to value? This question can be answered by listing unique instances in the world, namely, individual persons, works of art, predictions, gems, landscapes, and other singularities. Thus ultimately creative actualization can be either unique or normative, singular or general. Normative identities and principles can be creatively actualized but also singularities.

Different categories of value imply different values, that is, a plurality of values, and thereby provide an argument against reduction of values to value, that is, any supreme value or single one. Categories of value are normative generalities actualized in bearers that conveniently or conventionally identify irreducibly general levels of value. Such categories of value are generalized abstractions from instances, for example, beauty from beautiful instances (McDonald, 1997, ch. 4). Such values have value as individual bearers, which actualize a principle of a good of a kind. Categories are generic to inherently valuable bearers but with respect to their extrinsic normative relation, not their inherent value. Although the bearer is unique in some respects and its value is inherent to itself, it also creatively actualizes a principle which ties it normatively to others of a kind. Thus a bearer combines different relations and levels of value. At the level of an individual bearer value is inherent and unique; at a more general, normative level it is extrinsic and generic, if the bearer is not itself unique. Different principles may be actualized in the same instance.

Value is creative actualization in or by a bearer, whether an act, a function or a good, but may be located at a different level of value. Individuals have inherent value as such, but this may be as bearers of a principle, norm, or other general. One nail of a specific grade and size is pretty much as valuable as another, and no great value is placed on its individuality: which nail of this type is placed where in building a house is not terribly important. However, an individual plant of a rare species may carry unique characters which differentiate it even from all the others of its species. It may be doubly rare and thus doubly valuable. The question of locus concerns the value of individuals, of life, and of the instruments of life. For although many theories of value argue that all value is value for life, this position has many difficulties. If life as such has intrinsic value, which it would if value is value for life, taking any life would be morally objectionable. However, predation and the food chain require the taking of life (Callicott, 1989). The question then becomes where is the location of value in respect to or relation to life.

The mode of the creative actualization of self as intrinsically valuable is survival. To survive is to creatively actualize the self, self-actualization over time. Survival value can include instrumental value in the form of those creative actualizations which are a means to survival. Survival of an intrinsically valuable self, however, is good in itself, since the end is not preservation of another but one's own. A self is an intrinsic value which is actualized over a period of time by individual actions which support survival. Thus the actions come to an end through successful grounding in the world of the possible means to survival. However, value does not end survival like it ends action. By ending action value avoids futility but by aiding survival, value avoids death. Thus actions have instrumental value to selves who creatively actualize themselves through such action. The end of individual actions enables the self to commence other actions which can preserve it: self-actualization. Such biological survival or self-actualization is the ground for, although not the origin of culturally derived values. Both self-preservation and its activities are grounded in world by value. (However, survival should not be equated with "existence" or "standing out." Surviving involves self-actualization, that is, the intrinsic value of an agent. Thus in this sense it reflects an evaluation: that we ought to survive or that survival is worthwhile and thus we will creatively actualize the means. Survival is anything but duration without change, that is, "being." On the contrary, survival involves continual change.)

A value is defined with respect to other values and often in relation to other values, not in reference to or in relation to being, the subject, and so on. Such relations are independent of changes in metaphysical fashion. The inherent value is the contrary or opposite of extrinsic value; instrumental is in relation to intrinsic value, and so on with other value relations. Value is creative actualization, but this includes all value, or value in general. Inherent value belongs to all goods that agents have creatively actualized. Indeed, it includes goods of the world that are the basis of judgments of "existence": their actuality originated and is grounded in value. Intrinsic value belongs to the actions of self-actualizing agents, since the end is creative actualization of the agent who acts: creative actualization of self through actions of that same self. Such an agent is intrinsically valuable since the actions of the agent themselves have inherent value, and thereby creatively actualize their own conditions, the agent. Self-actualization means an agent creates itself through action and thereby is valuable in itself. Thus the "subject" is a type of self-actualization, also grounded in value. Instrumental value belongs to actions that aid survival. This formulation upholds but attenuates the distinction of actions valuable in themselves from those done for the sake of some extrinsic end.

Is there a good apart from humans? Is good assigned by humans to animals and other non-human life? Can good be considered apart from humans and their actions? Quality of life is not exclusive to humans, since some animals of the same species may be fairly appraised as living a healthier life than others or a luckier one, in which their conditions of survival are easier. Nor is life as a project confined to our species, since each movement an animal makes that is "successful" creatively actualizes its own life: self-actualization. The animal acts as an individual, aware of the possibility of predation, and that its own survival would be forfeit were it to be caught and killed. Otherwise it would not be constantly on the alert for predators nor resist predation. To call this an "instinct" is simply to cover our ignorance with a concept: we do not know how or what animals think. What is clear is that they can perceive as humans do, and react for the future to what they perceive, that is, act so as to avoid capture and predation in the future. Animals act for an end that is for itself: self-actualization.

Although successful movements to survive and preserve themselves as individuals involve functioning properly, these are not quite the same, since all the organic systems must function in addition to successful motion as the condition of it. The organisms must be functioning properly to move. Their movement is a successful functioning as well, but the strategy of movement is distinct from the movement itself. Each movement of a pursued gazelle must anticipate those of the pursuing cheetah or it will be caught and perish. Thus each movement is a creative actualization of both movement and the strategy of movement: a normative principle. Those that are successful result in survival, the goal of the activity. Thus they have instrumental value for survival: value as means. To say that this occurs by accident is a good argument against determination of responses by mechanical stimulus. The strategy of a prey—or for that matter a predator's motion—is successful or unsuccessful. To label this description as a "human interpretation" would apply to alternative theories as well, namely, "instinct," "stimulus-response," "mechanism," and so on. However, the alternative theories cannot explain non-human organisms acting for an end which lies in the future, nor one which is specific to a situation or context.

What is the justification for arguing for a wide scope for values? In one sense the universality of values is not a problem, since all theories of value agree on a wide scope for value terms. Even the most irrational subjective theories, emotivism and affectivism, would admit that there can be feelings toward any and all goods of the world. In this sense nothing is value-free. Even animals, plants, and minerals have instrumental value in anthropocentric theories and thus value. The basic question for value theory thus becomes how value can go beyond the subject-object distinction as the basis for it in creative actualization.

1) Values are involved in all judgments about what is human, natural, and real. As such they pervade all human activities. The generality of value lies in differentiation. Each instance or good is differentiated on some level from the general not by virtue of their identity, but their exhaustion of different normative possibilities, at least for the present. Any differentiation, whether a distinct good, a quantity, or some qualitative differentiation can have value in some sense, since if it is differentiated, or constituted as distinct, it is considered worth differentiating. Thus value is the ground of the differentiation, including all the differences in any classification or taxonomy. The justification for treating any differentiation as a "value" is value itself, that is, such differentiations are self-justifying if they contain inherent value. Such differentiations have been evaluated and judged as worthwhile qua value, that is, inherently worthy. Indeed the rubrics of the classification are products of critical evaluation and have survived a theoretical or hypothetical competition with alternatives. Thus the ultimate unit of measurement of sand as a "grain" of sand, as opposed to "just sand" is prescribed: a norm reflecting evaluation. A determinate mathematical quantity is prescribed (McDonald, 2004).

2) We are justified in bringing any human innovation and creation under value, for it is creatively actualized for its worth and different and novel from what was previously actualized in the world, a novel good. By extension, this applies to goods creatively actualized by other species, honey by bees, for example.

3) Further, anything which affects human life can be taken as a value, since it affects human life to some degree, whether for good or bad, or better, and so on.

4) We can bring all under values as different ends and purposes may conflict; to give each priority is to accord them equal worth; to subordinate one or another is to assign them different worth.

5) Values are distinct from ends or purposes as their ground. However, goals tend to become evaluated in a hierarchy of importance or ordered in time; one end is pursued over another as more worthwhile. Thus ends, goals, and purposes are pursued in terms of their value.

6) Identifying an instance involves a normative evaluation, that is, the best classification of the instance as a "token" of a norm. Ironically, this identification also differentiates the instance, placing an instance in a relation of identity and difference: a normative identity distinct from other normative identities. This flower is a lily not a tulip. Values ground such correlations as superior evaluations of correct identification, in normative judgments.

7) Only the value of differentiating or identifying, whether more formally or informally, can justify such a creation of differences and identities. Value justifies the "bringing under" a norm, that is, identification and corresponding differentiation from other norms, including judgments, which bring instances "under" a normative identity.

8) Bringing an instance under a category of value: economic goods, religious, social, artistic, and so on—must be adjudged as a type of evaluation and be judged in terms of its value. In this case value is assigned to the instance. Normative hierarchies also involve categorization, for example, bringing a species under a genus. To bring under a category of value is to bring under a value. To distinguish the value of an instance from these kinds of value or from previous normative identities is to distinguish its value.

9) To subject anything to evaluation is to bring it into human purview in the context of such categories, that is, as known, beautiful, useful, healthy, and so on. These are distinct kinds of value, not necessarily identical with other categories in worth, nor differentiated in terms of hierarchy. To categorize in this sense is to differentiate value at a specific level. An adequate theory of categories covers all instances in scope (McDonald, 1997, ch. 4).

10) The grounds of categorial differences in worth are evaluative. It may well be that one categorial good is rated higher than another is by persons or cultures, for example, religion. This ranking activity is also evaluation: differences in worth. Evaluations must be grounded in the world and thus the ability to evaluate differences in worth or identical worth involves bringing all under value. Everything can potentially be evaluated and then grounded in the world as value.

11) I am not arguing that bringing everything under value is to equate everything with value. To bring an instance under value requires justification. A tree is a tree. However, this normative identification must be justified. My argument is that the justification itself must always be evaluative. This involves scales, hierarchies, "rating," and the like for at one level the tree may be treated in one respect (for example, by biology) and in another context, in

another respect (no intellectualist hierarchy here as a priority, without evaluation. One character of an instance is worth as much as another. A diamond is as much a beautiful good as crystallized carbon.) The justification for such normative classifications, must be superiority to other possible classifications, thus involve evaluation. Even levels of classification such as family, genus, and species must be justified axiologically. Only its value or worth can justify such a classification. To attribute a character to something is only justifiable in terms of its fruitfulness, usefulness, and the like, that is, its value. Worth must be justified and demonstrated, not just asserted. Thus its ultimate ground is the value of the evaluation.

12) Ultimate justification must be in terms of value, for to question value as ultimate justification presumes value. Since philosophy is concerned with ultimate justification, it is a value discipline as its basis (McDonald, 2004). To choose some other basis for philosophy, for example, being, or the subject, is to evaluate it as more basic than value. However, this presumes justification, that is, the value of such an evaluation. Thus it presupposes values as the ultimate justification, that is, a basis in value.

13) Ultimately the question of whether we can bring everything under a value is a question of value. Values are self-justifying if they are basic. To question the value of values as such would be absurd. Value is presumed in the question.

14) The value world includes elements of the world that are not transformed precisely because of their value, for example, the Grand Canyon and like landscapes. They are retained as they are for their inherent good. Although value is tied to action, the Grand Canyon is beautiful without respect to human action or to time. It was not formed by past actions, although the "action" of the Colorado River in forming the Canyon is a metaphorical sort of action, distinctive as without agency or purpose. Again, world has value apart from human valuing and as the condition of human valuing. I will return to this point in a later chapter. The point here is that value and evaluation can include the entire world in their scope.

Creative actualization of a good is as inherent in the good. Value is located in the goods, for without creative actualization of the good there is no value, apart from inherently valuable actions and functions. Mere desire for value is not enough, since only a specific good will satisfy the desire, not other goods. Thus the special good inherent in the good is required. A flawed good is as useless in such a situation as a different good: a plane which will not fly when the desire to fly has been evaluated as a worthwhile goal is as valueless as having a gold watch but no plane. The good of a plane is required, not another good, and this kind of good is inherent in good planes, not other goods. Other goods have their own inherent good. The goodness must inhere in the good or it is only wishful: wishing or desiring that a watch could turn into a plane.

The creative actualization of possibilities may not belong to the present "reality," however, that delimited by "being," "existence," "states of affairs," and the like, since their creative actualization may lie in the future. This shows the limits of philosophies of "being," "existence," "states of affairs," or other substitutes and tokens for ontology, since they can neither account for changes and transformations out of "being," "existence," "states of affairs" as a ground: what is creatively actualized is new; nor include them in the inadequate and incomplete inventory of their very limited world. The world of "being," "existence," "states of affairs," and the like, is a small one, a slice of time capturing only one narrow range of possibility. Creative actualization, oriented toward multiple possibilities, is far larger in scope.

3. Modes of Value

What unites all values as values? What is the "one-in-many" of value if we distinguish instrumental values, intrinsic values, kinds, cultures, and personal values? What is the identity of the ultimate term of differentiation? As I argued in the previous chapter, value is identified by creative actualization in different modes and loci. Value as a descriptive term involves creative actualization, when potential worth has become inherent in goods. Modes of value may be described as the ways in which creativity is actualized, the different ways in which novel value is brought into world. Creative actualization is a distinctive mode, independent of previous metaphysical modes (I examined the modes of other theories of value in the Preface). Although creative actualization is a common character for all value, it may be brought about in different ways. Many of these have already been mentioned, for example, quality. In this section, the various modes will be briefly outlined.

The first distinction of modes has been utilized in the previous sections. This includes value, function, and good. As I use these terms, value is successful creative actualization, while good is any end-product creatively actualized. The first plane included successful action in the first creative actualization of flight, and also produced a good, the plane itself. The proof of its success was its continued functioning as a plane, a fact not evident to perception of the plane on the ground. While goods are distinct from value in the abstract, they are united in practice, since the functioning of the good reflects its value, its creative actualization as good and thereby as capable of successful functioning. A pot that is of good quality continues to function as a storage container and was creatively actualized with this functioning in mind. The success of the creative actualization endures as the good, since the good continues to achieve the success. Each use of the good involves repeated creative actualization, achieving new successes. The good is a means to new functioning, actions that have value. The use of the good as a tool is inseparable from the action of achieving new ends.

Some modes of value are universal, while some are confined to specific kinds. Universal values are defined in contrast to specific values. While they are normative, they are of a general sort, since norms can be specific or general. An example from above is inherent value, which all values have as creatively actualized. Inherent is defined by contrast with extrinsic, that is, value which is derived from relations or other values. Inherent value and extrinsic relations are universal, common to all kinds and individuals. Every creative actualization is of an inherent value with relations to other inherent values. Thus normativity of this sort is also universal in the sense that inherent value and extrinsic relations are universal.

Since such relations are universal, they are necessary and thus required. Required values can be actual or evaluative. Required values can move from potential to actual through action, another mode which will be covered below. Required values include moral values as obligations, that is, what is required for values as a condition. Such required values are universal conditions of other values, that is, necessary for their actualization. Moral values, I will argue, are the basis of any culture, that is, the ultimate values without which it could not continue. All other cultural and social values are possible because of moral values; they are thus inherent in cultures and societies. Required values are the condition of the products and modes of cultures and societies, such as ways of life, thought, and action. None of these products and modes of cultures and societies are value-free or independent of morals.

Does creative actualization require a fourth premise of necessity? Immanuel Kant would argue that since such values are necessary that they are unconditional or absolute. However, it does not follow from their necessity in relation to culture and society that they are absolute. For they are contingent in a sense upon such societies. Another requirement is that every good is a good of a kind, that is, actualizes a normative identity. Norms are universal by contrast with transformations by action, since actions give a new norm to a good. Transformation of an empty, urban lot into a garden gives a new normative identity to the land. The normative identity is creatively actualized by the transformation and thus can be distinguished from the transformation. Transformation by action may involve experimentation or trial and error to achieve the right effect, the origin of more formal experiments in science. Experimentation is a creative effort whose end is novel and thus may be uncertain. Duplication, by contrast, is of the tried and true, the repetition or copying of a previous success, and thus includes normativity. Values that require transformation are contingent or consequent; they depend on the action of transformation. Action itself may also be a distinct mode of creative actualization, when the value attaches to the action. This involves the distinction of inherent from instrumentally valuable actions. Repetition is a distinct mode of creative actualization and gives evidence of world: that goods can be creatively actualized as copies in our world. Goods can be creatively actualized as copies, identical instances of models. Thus norms and principles can be creatively actualized.

Some norms which are creatively actualized are universal, while others are general, but limited to specific kinds. In one sense, universality may be limited to a kind, for example, in the statement "all apples are red," red is universal to apples but not universal to all goods, since oranges are not red. Thus "universal" is an equivocal, that is, the concept can be ambiguous in meaning. As a specific kind an instance does not belong to universality in the sense of all actualizations but only specific kinds. However, generically specific kinds are normative and thus qua norm are universal: each instance creatively actualizes some norm and thus normativity as such, apart from any specific kind, is universal.

The instantiation of norms is their creative actualization; goods often combine identity and difference in one instance, that is, a distinct instance as belonging to a distinct kind. Instantiation of norms is a distinct mode of creative actualization, but is also universal, since all instances can be identified as belonging to a kind, species, or other norm. While goods may be singular and some art works exceptional, they nevertheless actualize normative identities at a specific or generic level. For example, Michelangelo's Sistine Chapel frescoes probably could never be duplicated and thus are singular. However, they are paintings, art works, and frescoes, and thereby actualize norms as well. Thus the instance may actualize creatively in more than one mode. A hammer instantiates the normative identity "hammer," perhaps a specific type such as "baldeen," "sledge," and so on; qualitative factors such as a good price, reliability, resistance to rust and decay, a sure-grip handle, and so on; and also its inherent value as an instrument. Further, the transformative method of creative actualization of a new instance of a kind is usually distinctive to that kind, for example, making hammers is different than making china—to speak nothing of making music.

The mode of actualization of such norms is not tied to any one "reality," that is, possible state of the world, since duty and other norms may be actualized by different ages, at different stages of historical development. Actualization of norms involves teleology however, since we aim at actualizing a duty in the future. This applies to evaluations in general: what is deemed worthy and made into a goal becomes the aim of action, and the world is transformed in some respect by such actualization.

Some values may not be physical goods, but actualized as ideas and standards. Whether a particular good meets a standard may not be apparent from its appearance. Rather, it may require examination, comparison, testing, and the like to see if it meets some precise standard. However, creative actualization may be in terms of some such standard that is creatively actualized. Creative actualization of standards is a distinct mode of creative actualization of norms. A specific type of creative actualization of standards is standardization, where standards are met by a series or each copy. Standards are not transcendent but historical and cultural, with a historical basis or ground. Each age has its standards for evaluation and attempts to creatively actualize them; each culture its norms which it enforces. However, social norms are of distinct types which will be discussed below.

Ideas may also be creatively actualized by the imagination. What is imaginary may become actual, for example, flying machines, constituting a distinct mode of creative actualization. In one sense creative actualization involving the imagination is paradigmatic, since all creative actualization involves imagining a goal in which what is imagined is creatively actualized. Similarly, creativity with regard to means involves imagination.[14] Ideas may be actualized in "non-physical" form as well. The mode of creative actualization of many kinds of intellectual and cultural values is non-physical, for example, poetry, philosophy, and scientific theories. Although these may be written down on a physical medium, a piece of paper, the good is not the paper but the ideas expressed on the paper; not the ink but the content. Again, this argues against "being, existence, states of affairs" since poems are not things or entities.

The mode of creative actualization which contrasts with universality is specificity. Goods may be divided into kinds and the evidence for kinds is that such a division is the presupposition of value theory: that there is one good that the different kinds have in common. We may argue that a diamond is an economic good and friendship a social one. These imply different categories of good, general kinds which cannot be reduced to one another and are creatively actualized in different ways, categorially distinct ways. Economic goods, for example, are made as a product, but social goods are not. Categories of value must include all kinds within their scope in order to constitute an adequate theory of categories (McDonald, 1997, ch. 4). Such categories are not categories "of" in a relation to a first, but different categories, delimiting the basic kinds of values and goods. While they require creative actualization as instances, and can be "predicated" of such instances, their mode of creative actualization is distinct from instances. Each category is inherently worthy and thus distinct from all the others; but the categories or kinds of good distinguish goods at another level than individual goods. They creatively actualize normative differences between kinds. Friendship is a relation, while a diamond is not. More, friendship cannot be reduced to an economic relation so is categorially distinct. As an old friend put it, "you can't buy health and you can't buy love." Again, scientific truth cannot be reduced to economic prosperity any more than prosperity can be reduced to truth. But these values are presumed by their respective activities as general norms over their field. They are independent of any one scientist or economic enterprise; variables independent of persons.

Subjective and psychological theories of value would argue that such goods are all based on desire, feeling, interests, and so on. However, such theories confuse personal esteem with inherent worth. A diamond has inherently valuable qualities whether any person or individual values it or not. Personal values are an independent variable, since the diamond may or may not be liked by different individuals. Usually the worth of some good is argued due to its inherent quality, for example, its usefulness or superiority to previous goods. The value is categorial, not personal, and this can be a mode of necessity, not choice, the rubric for personal values. One cannot mine coal by playing the violin, no matter how much one may will, desire, feel, or emote about it, take an interest in it, or have "correct attitudes" about the violin, for economic goods and aesthetic are categorially distinct. It is necessary to use specific methods to mine coal, and these are "categorial imperatives." True, new methods may be creatively actualized. But these are also independent of the whims of a subject.

Each category may have distinctive modes of creative actualization. These include production of goods, as I noted above, particularly in economic life, art, and entertainment. However, some goods in these fields are not produced, for example, the Grand Canyon as an aesthetic landscape. Others are merely gathered from the world, for example wild berries. Others may involve personal evaluations of beauty or economic worth, for example, willingness to pay for "pet rocks" or judgments of beauty regarding a weed on the part of certain individuals. While creativity is meliorist, deactualization of many landscapes would hardly be an improvement in the world. The actual is not limited to the novel, and includes previous actualizations, goods that have already been actualized, as well as goods of the world with inherent value. The Grand Canyon is an actuality with its own inherent value. Thus beauty may involve a categorially distinct mode of actualization: it can be purposive or random, by chance. Nevertheless, separating out the Grand Canyon as a distinct formation is its creative actualization as a normative identity.

Intellectual goods are not "physical" and may involve unique modes of creative actualization, for example, prediction. The prediction of the planet Neptune was a good prediction of a novelty which was, in effect, creatively actualized by the prediction, since the actuality of Neptune was not known prior to this time. From seven planets eight emerged, changing our view of the heavens. Similarly, Galileo's telescope expanded the number of "moons." Knowledge then may be of actualities or potential novelties; new knowledge is creatively actualized through good predictions while bad predictions fail to creatively actualize what was intended: fail to achieve their goal of expanding the novelties in the world. Scientific prediction is a distinct mode of creative actualization, categorially different from others. It does not involve production, but often creatively actualizes new instances or kinds (species) in a distinctive way. While it is not necessary that each category have distinct methods of creative actualization, they generally do. Similarly, categorial values are not the same as social and cultural institutions, but we can classify these institutions in accord with categorial values, for example, religious institutions, economic institutions, social institutions like the family, and so on.

All inherent values are either intrinsic or instrumental since these values belong inherently to instances. The bearer of intrinsic value is any self-actualizing agency, that is, any living thing, whether human or not.[15] Particular uniquely beautiful aesthetic values may also be intrinsic, although this may only be a unique form of inherent good: the irreplaceable. The beauty of especial landscapes and artworks is so unique and exquisite that they may approach intrinsic good. This beauty may be intrinsic not as self-actualizing but as an end valuable in itself. Such beauty as that of the Grand Canyon is worth viewing for its own sake.

While any value or good can be used instrumentally, instrumental values in the proper, more-narrow sense are not self-actualizing. Tools are useful inherently but not intrinsically, since they do not preserve themselves or attempt to remain alive and survive. Instrumental values have inherent value as instruments but also value for agents or value in relation to intrinsic value, extrinsic value. The value of continuing actualization attaches to both instruments and lives. We create strategies to prevent the decay of useful things (painting, refrigeration) and lives (medicine, nutrition). Thus they escape devaluation. However, such strategies must deal with changing circumstances. A tool left in the rain must be cared for; a plague must be fought.

Some instrumental values and goods are required while others are chosen. Those required for survival differ from those that are optional, for example, food and nutrition as opposed to cuisine. Survival values are "imperatives" or necessities given the intrinsic value of self-actualizing agents, that is, contingent upon the value of these agents. They are instruments of the continuation of life: preservation of life is the end of such values for self-actualizing agents. However, they are not relative or arbitrary: living things can eat definite kinds of food and will die if they do not. Rocks are dandy, but have no survival value as food, whether or not we emote at them. The actions of a self-actualizing agent always include survival or self-preservation; in turn the self continues to be grounded in the world by the series of actions which enable it to survive, but is not ended by them. Survival is a distinct mode of creative actualization, then, involving creative actualization of the necessities of life in different respects. These may involve gathering products like food but also the actions involved in defending against predators, that is, it may apply only to some actions, like fleeing. While life ends at death, survival is not creatively actualized by death. Thus creative actualization as value is involved in survival.

Instrumental values which are not required for survival may nevertheless enhance life and improve self-actualization. Such values are not necessary so entail some freedom of choice.[16] One argument for choice is cultural difference, for example, cultures with and without incest taboos. A limit on such freedom includes available materials and the historical state of the culture. We have access to modern technologies, science, and the like which ancient cultures did not. Again, corn was available to Native Americans prior to 1500, but not Europeans. As I have noted elsewhere, previous forces cannot determine such choices, since they create novelties in a future. Societies and cultures can make these choices or leave the decision to the discretion of individuals. Social actualization is distinct from individual since it reflects social forces; indeed, individual values are, initially, socially instilled. Incest taboos are not individual but social and cultural, and the individual who is repulsed by incest learns this value from their culture. (I am equating "a society" with a culture in this context but since society is an equivocal term, it has several other senses that are not equivalent to culture.) Social norms involve extrinsic relations of self-actualizing individuals, which are shared by the individuals within the culture and creatively actualized by them in their lives, a distinct mode of creative actualization.[17]

Creative actualization may also be individual and personal. One individual may imagine a project and creatively actualize it, reflecting personal values. An attempt may be made to creatively actualize personally valued inward states such as happiness or peace of mind. This is the proper realm for what the tradition has named under "subjective" values, those connected with feelings, moods, and inclinations, all of which must be creatively actualized within the person. Happiness is not only a matter of passivity but is actively pursued and its conditions creatively actualized. While it may be difficult to separate the purely individual from the social, at least some values and projects can be assigned to individuals, namely, those which are truly autonomous from society, "self-legislated"—and those which although they reflect social norms in some sense involve a novel approach to such norms. Indeed, "autonomy" itself is socially conditioned as are duties defined by a culture. The individual can often decide whether or not to follow such social rules, but this may itself reflect a loose society, which allows deviant behavior, and a strict one that does not. For example, fundamentalist Islamic regimes carry out such sentences as stoning of adulterers, execution of gays, and cutting off of the hands of thieves, unlike more relaxed regimes in the same countries which winked at such actions (the Shah; mullahs in Iran).

The idea of invention was socially approved in the early twentieth century, many inventions pursued by several individuals and reflecting social input. However, individual projects, for example, an air machine, had different forms of invention and success rates. Such projects were pursued out of the worth in which the individual held the creative actualization of the goal. Further, some experiences are shared but the permutation of all experiences in one individual is almost certainly different from that of another. Inner states are not shared, even if they can be communicated, which is precisely why value cannot be grounded in the subject. There is no way of knowing if two individuals are talking about the same inner state, since they cannot be compared.

Personal and individual values are grounded in the world, as the person is in the world. However, the inner constitutes a distinct realm that is not public, and thus such values are not creatively actualized in the same respect. They are indeed grounded: the project or state is creatively actualized or fails to be achieved as a goal. The goal is creative in the sense that historical circumstances are involved—Europeans could not personally value chocolate before 1500, and novelties involving chocolate, such as cakes and ice cream, had to be creatively actualized. The goals are actualized both in the sense that they are grounded in the person who is in a world, and that the goal is a value, one which has been personally evaluated. But although the goal is actualized and pursued because it is personally valued, it is not a mere wish or desire, since the agent acts upon the goal to actualize it. This may involve personal evaluations of quality both before and after successful creative actualization, including evaluation of different goals and later of results. Value is distinct, then, from mere feeling, desire, attitude, and other psychological states even on a personal level. However, some personal values remain private and inner, and thus autonomous from other modes of value.

Instruments have their own modes of value. These involve quantity, quality, functioning, and techniques. Quantity and quality are independent variables since a good may be quantitatively one but qualitatively superior to another and inferior to a third good. An individual good or instance of a good has been creatively actualized as a single quantitative unit. Quantity may constitute a unique good or an instance of a kind. This good is inherent in the instance apart from comparison to other goods, that is, extrinsic evaluations. Otherwise the instance is defective and creative actualization has not been achieved. Defects may be inherent not comparative: quantity of good is thus distinct from comparative good. Comparative good is evaluation of relative good, that is, of inherent goods in relation. However, such evaluations must be grounded in the value inherent in the good, not subjective whims, or they are ungrounded: groundless. Such comparisons are less a mode of value than based on some mode of value, then, although they can lead to improvements that are creatively actualized.

Quality is an equivocal and may refer to several distinct creative actualizations. In one sense all creative actualizations may be improvements and thus improvements in quality. Quality may then be the basis for projects of worth, that is, worth differentiating in terms of such qualities. However, as I noted above, quantitative differences in the form of instances of a kind are also creative actualizations but an independent variable. Creative actualization of a quality is a distinct mode of value, then, with several varieties.

Quality may attach to a part or a whole.[18] A whole good may be of inherently good quality, for example, the Sistine Chapel frescoes. It may also attach to a part, namely, some improvement in some part of a previously actualized good, for example, improvements in efficiency. A whole of quality, a good plane, is not an attribute of a substance or being, since it can be predicated of the whole. The quality is inherent in the whole and cannot be separated from the whole as a functioning whole. We may value green and prefer a green-colored car, the creative actualization of this color in a car as a bearer of value. However, we may value a whole as well, namely, a car as a whole, not just its color. While the color may be an attribute, the car is not and its quality attaches to the whole, not just the part, the color. Quality may also attach to or include creative actualization of a good that endures as a good over time and does not decay, if this is evaluated worthwhile.

Two other types of holistic value are the harmony of parts working together, and symbiosis as co-conditional value.[19] Neither are a denial of inherent value but a recognition of the value of the relation of parts of a whole, a distinct mode. Harmony is a metaphor from music, but works well as an overall description of mechanical relations as well, in that all the parts of the machine must be functioning well for the machine as a whole to work well. Similarly, organic bodies require the health of all the organs for overall health.

Quality may be a topic of evaluation, that is, personal, categorial or comparative evaluations utilizing critical standards. Ideals and standards may constitute goals for achievement of qualities in this sense and if the creative actualization of them in the project is successful, the quality is achieved. Quality is projected as a goal and if the standard is met, creatively actualized. Again, personal values may be generated by critical evaluations of present actualities and this may generate new goals which if creatively actualized constitute some sort of qualitative improvement.

Value may also attach to qualities that involve abstractions such as justice. Judicial decisions which are tightly argued, consider all the evidence and case law carefully, and come to a fair decision are quality judgments, in which "justice has been served." These qualities do not attach to "substances," since there is no "matter" or "substratum" to a decision, except by far-fetched metaphors, for example, the content of a case as its "matter." Thus qualitative evaluations and quality itself are independent of metaphysics and a distinct mode of creative actualization. Indeed, quality is often the ground of the "real," since what is "realized" is the result of projects that attempt to creatively actualize a quality of some sort.

Instruments which "function well" constitute another mode of creative actualization. These are limited to active goods or their use, since a painting can hardly be described as "functioning well." However, instruments that are used as tools may function well while in use, but, like a painting, are left "standing" and "waiting" if not called for. Life too may "function well" as Aristotle believed, but perhaps this comparison is a metaphor. Certainly biological organs function well or badly but applying the epithet to the whole may be a stretch. While a person may be in good health on the whole, this is generally the resultant of the sum total of the health of the organs, and thus the health of them collectively. Means certainly function well or badly but can ends be said to function well? In any case, life uses strategies for survival which involve functioning well. Functioning well is relative to the purpose and if a strategy enables survival, it has achieved its purpose. Also, functioning well may be another variety of quality or involve quality since mere functioning may be distinguished from good and better functioning. While this involves evaluation, evaluation may be grounded in an inherent quality of the good. Functioning well frequently unites action and a good in one activity, in the creative actualization of a goal.

Last but not least are techniques. Techniques are a mode of creative actualization that can be separated from their products, since the etymology of technique is from a word meaning "making." The making itself is a form of creative actualization distinct from the product, the good produced. Techniques are wide in scope and also include surrogate "makings" for example, machine "manu"-facture, and also Taylorist assemblies, that is, breaking down making into limited segments for efficiency.

The elements of any creative actualization can also be described. Creative actualization is the successful grounding of a project, which is evaluated worthwhile. Value is distinct from evaluation since it grounds the potential goods in the world, that is, the evaluation that a goal is worthwhile leads to its creative actualization in the world. As I noted in the last chapter, creativity is novel in some respect while actuality ties action to world as value in the world. Action may involve more than one action as a means or as an end, for example, in building a plane; or complex modes of action in which they overlap. Action normally involves the freedom to act in the sense that we encounter no impediments to action. Thus evaluation, creativity, actualization, action, freedom, and a world are elements of any creative actualization.

The goal is another element which is evaluated worthwhile, a project of worth. The goal of creative actualization is grounding a potential good or value that has been evaluated worthwhile in the world. This involves causality of means to ends but also the unique grounding relation of teleology discussed in chapter one. A ground is both a projection of a critically evaluated goal in the future and the objective or worthwhile purpose of action(s). However, agents actualize these projects in the world only through value, that is, successfully grounding in the world. Generally, agents undertake projects in a specific situation or context, and evaluate the time at which they begin them for the optimum time if necessity is not a factor.

If creative actualization is successful, it brings about complex relations of instances to norms, since each good is a good of a kind, at least at the generic level, for example, a unique painting. Thus it may creatively actualize one or more norms, whether these are simple identities, standards, ideas, principles, or other "generals." The inventory of goods of the world is increased, since novel goods are creatively actualized. Ideally, the diverse goods are compossible, that is, more good is achieved, not less or "breaking even." The condition of compossibility is reciprocity and amelioration, a point I will examine in a later chapter. The goal is evaluated as improving the world in some respect, that is, amelioration.

The world is an element that is transformed by this practical activity but may also be changed in a special mode of creative actualization, theory. Theory normatively prescribes what are the contents or inventory of the world and we can evaluate competing theories for relative worth. Thus the planets are gods in one view (Jupiter, Pluto etc.), and material in another; the "heavenly bodies" as holes in a fifth element allowing God's light to shine through in one theory, and flaming masses transforming hydrogen into helium in another. Successful theories give the elements of the world normative identities in line with their elements. Theories may transform the normative descriptions of the world and thus transform the world in a special way, for example, from when stars were viewed as giant fires to the Einsteinian revolution in which they are undergoing nuclear fusion. Again, breaking off a new species from an old is the theoretical transformation of the world which does not change the quantity of its inventory. In both cases the world is transformed by creative actualization, whether by practices or theories. Creative actualization of a theory is at the same time a transformation in our normative descriptions of the world.

Generically, we can distinguish three modes of actualization. The first is imaginary, in which fantasies, projects, even the impossible (imagining Casper the Ghost) are in the mind, but not in the outer world. Ideas "spring to mind" as imaginative constructions and thus they are only actual in the sense of actually ideas in the imagination. An imaginary \$100 is not an actual \$100. The imagination and its creative ideas have instrumental value, as proposals to be evaluated and possible values that require creative actualization. Evaluation of such ideas to judge if they are worth actualization is required before bringing such creations to the world through action. This mode is regulated by evaluation: which of these is worthwhile dwelling on as opposed to a will of the wisp.

The second is creative actualization: bringing about a good or project in the public (outer) world. This mode is regulated by value, both in the sense of feasibility and worth. Finally, there are norms, which are implicit in creative actualization. Indeed, creative actualization as the general character of value is itself normative. Any project or good is an instance of a principle even as it actualizes that principle. Moreover, principles connect past, present, and future change. Insofar as principles are creatively actualized, they are ruled by standards. Good must be generic; it cannot be specific if all the different kinds (pluralism) can be evaluated and judged with axiological predicates. (Compare transcendent attributes/predicates in the scholastics). However, this generality is normative not ontic. As I noted above, the first case of creatively actualizing any good is at the same time the origin of a norm: a principle in both senses of an originary instance and the archetype for future instances.

In sum, the elements of value are interconnected since value is grounding in the world and the world is its condition, but a world is transformed by creative actualization and thus value is the ground of such changes. Action is generally the means of creative actualization but is grounded in worthwhile goals whose successful completion in the world ends and justifies action. Action also transforms the world but is based in the world. Creation of goals involves imagination but these are creatively actualized as values, functions, and goods. The other elements stand in similar interrelated connections.

4. Evaluation and Norms

I have already distinguished evaluation from values.[20] Evaluation regulates possible values: should a potential good be actualized? Would it be worth pursuing as a project? Would such goods constitute an improvement? Value is the creative actualization of such potential actions and goods as actualities in the world. If they are not actualizable they are without value, although they may retain potential worth. The attempt to actualize them is "no good," whether it falls short, is defective or is impossible. Possible values are in the realm of evaluation: some future outcome is evaluated better based on some standard, possible improvement, and so on. The evaluation is prior to grounding in the world as a creative actualization. However, it has not yet been creatively actualized as a value, that is, successfully grounded in the world. The evaluation is based on this possibility of grounding: of actualizing a value as an instance which is distinctly valuable or good, different in some respect from other, previous creative actualizations. The potentiality becomes inherent in a good only when it is grounded, that is, creatively actualized. Evaluation of potential goods as such is not grounded.

Evaluation includes the judgment that an improvement in some respect will be creatively actualized by a project. Thus evaluation is generally comparative, since even creative actualization of a novelty is evaluated as an improvement of the world. Evaluation is thereby critical, since it is not complaisant with all previous creative actualizations, with the world as of that time, the world of "being." Even after a successful actualization, the critical, if speculative question remains: has true improvement been achieved? Was the project worthwhile? The car works, but what of the pollution? Does making all goods the same on an assembly line overlook quality? Does it capture only economic value and not esthetic—that is, is it good enough? Such critical evaluations may constitute grounds for further improvements.

We should also consider more ultimate critical evaluations, for example, are all goods and values good for someone, in some way? Is the good for humans at issue or should all of life be considered? Do competing versions of what is good for humans (pleasure, self-realization, and so on) indicate some problem here? Does this imply diversity or relativity? Is what is good an open issue, unresolved? What is ultimately valuable for humans? For all of life? In critically evaluating the actual we often make reference to standards, either those which launched the project or those which have become established through custom, which are a species of the latter. Standards are often measurements for ends.[21] However, such standards remain in the realm of evaluation: they are critical tools, not actual as value or goods, unless creatively actualized. Standards are not values, although they can be brought about as values.[22] The latter is an independent variable, however. Standards may be more constant than values, unchanging standards. However, new standards may supercede old ones as a melioristic process.

Projects are the attempt to creatively actualize an end or goal. Thus ends that are goals are not yet a part of world as creatively actualized. Ends that have been evaluated worthwhile as yet have no value. However, ends are those projects that are evaluated worthwhile. Evaluation includes a critique of possible ends to determine which is worthy of activity, of a project. Not all ends are worthwhile, whether due to their worth or their practicability. But some projects may turn out to be possible or worthwhile after all, which at first skeptics judged as folly. Value is independent of evaluation. But ends are evaluated for worth prior to the activity to creatively actualize them. Evaluation is in part a critique of ends.

Evaluation is the logic of freedom since novel creative actualizations transform the world. As novel, these cannot be determined by past conditions; critical evaluation of their worth rules out previous determinations in the process of comparison. The differing possibilities rule out pre-determination of which shall be actualized, since none has yet been and deciding which possibilities are worth actualization requires critical evaluation. Indeed, determining factors must be creatively actualized in the world, since they are without actuality unless they have been brought about in the world. This requires their evaluation as worth bringing about: as superior projects in comparison with other, inferior possibilities. Different possibilities can be actualized using the same time; evaluation of these has regard for the best use of future time not past time. Freedom is the basis for change, for transformation of the world as an improvement. Evaluation, the logic of freedom, precedes actuality.

What then of values which are worthy but not yet actualized? valuable but not yet enacted? Prior to the Wright Brothers success, planes might have been considered Utopian; other projects might be considered worthy but not feasible, for example, universal justice. If they are not pursued, they have not been evaluated as worthwhile: evaluation results in action. However, some projects may be premature and require further work. Others may be ongoing efforts with successes and failures along the way, peaks and valleys, for example, bringing justice to everyone. Standards and ideals that have not been creatively actualized may still be worthy and have potential value. Their evaluation as worthwhile goals for the future remains as an "imperative" or as a project. The potential value of such ideas remains in the realm of evaluation. However, value is creative actualization and bringing about standards and ideals may require as much trial and error as any other good.

Norm is a generic terms for most "generals," which includes universals, principles, rules, laws, kinds, and other identities. Norms also include the deontological realm of standards, ideas, and the "ought" insofar as these have not yet been creatively actualized. Generals are norms since they provide an identity of all instances of which they can be evaluated correctly. These are partly transcendent of their instances, since they apply to future instances as well. Not all instances of a general may have been creatively actualized, for example, future polar bears or their instantiation of the law of gravity. Some norms may never be actualized, for example, standards of perfection. Ideas in the sense of theories, or abstractions from multiple instances are norms, both as generals and thus as providing a normative identity to which instances ought to conform.

Norms are in the realm of evaluation since if they are not creatively actualized they remain mere possibilities. What ought to be predicts what should be but may not occur. Theories tell us what ought to occur, but may not. If it does not accord with the prediction, the theory may have to be altered or revised. The distinction between a theory and deontological norms is that an agent will creatively actualize the rule or principle in practice. Norms are brought about in action and activity that aims at their creative actualization in the world. An act according to principle creatively actualizes the principle, just as an air machine creatively actualizes the principles of flight.

Both generals and deontological norms, which are closely related, are in the realm of evaluation, since they are better or worse covering principles, which may or may not have value in practice, and because they are at least partly potential. Practice is the condition of creative actualization of the norm, that is, the "objectification" or grounding of it as a good. Norms are not fixed in stone, since superior norms can be created. New and better ideas, theories, laws, principles, standards may also supercede them. Duration is not a virtue of a norm.

What is the relation of norms and values? Value is the cause or ground of a norm, idea, end or other practical possibility becoming actual as well as the moment at which it is actualized in the world. The historical moment at which the ideal merges with the world and thereby transforms it is one mode of value, creative actualization of norms. Value is achieved at the point of success where what "is not" is changed to what "is" or more strictly where what "was not" becomes actual through creative means. Similarly, principles are manifested in the world through creative means in which creative actualization of the principle becomes the goal. Thus there is a close relation between ends and norms in practice, namely, that all norms require ends while all ends presuppose norms. The use of rules and principles aims at some end by which it is governed. Yet the end is of a kind and thus normative.

Values and goods can be recreated as copies identical in most respects to an original. The precision with which this can be accomplished has increased with time. Thus it might appear that creative actualization belongs to value theories which stress identity over difference. However, the creative actualization of new kinds results in plurality, not identity. Duplication requires plurality and thus is secondary, since an initial creative actualization of a kind is necessary for its reduplication. The kinds are not instances of a universal, of an identity, but plural. Universality and generality are of the character of norms, instead of values, which are manifest as inherent in bearers. Thus norms that have not been creatively actualized are in the sphere of evaluation, not value: subject to critical appraisal to determine if actualization is worthwhile. Their creative actualization is as a unique novelty, distinct from other values and thus adding to the different kinds, plural goods. Norms are actualized in the world as inherent in an instance.

However, their creative actualization is distinct from their potential worth as norms. An idea prior to application has only hypothetical or potential value. Theoretical ideas are evaluated as "covering laws" for a relevant set or field. They are judged better or worse than competing ideas in terms of accuracy, scope, prediction, fruitfulness, and other standards. Ideas in the sense of projects are creative but not yet actualized: the imagination has not yet been made actual. The idea has not been tried in the world, so it is in the realm of evaluation, not value. The idea would be valuable if it could be creatively actualized and its potential worth may regulate our activity of trying to bring it about, which makes the idea normative. If someone tries it and it works it has value. However, the idea may or may not have personal value to individuals who work to bring the idea about, as a pet idea, no matter how many failures they are faced with. Further, the confusion of norms and values, that values are deontological, or what ought to be, is akin to the confusion of value and evaluation.[23] Values mediate what "is" and what ought to be as actualities which may manifest or bring about norms in one mode. Value is achievement of the principle in the world, not a possible principle. Creative actualization is the practical achievement of the norm: the "objectification" or "realization" of it. Due to creative actualization, norms that "were not" now "are" actual.

When we act upon an "ought" we actualize it. Thus oughts are not only some transcendent standard, for they may be creatively actualized over time and intervene in history. Rights, justice, truth, and beauty are not beyond our grasp but can be creatively actualized in the world, transforming the world. To enact norms by creative actualization through pragmatic means is to change the world. Standards and principles can be brought to earth and grounded: creatively actualized in practice. Values are grounds of action which alter the world for the better, since they actualize what ought to have occurred or should occur. In turn, value is the measure of which norms are practicable, at least at a point in history. Standards and other norms must prove their worth through actualization. "Oughts" which are actualizable become values, that is, creative actualizations. To creatively actualize a principle or other norm is one mode of value. As I already noted, those that cannot be actualized may be Utopian. They may also be worthless although they may remain as critical standards. There are also good ideas (flying machines) and bad ideas (time machines) whose test is whether they work out in practice. However, it is impossible to rule out such ideas altogether, since at some point in the future they may be actualized.

Since norms in the form of "oughts," deontological principles or directives, can be creatively actualized, or brought about in the world, the distinction of "ought" from the actual is attenuated. The distinction of standards from facts is also attenuated and with it the standards of philosophy and evaluation from the facts of science. Knowledge is the creative actualization of facts, or critical knowledge that has been proven, or at least for which there is a great preponderance of evidence, in accord with the standard or norm of truth. Philosophy dwells in both, as does science. But philosophy concerns the whole not just particular fields or even just knowledge (art, psychology, etc.; McDonald 2004). The actual may reflect previous norms that have been brought into the world through action. If we can actualize norms, norms can be creatively actualized and "ought" has become "is." If we say, "the lawn needs mowing," so "I should mow it," and then do so, then the lawn has changed from unmowed to mowed. We have moved from an "is" to an "ought" back to an "is". Creative actualization means that what is creatively actualized functions as it should function: the norm and the instance, "ought and is" are combined as actual. The argument is not that there is no distinction of norms and actualities, but that the two can form a composite or synthesis if norms are creatively actualized in the world. Thus creative actualization constitutes an argument against the absolutizing of the distinction, as if the actual and the normative were in two different universes, one natural and one "non-natural." If they were, logic and math, both normative, could not be applied. They would be useless. "Oughts" and other norms regulate practice, where some good is put into practice which was not but can be. If this good is successfully brought about or creatively actualized, then it is good. An "is" is made of an "ought".[24]

Creative actualization, then, is neither idealist nor realist, for it dissolves the distinction. The "real" reflects previous creative actualizations and one mode of such values is creative actualizations of norms in the form of ideals. Following pragmatism, creative actualization seeks a middle road: pursuing the possible with an eye on the actual, but the heights of actuality not the depths. Mere idealism, which includes speculative rationalism, can ignore the actual in favor of the "counterfactual" and thus the value of what has been achieved. While idealism is correct to emphasize higher values, it judges the actual successes too harshly and often without a fair appraisal.[25] Mere realism, which includes empiricism, wallows in the actual without trying to improve upon it. While realism and empiricism correctly emphasize the truth of the actual, of what values have actually been pursued, they ignore the developments leading up to the actual. The ideal is made real by creative actualization, attenuating the distinction.

Creative actualization transcends the actual by creation of novelties; values are not ideals for they are immanent, and without creative actualization no ideals would or could be part of the world. Creative actualization overcomes the "opposition" of transcendence and immanence by actualization of transcendent norms, ideals, hypotheses, and principles. Opposition is superceded by revaluation, in which the transcendent is judged worthwhile and practiced as immanent. Creative actualization revalues and transcends the actual with novel transcendental principles and actualized norms. Reality is remade, revaluing realism by demoting it to emergence and consequence.

I argued above that the goal of duplication creatively actualizes a norm or principle. An identical kind is sought as a project that copies the relevant features of the original. Thus normativity is worldly not only "mental," that is, imaginative. The question remains about whether creative actualization is concerned with the type or the individual act, identity or difference. Creative actualization in one mode can be the actualization of a kind as an instance, the actualization of norms in an individual as a composite of the identity in the distinct instance, different from others qua individual. Value as the most general term of difference includes creative actualization of differences in kinds but in single bearers of an identical principle. Action and activity can embody generals as copies. They are individuated by action: dependent upon different acts over time for their actualization. But acts also fall into a pattern of kinds, embodying an identity as a kind. Types of action are categorially defined as of one type or another, whether in terms of general activities, such as "economic," and "political"; or functions, such as climbing stairs, or cooking. Thus the opposition between identity and difference is also attenuated, a point I will examine in more detail below. The question remains whether creative actualization is confined to bringing about a novel type or can be concerned with a unique individual. While the latter is rarer, it does occur, since art works, experiments, and predictions can be singular. These are distinct modes of creative actualization. What is also singular is the time of each creative actualization: values ultimately differentiate in time, transforming the world over time. Finally there is the individuality of the bearer.

Identity refers to both the instance and the kind as their relation. Identity in one sense means that an instance is itself or identical with itself. In another sense it means the identity that an instance has as a kind, which identifies its species. Both of these senses also involve differentiations, however, since to identify an instance itself differentiates it from other instances, and to identify a kind is to distinguish different kinds. I am arguing here for the correlativity of identity and difference: to identify is to differentiate; to differentiate is to identify, since in the latter case one is differentiating out an identity. Correlativity is primarily in the normative dimension, since such identities are norms, that is, they apply properly to all instances. However, there is also the relation of such norms to the actual as instances of principles, and other generals.

The question for our theory is, which is value, an identity or a difference? The answer is both and neither due to the correlativity of identity and difference. In some respects, value is the identity element of all creative actualizations, since they all create value in some respect. Again, creative actualization of a kind, a principle, or a copy is creative actualization of a normative identity. The kind is different from other kinds; the value distinctive. In another sense, creative actualization differentiates the instance as novel and as distinctive, and thus as different. Value as creative actualization has elements of both identity and difference. However, as I shall show, value as the ground of identity and difference means that value is ultimately beyond either, since their actuality is based on their value. In a sense, value as the most general term of differentiation is an identity-difference.

Creative actualization is also beyond the identity-difference opposition since the latter is a normative requirement. Goods of the world combine identity and difference insofar as they combine such contraries in a distinct instance of a kind. The opposition is, again, in the realm of evaluation, not in the world of values. Thus a good is beyond identity defined as the contrary of difference, the normative realm, since it has been creatively actualized in the world, changing the world. I am not arguing that contraries, for example, "cold-hot" can be creatively actualized as such, as an instance. However, value is to some degree the combination of identity and difference, and thus overcomes their "opposition."

I argued above that value is the most general term of differentiation. Creative actualization is the creation of actual novelties, which previously were not part of the world's inventory. To bring about a flying machine in the world is to create a novelty and this makes the world different than before: changed, if you will. There is a new machine in a new historical time and this marks one actualizing of the difference of the new historical point from previous circumstances. Value as the most general term of differentiation is the historical point of differentiation, in which a difference is creatively actualized.

Again, values manifesting norms as principles, bringing them about, creatively actualizes new kinds as distinct. Copies of planes are made which were not actual prior to that point and which are different than prior historical times. Creative actualization of flying machines in general is their differentiation in another sense then, as a category of different. A whole new kind has been creatively actualized which is different from all previous kinds. This category is finite as different since this one is distinct from other categories, that is, it does not encompass the whole, but a limited part in relation to other parts as different from them. In this respect, "values" in the plural imply differentiation, since otherwise there would only be value. These differences are not different attributes of an underlying being as substance, since then they would not constitute differences. There is no *pros hen* to a substance of the different qua different.

However, the machine has been identified as a flying machine prior to its actualization, since this delimits the goal. Creative actualization of a distinct instance of a distinct kind also involves its identification as a distinct kind, the normative identity, "flying machine," "airplane," and other synonyms. Reduplication further identifies it as a kind and creatively actualizes a principle, which is a species of normative identity. Values can be recreated, which brings a general to actuality. Further, value as the identity of all individually or categorially distinct goods is necessarily general and abstract. Value itself as a generic term identifies what is common or identical to all such goods, principles, and kinds in some respect. Since creative actualization is a generic characterization of value, it is itself a principle or general of a sort, although a principle of all principles which can be added to or part of the world, that is, regulative or normative over transformation for the better through action. Values are principles by which we choose and act in another sense, that is, evaluation and creative actualization in general. As principles, they are normative, since principles are a species of norm.

Since the creative actualization of novelties is both differentiation and identity, the problematic of identity and difference is a dead end, at least with respect to value, since they are correlative. The instance as different from others is itself, self-identical. Differentiation of a kind is also of its distinctive identity. Neither grounds the other, as they mutually imply each other. There is no sufficient ground for distinguishing identity from difference based on identity and difference, since each identity is a differentiation of a sort and each difference an identity. Thus to approach the problem from a logical perspective, that is, from the normative dimension, is a dead end, since neither are ultimate nor grounds.

I would argue that correlativity is required for identity and difference as normative. Identity and difference are in the normative dimension, which requires its own rules, since it is not necessarily actual. Their actuality is subject to creative actualization. A flying machine as a type would be different from one that did not fly whether or not the machine had been actualized, that is, in "theory," or in "principle," the normative realm. Similarly, truth and falsity are truth-values in logic apart from their content. (Peirce, F. C. S. Schiller, and others have argued that logic is a normative science). A normative realm of pure identity would not encompass differences, and could not account for different norms and different identities.[26] Similarly, a realm of pure difference could have no kinds, no principles, no norms, and no concepts. Thus it would not be part of any normative dimension.

I argued above that value includes creative actualization of norms as one mode of value. Thus the actuality of the normative is grounded in and by creative actualization. Value includes creative actualization of norms as a composite of identity and difference. Value provides the ground upon which identity and difference meet as an instantiated norm, an actualized identity. Value is the point at which the relation is actualized in the world as an instance. The composite is brought about as actual: identity combined with the different. Although this sounds paradoxical, since identity is the contrary in some respect of identity to what is different, the paradox is in one respect the value of value, since it can combine contraries in an actual instance. Identity and difference are never combined in the same exact respect, except insofar as any identity is different than another. But an instance of a norm, identity, principle, or standard combines a particular, specific or generic identity with a distinct, different instance. A plane is both one individual, different from every other, and the same as itself as such an instance. The aircraft is also the creative actualization of the normative identity "flying machine," which differentiates it from other kinds of machines. Value is not identity or difference as such but grounds their relation. Only those correlations of identity and difference that prove feasible have value and can be actualized in the world. As a project, the norm is the identity element and achievement of the goal is the differentiating element. Since values as the ground of the relation of identity and difference establish their relation, value is neither identity nor difference, but the ground of their actualization as a composite in the world: the instance as a kind. Each such correlation that proves feasible is valuable, or a good in the world, including actualized norms, oughts, principles, and concepts. Since qualities as attributes are principles as general, they are in the realm of norms, not value. Thus values establish relations between seeming incompossibles, identity and difference, as novel actualities. That a norm can be actualized as an instance is to the good: good that they are compossible in world; wonderful that the normative can be actualized.

Nevertheless, in one sense value is an identity, as I noted above. Value was also described as the most general term of differentiation. Which has priority in relation to value? Since value is creative actualization, priority must be given to difference. This evaluation can be made on several grounds. The first is that creative actualization involves the bringing about of the different. True, actualization also instantiates some norm. But the point at which it successfully emerges is creative actualization. And as a type its value is a novelty in the world. Thus it changes the world, since the world is thereby made different through creative actualization. A novel value is differentiated from the whole, the world. This emergence of the novel is a historical differentiation. The novel has emerged as a revelation. Subsequent actualizations, that is, of principles, are not as novel or different although they are novel in some respects. They are not different from the first breakthrough in kind nor as unique events in history. Value marks a historical difference, the successful actualization of what "was not," and thereby historical change, grounding novel identities in an instance.

Second, because identity and difference are correlative normatively, their conjunction in the world is a determination of the actual, that is, both an actualized normative identity and a historical determination. The world is given specificity in the changing actual. History is determined in a choice of what possibilities have been evaluated worthwhile. Value as the ground of identity, including principles, justifies its worth. The worth of the actualized normative identity avoids the arbitrariness of its constitution as a compossible. Creative actualization of the identity is justified if it is differentiated in the world as a principle of good. Value is the ultimate term of differentiation, then, since it justifies such differentiations of the world. Value marks worthwhile as opposed to arbitrary differentiation.[27] Value difference marks an improvement in all categories, since what has been differentiated is good, regardless of the kind.

Third, value is the ground of such differentiation as a project. Value is also the ground in the identification of the proposal as worthwhile, as an identity worth actualizing. The ground of the differentiation of a new normative identity is its value, thus value is the ground of differentiation. However, its successful achievement is as an instance, including a quality as a mark of value. The same actual quality is both an instance of the quality, its mark as the normative instance, and the ground for applying the norm, that is, the basis for the identification as a norm. A normative identity is both instantiated as a value and the mark of the norm qua instance of the identity.

Some aspect of reality is differentiated and considered worth differentiating by normative identities, that is, "concepts." Instances are clusters of normative identities, that is, multiple predications can be made of any individual, whether of size, shape, relation, or other "predicates." Normative identities can be "predicated" of bearers in categorical propositions, for example, "this machine is an economic good," or "friendship is a social good." "Facts" are selective normative relations, for example, "green is a color" brings the normative identity "green" under the normative identity "color" using norms of syntax (copula, article, distribution, and so on). However, these conceptual distinctions must also be grounded in the world, that is, justified in terms of its superior value as "correct." "Meaning" intends a world in elucidation and evaluation of suitable "concepts," which are a species of norm, that is, normative identities.

The "habitual" or "general," to use Charles Peirce's terms, follows upon the creative or unique, the initial creative actualization of some new value or good. Thus the latter is the *sine qua non* of the former and this provides a norm for evaluation of the relation of identity and difference. Generality or identity must be derivative, for generality is manifested, if at all, over time, after creative actualization of the unique, the different. (This principle is consistent with Darwin's notion of the origin of species by natural selection. Favorable variations become generalized in a species.) While a general element is needed for creative actualization, since an activity is specific and thus determinate as a kind of activity ("economic", "religious,")—yet such categories also differentiate the activity from other activities identified normatively.

Values, as I noted above, implies different values, at least in some respects, since there are more than one. The unity of the world is differentiated into a plurality of different goods. Thus creative actualization of different goods of the world is pluralistic, and inherently opposed to any monism of value. No value philosophy could be monistic, for different values are entailed by "values." One value implies others, forming a plurality. Again, any scale or grade of values, and hierarchy involved in evaluation implies differences in value. The emergence of novelties implies differences in value. Value philosophy is a philosophy of difference and thereby plurality. Monism of value includes all hierarchies of value, that is, all *summum bonum* theories, a point I will cover in chapter five. (Participation, foundation, and *pros hen* are the metaphysical models for the *summum bonum* theories). Monism cannot be a principle of values, but only of "value" since if each thing was either ontologically, functionally or axiologically attached to, related to or partook of "one principle," it could not manifest another principle or norm, that is, its distinct value. The value of food is different than the value of clothing. If their value was the same, or identical, we could neither eat nor be warm. This differentiation of value is the reciprocal relation of distinct values. Thus their differentiation is neither a foundation of value nor the basis for a hierarchy. As different they are plural, and this cannot of itself provide a ground for subsuming them under an identity. Plurality of value is an argument against identity of value, since value is differentiated. Thus a hierarchy of values cannot be their ground, since such a hierarchy would be an identity of all value. What values have in common is creative actualization and meliorism: plural improvements—but these characteristics are normative, not actual except as creatively actualized, thus as different. The world can be improved in many ways through creative actualization. However, "values" is value taken collectively, the sum of all instances of value. Value as a principle is not distinct as an actuality from any instance. Pure principles as identity patterns are not "ontological." Indeed, value is required for the actuality of novel principles. Thus fundamental ontology cannot account for principles, since they have no "reality" without value, that is, as embodied as instances (McDonald, 2009).

Creative actualization differentiates in some respect, if only as a unique copy: unique in the place it occupies, the time it is made, the ingredients, and so on. Improvement through difference is not equivalent to identity through difference, their correlativity. For the world is not identical to what it was before, but different, even if the difference constitutes a novel identity. Making is covered as well as theoretical differences and hypothetical differences, since to make is to differentiate.

Summary: In this chapter I have discussed inherent value as a character of actions and goods. The distinction of inherent value from the instrumental-intrinsic value distinction was made and the separation of the problematic of value from being. Value is not the "attribute" of a substance or a being, since creative actualization can be the cause of the entire good, not just one attribute of it. This is the case in technology, for example, as well as scientific predictions, such as the actuality of Neptune. The actual is created not ontological. Again, the instance may itself be grounded as a functioning actuality that is not necessarily perfect. Thus the instance may be creatively actualized as a whole, for example, plastic devices, and thus not add an attribute to a substance as its perfection. I also distinguished the bearer, locus, and scope of value, and I argued that individual bearers of value could nevertheless creatively actualize norms and other generals. The locus of intrinsic value lies in life, although life is not thereby intrinsically valuable or the character of all value. The scope of value includes potentially everything since value goes beyond the actual to the potential.

However, this means that value is sometimes hidden, sometimes manifest. Different modes and elements of value were covered. Since creative actualization can bring about what was only potential, value marks the successful achievement of norms considered as generic: including ideals, principles, rules, obligations, standards, and other modes of potential generalities. Value differentiates them as instances that have been creatively actualized. In the next chapter, justification for creative actualization will be examined.

Three

MORAL JUSTIFICATION

A problem for all theories of value is that the good may be used for evil purposes.[1] In the twentieth century, concentration camp guards, administrators, and others associated with such evils have even tainted duty: the excuse they offered was that they were "doing their duty." The courage of the SS in battle, and the loyalty they showed to the Party were in the service of outrage: virtues can be turned to evil purposes, as Kant knew. Other theories of value are *a fortiori* subject to moral abuse, whether a *desire* to rape, the *happiness* of the polluter who has not been discovered, the *self-interest* of a corporate criminal, or any other so-called "values" used to justify immorality. The relation of non-moral values to morality is a major problem for any theory of value. Indeed, this problem has frequently been used to classify ethical systems: those which define moral good in terms of some other good are contrasted with those that argue for morals as the condition of other goods, that is, "deontological" and "teleological" ethics.[2]

The basis for all practice is at issue in this divide, that is, whether some acts fall outside of morals or whether all must be within moral limits if they are to have value, including collective actions, economics, politics, and even scientific practice.[3] In this chapter I will examine the relation of value and obligation with respect to creative actualization as a theory of value. I will begin with a distinction between evaluation and value, and the use of evaluation as a critical notion. I will argue that creative actualization requires moral values as the condition of other values and thus as the ultimate justification for values. I will then discuss the reciprocal structure of duty, justice, and other moral norms as recognizing the other. I will conclude with the qualification that morals provide limits to other values, not their content. I will use the terms "ethics" and "morals" interchangeably in what follows. Although the two words have different etymological roots, no one has made a strong case to separate them.

1. Creative Actualization and Morals

Moral Values and Moral Evaluation: the problem of the morality of creative actualization is valid, since with technology humans can and have creatively actualized the means to destroy the entire planet: the "omnicide" of "total" nuclear war. A more subtle form of omnicide is the slow poisoning of the biosphere by the irresponsible use of toxic chemicals, their release into the atmosphere, soil or the water table, and thereby the gradual destruction of habitats for all life. Creative actualization of nuclear weapons is "good" in the limited sense that the weapons work or function properly, but is hardly good from the point of view of living organisms. Synthetic chemicals may well destroy insect pests, and so work well; but their effect upon the rest of life as long-term elements in the environment is problematic, to speak nothing of their destruction of beneficial insects and predators. Creative actualization may ultimately have bad results, which were not originally foreseen. Worse, we may creatively actualize evil. As with pleasures, not all creative actualizations are morally good. In order to deal with this problem in theory, moralists have distinguished right from good. Similarly, obligation must be defined apart from value. What is good may not be right: Socrates argued that those who do the seeming right might be ignorant of what is truly good (*Republic* I). The relation of value and obligation should be formulated such that there is no doubt in principle about what actions are right and what wrong, except perhaps in the hardest cases.

Moral theory describes what should happen in principle. The hypothesis is that moral practice will follow and can follow moral theory, moral principles. What ought to happen will come about through action. Moral rules describe the "ought"; moral practice, what actually happens. This distinction allows for moral evaluation, the careful consideration of what we should do prior to acting upon our decision. We can morally evaluate different possible courses of action, and their merits and problems. The one we evaluate as best is the course of action that we actually pursue or should pursue, constituting our choice. The action that will creatively actualize our decision lies in the future if we are evaluating it, it is not yet actual. Thus the action is not an attribute of a substance. Evaluation of what "should be" is prior to what "is," its condition, and its ground. Goals lie in the future but are the present ground of what will "be" or more strictly what changes will occur.

Evaluation is distinct from practice and prior to practice. Practice is temporal and our activities are justified over time. We evaluate multiple possibilities, but chose only one at a time. Moral evaluation is distinct from moral practice, then, in several respects. Evaluation is an activity, but not yet an action. It precedes the practice that implements the result of the evaluation. With evaluation agents consider several possibilities: evaluation dwells in possibilities not actuality, and is the condition of principled moral action. Evaluation also differs from actual value in that multiple values are compared. Value has to do with inherent value, the good of an individual instance or bearer of value, especially its creation. Evaluation compares different values and thus assigns value to actual differences in value.

Another major difference has a bearing on the problem of the relation of value and obligation. Creative actualization does not of itself exclude values. Any good or value that can be creatively actualized has merit as a value; no good is excluded by creative actualization as such, so long as it functions well, meets a standard or some other practical test. But critical notions arise with the question posed by Nietzsche of the value of a value. Creative actualization does not consider some critical questions: the question is not whether we can but whether we should creatively actualize some possible good, whether a novel synthetic chemical, a clone, a genetic "enhancement," a suburban "development," and so on. We cannot answer the question of "ought" based upon "can" alone. Moral evaluation is required to critically appraise any possibility. Evaluation in general is required as well for future actions in consideration of a host of factors, such as will a particular creative actualization achieve its intended goal (pragmatic considerations)? Is it culturally suitable? Is it personally fulfilling? Is it the best choice of those available? Evaluation consists in critical appraisal of possible choices prior to creative actualization.

The critical question for evaluation begins with the query, "what good will that do?" What good will be creatively actualized by an act? What future will ensue by means of such activities and what kinds of values will be actualized by them? These questions are distinct from comparative evaluations, that is, the relative merit of different courses of action, results or possibilities, and superior alternatives. Comparative evaluation also includes conflicts of values, for example, those raised by conflicts of novel creations with values in the world, whether by chance or design (Lake Powell vs. Glen Canyon). The esthetic of nature has here come into conflict with economic and other values, including the esthetic of a created body of water. The value of the world as it "is" in conflict with the value of what it should be. Evaluation can be comparative, deciding between different possible courses of action; or critical: appraisal of a single good, a particular goal, or even a whole order. The critical question can arise either before or after comparative evaluation, and uses different standards. Critical evaluation inquires into the worth of the possibility, whether immediate or ultimate—but also its justification, whether it constitutes an amelioration, questions of quality, and conformity to standards. Asking the question of the value of a value, or "is it good?" concerns value as a critical notion, evaluation, not creative actualization.

Critical evaluations raise the question of the value of a value, that is, its ultimate worth. Such questions arise not only in evaluation of choice but because different individuals may come to different conclusions about what would be good or right to do in a situation. What is the right or best action to take is then problematic. Competing versions of good and right means that they are at issue: does this imply all goods are relative to someone or that the issue is unresolved? Is what is good or right then an open issue? If good and right are problematic, then their value, whether immediate or ultimate, requires critical evaluation. To ask the question of the worth of a value also raises pragmatic questions of whether its creative actualization would indeed be an improvement (meliorism) as well as the practical question of the best means of creative actualization: is it actualizable at all? Asking, "what good would that do?" is critical in terms of ultimate improvement, a melioristic evaluation. Critical evaluation is in part pragmatic.

Evaluation concerns potential creative actualizations: in the realm of the possible.[4] However, such possibilities may become actualized if evaluation is favorable and they are acted upon; they can be creatively actualized in the world. Since evaluation involves appraisal of possibilities, evaluation no more starts from the "given" than morals. On the contrary, evaluation is frequently critical of "the given" and raises the issue of the value of the given. Evaluation considers competing possibilities, in which the "given" one counts only for one and may have given rise to the problematic that preceded evaluation. The "given" may require critical evaluation: raising the issue of the value of such a creative actualization in the past that is now present as a problem. The worth of the potentially valuable ought to be evaluated, since values are not conditional on time. The critical issue is whether they are worthwhile: whether they are worth actualization.

All kinds of potential values can be critically evaluated, from moral to personal, including appraisal of different modes: categorial, inherent, intrinsic, and so on. In this sense value has much wider scope than actual creations, since we can raise the question of the worth of anything, including chance formations which were not creatively actualized by agents. The question, "Is a wild landscape beautiful?" is one for evaluation. The question of whether it should be creatively actualized cannot arise with a landscape that preceded human agents, but the question of its ultimate, comparative or inherent value can always be evaluated critically. However, actuality, actual goods, are irrelevant to evaluation, since we can raise the issue of the worth of goals that have not yet been actualized and critically evaluate the worth of such goals, as well as the worth of actualities. The inherent value of a good is valuable but not ultimate. Its value can still be critically evaluated in terms of melioristic improvements, ultimate worth. What is good in itself demands critical evaluation. Categories and kinds of value require periodic appraisal as well. Are the kinds valid? Is health distinct from pleasure as a kind of goal?

When a conflict arises over what is right and the moral choice is problematic, critical evaluation of alternatives is called for. Once a moral problem is resolved, creative actualization is required, but although moral issues and principles are part of evaluation, they must also be critically evaluated. Moral problems are also raised by "hard cases" where moral principles apparently conflict. While war is wrong, is a defensive war moral? Do the rights of society, the greatest number, or the individual have priority? Does the good of society justify the sacrifice of individuals in war? If individuals have rights is famine relief required? Or is a "lifeboat ethics" in which the limitations of the carrying capacity of the planet recognized the right course? Personal values are also problematic: is the lifestyle of a redneck worth destruction of a redwood?

Moral evaluation is a specific kind of evaluation that nevertheless has universal scope: the moral value of any instance can be raised as issue, the moral worth of any possibility a question. We may evaluate something worthwhile but also raise the issue of the moral worth of the evaluation: is it moral? As with evaluation in general, actuality is irrelevant to moral evaluation as such, since moral principles remain valuable in principle regardless of whether they have been creatively actualized. The evaluative realm of possibilities has its own norms. Actual circumstances may not be irrelevant to us of course, especially as practical agents, but they may be to the value of a value in evaluation, as opposed to the actual world.

One objection that might be made is that moral values are supreme in this view and this creates a moral hierarchy of values. While hierarchies are part of evaluation, the creative actualization of a choice gives the result its own value, which is not contingent on a summum bonum. Morals are not supreme as values then, although moral choices are the appropriate choice or alternative in specific circumstances, and may rule some out in critical evaluation. A moral choice is a better choice than alternatives, which would be bad choices. Theft is wrong because a moral choice is more suitable, right, than the purely economic choice of theft. This decision is less moral "over" economic values so much as moral alternatives actualized instead of economic alternatives (theft). Evaluation of alternative actualizations has as a condition the creative actualization of reciprocal values in economics. Moral relations are reciprocal; but evaluation and value are distinct.

The problem of practice still lingers, however. The main problem is the gap between theory and practice, that defining what is right in theory may not affect actual practice. Our evaluations of what is right may not coincide with our behavior. There is also the problem of moral competence and luck, whether our intent to do right is enough even if the practical consequences are not in line with the standards set by theory. The issue is not only why should we follow theory and act according to moral principles, but also how can obligation be actualized. I will address these problems in respect to creative actualization in this chapter.

Creative Actualization of Moral Values: moral value is distinct from moral evaluation. Moral value is a species of creative actualization in which agents creatively actualize moral principles, virtues, and cultural norms in specific acts. Generally, moral evaluation precedes action and involves the critical evaluation of various alternatives in the light of moral principles. Moral evaluation is distinct from acting and need not result in moral value, both because an immoral decision may be reached and also because moral rules may forebear or constrain some actions, rather than require them. (This constraint is in a loose sense an action, since some have argued that inaction is, in conditions of choice, a type of action. In this case, the inaction is the creative actualization of moral constraints.) Moral evaluation of choices is evaluative since it involves consideration of possibilities, not their creative actualization. Prior to actualization a decision must be made among different possibilities, evaluation. One consideration in evaluation is the potential for creative actualization, that is, that the possibility evaluated best is also practicable. The potential for value is presumed, namely, that we can creatively actualize our moral evaluations. Evaluation of what is morally permissible or impermissible, another factor, is not value, since it dwells in the realm of potentiality.

Values that have been creatively actualized may or may not be ethical. The distinction of moral evaluations from evaluations is that moral evaluations are evaluations that we are required to creatively actualize. Moral evaluations are practically necessary and what is practically necessary is obligatory.[5] Agents are morally required to bring moral values about, implementing them as moral values. This requirement answers the problem of the amorality of value theory as such. While we can creatively actualize what is evil, we should creatively actualize what is moral. Creative actualization as a theory of value addresses the problem of morals fundamentally, since morals live as creatively actualized. The question of morals then is one of moral evaluation, and not of theory of value, of whether we choose to follow obligation. Moral values also require creative actualization, like other values. Thus creative actualization, value, is required for morals. Creative actualization as a theory of values is morally valid, since we are required to bring about moral principles, and what is required is obligatory. Creative actualization is morally justified, since it grounds moral principles in, through, and by practice.

This requirement means that norms, moral principles, are not the only content of ethics. Intending to implement moral principles is not enough: we must carry through our evaluation of what is best by its creative actualization. Ethics should be manifest in action, not intent, or it is impotent. An ethic must be based on actions, not intentions, or its principles remain utopian, not to say hypocritical. A person's actions reveal their values; their intent may differ from appearances, that is, the meaning of the action. But it cannot *contradict* the action. Generosity may be a façade for self-interest, but the act of generosity remains. We may be skeptical of a person who claims they love books, but owns or reads none. This follows from a pragmatic(ist) criterion of values, that values are revealed in actions. A pragmaticist criterion of value would not do away with intent, but would subordinate it to action.[6]

Ethics is subject to verification, then, as revealed in behavior, in the original sense. The world of "oughts" or duties is not in another universe from the world of "is"—the possible removed from the actual—since they are brought together in moral acts, whether virtuous, dutiful, or consistent with morality. Ethics also has the office of connecting what ought to occur with what does occur: putting the principles into practice, through creative actualization of moral norms. When we act upon an "ought" we creatively actualize it in the world: the norm or principle is not some transcendent standard but can be practiced. The world is thereby changed, since novelties are enacted by pragmatic means. Thus creative actualization can connect mind in the form of imagination to matter: ideals actualized in the world. As such, values create what "is" or more precisely a new world changed from an older version. Value is the mark of a world of *change*, not of coming to *be*. Values also intersect facts, since creative actualization creates "facts" while the facts of a situation may generate efforts to change the facts, that is, creatively actualize novelties.

Since moral values are practically necessary, we are required to creatively actualize them. (The contrary holds as well, but not in the logical sense: not all necessary values are moral values, since some may have to do with survival. While survival may be morally required—consider injunctions against suicide—it is not strictly moral but biological.) Duties are potential moral values, which we are obligated to enact. Moral principles and norms are not moral values unless they have been creatively actualized. Acting on a principle is enacting a norm as a moral value. Morals are normative but can be practical through creative actualization of moral norms in actual practice: where ideals, standards, and principles are actualized as moral values. The "ought" is creatively actualized. The distinction of what ought to be would not arise unless it were not actual, but it does not follow that it cannot be brought about. As Kant argued, ought implies can. Obligations are not in another world, but require creative actualization in the world in which they have not yet been enacted. Moral norms are still creatively actualized as moral values, norms brought from ideals to actuality, but they are evaluated superior to non-moral alternatives.

Because different possibilities can be actualized, conflicts can arise over which values ought to be actualized. Moral values require actualization, while other values do not. Moral rules must be in accordance with life, since principles that put an end to life would soon end all moral agency. The necessities of life are frequently held up as an example of a moral dilemma: should we steal food to save the lives of our family? But if food is required it is practically necessary, and can be evaluated higher than other goods that are not necessary. Moral evaluation regulates what values should be actualized in any conflict over alternative possibilities.

What ought to be has priority over what "is" given competing possibilities, alternative choices about the future shape of the world. Moral values have to do above all with the future: with what world we shall bring about through creative actualization. Moral norms, including consistency and the like are obligatory evaluations prior to any creative actualization, but with reference to the future. What now is actual cannot control our destiny, although it can influence it. Insofar as our institutions have any worth, it is because they were a rejection of the actual for the newly possible. Rules and other norms that have not been actualized remain as principles and standards for the future. However, without creative actualization in moral acts, they are without value.

The distinction of value and obligation is attenuated in evaluation, however, since the best alternative is the choice we ought to make as best. If we creatively actualize the best choice, as we should, then the normative choice is actualized in the world and what ought to occur becomes part of the world. Correct evaluations are actualized as moral goods. Our present level of advances reflects past evaluations of what should be creatively actualized. Can obligation then be reduced to value? The relation of the normative, of which obligation is a species, to creative actualization is complex. Moral norms are creatively actualized in acts with moral value, or creative actualizations within the limits of moral principles and obligations. Creative actualization of norms brings them from mere "oughts" to actuality. However, creative actualization of the norm is the goal of the act, its purpose and thus cause. The act would not be constituted as moral without moral norms. Moral norms require creative actualization, but creative actualization requires a moral condition. Their relation is reciprocal, for both elements are required for moral value and their combination constitutes the act as moral. Such reciprocity is the relation of all the elements of value, action, telos, norms, and evaluation. I will return to this point in a later section.

The Relation of Values and Obligation: What is the relation of creative actualization to obligation? Do values ground ethics as in "consequentialist" theories or do morals condition values, as in "deontological?" Moral acts require justification. What justifies them? Intrinsic values are used by consequentialist theories to justify duties. The problem is that there is no logical relation between values and obligations. Take happiness: are we obliged to creatively actualize it? It is difficult to see why we are obliged to act on duties based on happiness: how does a feeling create or warrant an obligation? Are we obliged by such a value: are acts of duty required, that is, necessary given such values? Why is the obligation required, or necessary: why does it follow logically? Is the act I should perform grounded in intrinsic value? Or is it only worthwhile? Are we obliged if we do not feel happy? Or if such obligations end our happiness? Isn't a minor needed to derive a (social) obligation from an (individual) feeling?[7] But could such a minor logically connect value and obligation? There is also the consideration of the relation of logical necessity and practical necessity.

The issue, then, is why is anyone obliged in terms of an intrinsic value? How does intrinsic value logically entail obligation? Philosophers have proposed several answers to this question during the history of philosophy but at this point I will note that there is no logically necessary relation. It would not be logically contradictory to argue that something is good, worthwhile, or even intrinsically valuable and also that we do not have any obligations that follow from such goods. Even if intrinsic value is necessary for duty, is it sufficient? Still another problem is that if the relation were necessary, specific obligations would be generated whose consequences would be either impossible or problematic. If intrinsic values automatically entailed or necessarily entailed moral obligations, would we be required to actualize every intrinsic good? If this consequence were entailed it would mean, in human terms, that we would be required to have as many children as possible. Even so conservative an organization as the Roman Catholic Church does not draw this conclusion. In non-anthropocentric terms, we would be required to preserve every life. This would be pragmatically impossible, given predation (J. Baird Callicott, 1989). The relation of intrinsic value and moral obligation is more complicated.

The question is whether intrinsic value provides sufficient warrant, that is, necessarily implies moral considerability which in turn necessarily implies moral obligation. Logical inference is one possible relation of intrinsic value to moral considerability: that moral standing, and thereby moral obligation, follows logically from intrinsic value. Logical inference is only valid or conceivable as somehow a moral intrinsic value, that is if intrinsic value is equivalent to or implies moral value. Intrinsic value would be an end that alone can justify means. As a moral end, we ought to pursue it. Thus justified means follow from some one consequence or ought to be practiced. Intrinsic value is defined morally in such an inference. However, this warrant is not logical without circularity, since intrinsic value is defined morally, the warrant of what must be justified is itself. If this inference is dubious, as I argued above, why is anyone obliged in terms of an intrinsic value?

Both Immanuel Kant and the Utilitarians, who are about as far apart in ethics as can be while still remaining ethical, agree in denying the identity of value and duty. For Kant, we do not have a duty to good wills and especially not to instrumental values, but to universalize our own will, act for the sake of duty, and treat persons as ends in themselves. Similarly, the greatest good of the greatest number entails no obligation to individual pleasures, or even the happiness of particular individuals. Justification or moral warrant must be axiological in some sense, then, as ends justify means because of the value of the ends, not because of the moral value of the ends themselves. This factor shows the importance of Kant's deontological ethics and its transcendental arguments from the condition of values. Morals are independent of values, although a condition of values. Morals cannot be defined wholly in terms of values.

I will argue below that creative actualization as a theory of value is compatible in some respects with consequentialism, intrinsic moral value, deontology, and pragmatism. I will argue against any *summum bonum* or *pros hen* relation as a moral warrant, however, and thus any justifying foundation. I will also argue against the strong separation of obligation from consequences that marks much of contemporary ethics. Creative actualization as a theory of value has been articulated in part in response to the theoretical questions surrounding the relation of value and obligation that I outlined above. However, there is also the very pressing motive of environmental damage and the consequences of unchecked human encroachment on the environment. Since such consequences for the environment must be a concern of any serious contemporary ethic—indeed may be the criterion of a serious ethic—consequences must also play a role in morals. Since John Dewey's ethic considered both, the model provided by instrumental and pragmatic theories will also be given their place.

Environmental consequences separate this ethic from the anthropocentric, subjective ethics of the past, as well as the rights-based politics of neutrality that is its offshoot. The argument that rights "trump" other considerations, a hierarchical argument, can no longer be taken seriously. Rights are not absolute. The "right" of individuals to have as many children as they want, the basis for the population problem, not only impacts others, in the form of a claim upon them to provide food, clothing, and other amenities for their offspring. It also has the potential to destroy the environment in a futile effort to stretch resources to the maximum in order to meet such hypothetical claims. The environment is the condition of all such rights, for without suitable air, water, and food, no rights of any kind could be exercised. Such a "right" must then be limited by moral consideration of environmental consequences. Liberty is meaningless in a dead world. Such anthropocentric notions of rights can, from this perspective, be seen for what they are, namely, special interests masquerading as ethics. Similarly, governments cannot be neutral as to life-styles, plans, and goals. Environmental virtues are essential and morally required. A politics that refuses to draw or recognize such consequences is fanatical and ideological in the strict sense: the part masquerading as the whole. Individual "rights" are relative to circumstances and especially moral consequences for others. Moral progress means the inclusion of the environment as the hidden other, the consequences for which Western anthropocentric ethics has been continually blind.

Creative actualization is compatible with deontological theories so long as the theory of value is not confined to the ability of one species, humans, and consequences for non-human life are considered in moral evaluation. I will also argue that the non-living environment, the air, water, and soil are necessary conditions of human life. Microorganisms, plants, other animal species and their ecological niches are also necessary. Since they are necessary, they are morally considerable in some respect; more, we have obligations in some form to our conditions of life and survival. Thus transcendental ethics stands in need of a supplement: reform of its anthropocentric theory of value, and its subordination of considering environmental consequences. I hope the myth of the "empty formalism" of the Categorical Imperative has been addressed sufficiently by modern Kant scholars to have died a deserved death. However, although the Categorical

Imperative provided a person oriented theory, it is speciesist, and requires supplementation in the light of moral progress.

I will argue that such a reform entails consideration of consequences in moral evaluations. From the good of any bearer of intrinsic value, we can derive conditions of its possibility. These must include conditions of its life as necessary, since bearers of intrinsic value require life: the moral condition. I also hope to show axiological conditions, namely, meliorism in respect of consequences. The conditions of all life are the non-living factors necessary for life as having instrumental value and the holistic relation of organisms. Thus their rationality is not required for their moral considerability, since they are required for human life. Consideration of consequences for these collective elements of the environment is thereby required in moral evaluation. The separation of transcendent grounds for moral obligation and consequences is untenable. Outcomes are a moral factor. This argument also applies to consequences for other persons, that is, social consequences. Pragmatic conditions of possibility include consequences in moral evaluation, namely, what will the consequences of creative actualization in a specific situation or context be?[8] Is a good worth actualizing? Thus amelioration is one factor in evaluation.

The distinction of moral from non-moral values, as I argued above, is that only the moral values require creative actualization as a consequence of our action. They are universal as a factor in all evaluation: either as a per se requirement, ruling out bad alternatives, or as a limit on other creative actualizations. Thus not all creative actualizations are required: some may remain possibilities for choice. In Kantian language, their creative actualization is "hypothetical," not categorical (1785/1964). Their creative actualization is contingent then upon reciprocal moral limits and superiority over alternative possibilities, whether acts or goods: meliorism. We are not required to creatively actualize every possible inherent value, or for that matter novel intrinsic values, but only moral values. Moral values are exhibited in acts of duty and virtue that are undertaken as choices that are alternatives to creative actualization of non-moral alternatives. Morals require creative actualization as morally valuable acts.

The necessity of creative actualization of moral norms as values is important, since although morals may be an independent variable in evaluation of what should be actualized, the *value* issue is independent of moral evaluation. Value considered strictly per se, in abstract isolation and as such, does not have a moral warrant. But actions do, since they must be evaluated and their consequences for society, the environment, and so on must be considered. Whether an H-bomb is good is, then, is in one respect independent of its moral valuation. Instances are good if they have been actualized without defects. But this does not tell us if we should *use* it (Hiroshima, India and Pakistan) or even *build* it, which are questions for moral evaluation, not value theory as such. I agree with Kant (*ibid.*) that moral considerations are independent of value, that is, some actions are right apart from valued ends. But this belongs to evaluation, since we deliberate about morals, that is, evaluate different courses of action prior to enacting them. (Kant too confused evaluation and value.) Morals and values are independent variables then. But the results of moral evaluations require creative actualization, including moral evaluation of the worth of any possible value. Practice attenuates the distinction.

The theory of value as creative actualization alleviates the theoretical problem of value theory: non-moral consequences of the value. What if genocide makes some people happy: does this morally compromise the theory that happiness has value? Creative actualization of moral norms is required and therefore creative actualization cannot *justify* immoral evaluations, since they violate required creative actualizations as limits on what should be actualized. Creative actualization does not and cannot justify evil, since the decision to go against moral principles and obligations is made in evaluation. Acting immorally is evil or defective creative actualization, that is, without justification, since obligations should either have been creatively actualized, or should limit other creative actualizations. Any commitment to values generates obligations in the sense of duties, since values imply duty.

There is another sense in which creative actualization gets around the problem of immoral "values," disvalues or evil, because it also attenuates the problem of the relation of value and obligation. Without creative actualization, no moral norms are possible. The relation is reciprocal since moral norms require creative actualization as much as we are required to creatively actualize them. The relation is complex, since it involves a number of problems, but value should not be confused with evaluation, as it has been by almost all the major theories of value.[9] The idealists in general have argued that value consists in a normative, transcendent standard. Such a standard remains a mere possibility unless creatively actualized; and thus in the realm of evaluation, not value. The psychological school, early and late, has attached value primarily or even exclusively to states of the subject or psychological states that are used to evaluate all values. This confusion may indicate that the problem is not axiological, but valuetive. The theory of value as creative actualization, however, solves several aspects of the problem, namely, that value is a reciprocal ground of obligation, since obligation must be creatively actualized in order to take effect as an actual moral norm. But obligation in turn constrains creative actualization, at least as a limit to "the permitted," a point I will examine below. Since their relation is reciprocal, they are conjointly required, and the logic of the relation of value and obligation is superceded. Obligation requires value, that is, creative actualization as moral value. In turn, creative actualization requires moral constraints. Determining moral norms is a matter of evaluation of the best course of action, and moral "deliberation" belongs to evaluation. This requirement cannot invalidate creative actualization as a theory of value, then, even though some actions might creatively actualize evil. A hydrogen bomb that still functions properly and is in good working order is good, apart from moral concerns or moral evaluations of its use, since the latter belong to evaluation, although creative actualization of the bomb itself is also subject to moral evaluation prior to its creative actualization. The success that is a mark of creative actualization is a norm not affected directly by moral evaluation. Evaluation of whether we should set off a bomb presumes a "good" bomb, that is, one that will work. Indeed, creative actualization is a distinct requirement even for moral norms. Moral evaluation is indeed required prior to any creative actualization as the ultimate justification of its worth, but creative actualization of such norms is required as a mark of value.

The moral theory of value in Kant and the Stoics, that value always has a moral condition, is based on a confusion of the relation of value and obligation. This relation is not a one-sided condition; instead, the conditions are reciprocal, in that creative actualization is required for all moral value as much as moral evaluation and limits are required for moral worth. If moral evaluations are not creatively actualized they remain at best Utopian—at worst, hypocritical: saying one thing but doing another. Morals without action are practically an oxymoron. Morals must be practiced, that is, creatively actualized for situations. Creative actualization is the practical condition of moral principles. Nevertheless, moral evaluation of future acts is also required for moral worth, as Kant implicitly argued: separating out moral principles and using them to critically evaluate proposed actions. Moral evaluation is the precondition of moral actions, that actions are critically evaluated as required or acceptable, that is, within moral limits. Moral evaluation requires creative actualization and creative actualization should be within moral limits, that is, requires moral evaluation. Creative actualization of moral norms is required or they are no good, they remain a mere wish or possibility. Values are the condition of possibility of moral norms, since their creative actualization in appropriate situations is required.

Thus creative actualization, unlike all the subjective, self-interest theories, is justified as a theory of value, despite the potential for bad and even evil creative actualizations. Evaluation governs actions, or what we should creatively actualize, but not the value of the actions, or the products and results of actions in abstraction from evaluation, only considered in terms of its inherent value. Inherent value may belong to a particular case of creative actualization despite immorality. Stolen gold or money retains its economic value, or robbery would never take place. Pollution is evil in terms of moral evaluation, not economic value as such, the inherent value of what has been creatively actualized—for externalizing costs makes short-term economic sense. For moral norms also have a value condition, namely, creative actualization. The judgment that moral and not immoral ends should be creatively actualized is an evaluation of possibilities. Such moral norms do affect the value of action, however, or what we ought to creatively actualize.

Morals form a limit to creative actualization that themselves require creative actualization. Creative actualization is the practically necessary condition of moral norms and principles. In a practical sense, moral norms are conditional on values: an axiological ethic. Otherwise, such norms and standards are Utopian. An axiological ethic is not a normative ethic. Norms have values as a condition, but not in the sense of a warrant of value as such. As Kant correctly argued, ethical norms should be *evaluated* apart from values, since the morality of the value is distinct from practice and can never be derived from practice alone. Moral norms and creative actualization are distinct.

Since moral norms receive their warrant from a different source than value, the relation is not circular. My argument is not that value is grounded in obligation and obligation in value, since values involve successful creative actualization as their warrant, inherent value; while evaluation of moral worth does not. Duty is not derived from value, nor is value derived from duty. Their relation is not ground and consequent, but co-conditional or conjointly required. While a condition of moral values is creative actualization, and creative actualization is a consideration in moral evaluation, moral worth is not defined in terms of practical success. Nor is creative actualization the condition of their obligation, that is their practical necessity. However, if creative actualization could be shown to ground ethics, an axiological ethic, then ethics would also be independent of metaphysics.

The "logic" of inherent value requires reciprocal relations, not only among agents, but for all values, including moral values. Moral principles require creative actualization and creative actualization requires moral principles as a limit, if the value of the creative action is to be justified. The condition of intrinsic values, which as intrinsic cannot be subordinate, is a reciprocal relation in which their intrinsic difference is not subordinated as an instrument, that is, the duty of non-subordination. The condition of duty is the creative actualization of obligations, with the result of a reciprocal relation of intrinsically valuable agents. Subordination of either right to good or value to duty is unjustified.

"Values" imply multiplicity; inherent value implies that the value of such values does not derive from an extrinsic source, but belongs to the bearer. Extrinsic value is instrumental value for an intrinsically valuable bearer, but inherent value cannot be reduced to extrinsic value. Reciprocity is the principle proper to inherent value bearers whose value is not extrinsic. Reciprocity is the moral norm whose creative actualization as the relation of intrinsically valuable agents is required. Reciprocity creatively actualized as a moral value would constitute a society of moral equals. Its creative actualization is a duty as the condition of all intrinsic value, all self-actualization. Moral values are the condition of the intrinsic value of equal agents, and require the creative actualization of moral norms in action. Since self-actualization constitutes intrinsic value and such value cannot derive from an extrinsic source, creative actualization is incompatible with hierarchical theories in the form of a *summum bonum*, a supreme value that justifies the many values. Self actualization is reciprocal, thus not extrinsic. Creative actualization of reciprocity as a moral value is the reciprocal relation of intrinsically valuable agents, a co-conditional relation of norms and values. Thus value as creative actualization is not hierarchical, since a reciprocal relation of equal values is required.

A hierarchy of obligations, graded in proportion to value, also reflects evaluation, not value. Evaluation of obligations and the creation of a hierarchy is distinct from their creative actualization. Each obligation creatively actualized has distinct value and does not derive its value from the hierarchy. A highest good cannot of itself create obligation, since however good it is, it can only give necessity to value, not to obligation. Obligation is also to an other; a *summum bonum* creates obligations only to the highest, not the other.

Similarly, inherent value does not refer back to intrinsic value as its justification, since inherent value is not equivalent to extrinsic value.[10] The value of self-actualizing agents is intrinsic, and thereby autonomous. This norm of intrinsic value includes larger actualities that self-actualize over a longer period, for example, cultures, species, and habitats. Thus obligation is not derived from a highest good or a single set of intrinsically valuable bearers.

As for intrinsic value justifying subordination of inherent value as extrinsic value: this constitutes a devaluation of the inherent value of the bearer to the status of an instrument of another. Inherent value as such cannot be subordinate; nor can its value be devalued without its destruction, in which a bearer no longer bears inherent value. Nor can any obligation for such subordination be derived from intrinsic value. Creative actualization involves a distinct relation of value and obligation, in which value and obligation are independent, although in a relation. The question remains whether there is any obligation to all the many kinds of values, or whether they have any moral considerability.

A third model for ethics, that in a sense combines *summum bonum* and the intrinsic-instrumental relation, is that obligation is derived from intrinsic value and intrinsic value belongs exclusively to states of the subject, whether happiness, self-interest, conation, feeling, and so on. This model is reductionist: value belongs exclusively to particular psychological states. Dewey is right to argue that atomic, nominalistic models for ethics, based on individual factors like individual desires, self-interest, and so on are anti-social. We might add that they are amoral, since they attempt to derive obligation, duty, and similar other-regarding norms from egoistic, self-centered values. This impossible project has as its motive the attempt to unconditionally justify self-interest and give it moral respectability—the fox in charge of the henhouse, if not the inmates taking over the asylum. To attempt to give morals a warrant in terms of an individual self-interest is to undermine morals, since morals are properly a restraint on self-interest, desire, excesses of feeling, "attitudes," and the like. Morals are primarily social and duty is primarily to others. Obligations to self are derivative from the universality of principle, since one cannot be one self with obligations to another self, despite Kant's arguments connecting moral norms to the self in the *Groundwork*, which he acknowledges are "imperfect" duties

Egoistic "ethics" of all types, whether based on desire, feeling, self-interest, will or other subjective states, ignore the source of ethics and moral development in individuals, that is, the learning of moral norms, virtues, and other principles of a culture in childhood. The ego implied as a model in such projects is taken from a fully formed adult without the requirements of a moral upbringing: as if such an adult sprang like Athena from the head of Zeus. This is comedy, not morality. Moral agents can no more derive ethics from egoism, self-interest or individual psychology than scientific principles from witchcraft. Society is the condition of the individual, if only in the parental relationship. "Desire," "self-interest," "feeling," and other egoistic sources of "value" require ethical *restraints* not enhancement by egoistic formulations of duty. Such self-regarding psychological states must be evaluated in terms of moral worth and thereby cannot be the source of moral worth. The primary function of ethics is not a rationalization of self-indulgence but to limit individuals and acts motivated by desire, self-interest, and other reductive schemes for egoism.

I argued above that consideration of consequences is required for ethics. Several grounds can be adduced for this requirement, among which is that creative actualization of the norm in the world, its achievement, is a mark of moral value. Duty is what we are required to creatively actualize as a consequence of our action: duty must be creatively actualized in the world, not only as a hypothetical intent. Duty, therefore, is not as distinct from value as a consequence, as the purported separation of deontological and consequentialist ethics implies. Duty consists in norms of moral worth that have axiological priority in evaluation, and practical priority in action. Kant's argument that we cannot be sure of actualizing our moral goals in the world is not morally relevant. Even if we are sure of very little, which is dubious in practice, we must keep striving to implement moral values. Otherwise, as Dewey remarked, we are "evading responsibility for consequences." The distinction of the ethics of duty and of consequences is misleading. While moral worth cannot be formulated in terms of non-moral consequences, the idea that we should not consider consequences of our actions in moral evaluation is unworldly. On the contrary duty could almost be defined in terms of the consequences on others of our actions.[11] Although the rubrics for contemporary ethics sharply divide deontological and consequential ethical theories, the sharp division of these two approaches may hide a more complex relation. Similarly, the distinction of deontological and "virtue" ethics, that is, the ethics of perfection of character, has been exaggerated. Plato, like Kong fu-tse (Confucius), argues against any strong distinction of what is right and what is virtuous, while Kant treats pursuit of virtue as a duty.

Another issue is whether the norm of action done "for its own sake" (Aristotle, *Nichomachean Ethics*) makes sense, if interpreted entirely apart from consequences. If "liberality" is a virtue, surely its consequences for others are a factor? Would the liberality of a hermit hold any ethical meaning? Aristotle also argues that happiness (*eudaimonia*) is a consequence of a virtuous life. Moreover, it is dubious whether an action can ever be separated from its consequences. Isn't a courageous act the very process of aiming at courage?

Since consequences for social others must be considered in moral life, the rejection of consequentialism in ethics is untenable. Consequences are a moral consideration for intrinsically valuable agents that creatively actualize themselves, self-actualization, since this activity has their own life and survival as a consequence. Creative actualization of inherently valuable goods also is a consequence of action. In sum, consequentialism is a necessary consideration for creative actualization as a theory of value, since creative actualization changes the world: it always has consequences. Moreover, creative actualization of moral values as such has consequences. Practice creatively actualizes what ought to be as consequences of moral acts. Creative actualization of duty is a consequence of acting on duty. Otherwise, no duty is feasible. Indeed duty can be defined as what we are required to creatively actualize. Since such consequences can be evaluated morally prior to creative actualization, they should be. Creative actualization is thus the basis for an ethic that includes consideration of consequences, namely, whether the consequences are better. Are the consequences we choose to creatively actualize more moral than alternative possibilities? As soon as we are responsible for our actions, and not only for our intent, we must consider consequences of actions.

The requirement that consequences should be a factor in moral evaluation is a rejection of "pure" deontological ethics, for which duty is foremost and which is formulated without regard for or consideration of consequences. However, it is not a rejection of a weaker form of the ethics of duty in which consideration of consequences is reformulated in terms of duty. In this weaker form, only moral norms can determine moral worth, not non-moral consequences like "self-interest." However, the moral action should be determined in part by considering specified outcomes, namely, the preservation of the environment, of life and other species, of society, and individuals; as well as the promotion of plural values, meliorism, and virtue. Murdering another person is wrong partly because of the consequences for the victim themselves, their family and loved ones, the society, and the life that the person would have had. The duty of moral agents is attempting to improve the world by our acts over time, amelioration, by their consequences in the world and for others. Creative actualization is compatible with consequentialism so long as moral worth and moral principles are foremost in evaluation of action, intrinsic value is not speciesist or monistic but truly universal, and evaluation includes consideration of all inherently valuable loci, that is, not just living things but habitats, beautiful landscapes, and other environmental factors, such as air, soils, and water. Two issues should be considered, namely, what consequences should be morally considered and what is the relation of these consequences to obligation.

Environmental factors in the world are a condition of all life, the bearer of intrinsic value, and the consequences of action on the environment should always be considered (Callicott, 1989, Leopold, 1949/1968, Rolston, 1988). The first norm for consideration of consequences is environmental, since we are "in and of" (Dewey) an environment: we are not detached Cartesian subjects viewing the environment as an external object, but biological creatures within an environment that surrounds us and enters within our bodies, our selves. Since air, water, and food (and therefore soil) are necessities for self-actualization, their destruction constitutes self-destruction, and destruction of others. Consideration of consequences for our environment is required in a practical sense as the condition for all of life. The environment includes all of the factors required for a viable habitat, including not only pure air and water but also a healthy soil, the condition of plant life, which is the basis of the food chain.

Similarly, the environment includes other species necessary for its functioning as an environment. These species have a status independent of their role in the habitat, however, for otherwise their role would determine their value and their inherent value reduced to functioning, that is instrumentality or extrinsic value. Since species creatively actualize themselves over generations as individual instances, moral consideration of individuals of rare species is required, as the condition of preservation of species. Consideration of consequences of action on species is required to attain actual universality, as opposed to universality for one species, specious "universality." Universality for all self-actualizing agents is the standard for moral worth, not only for one species, humans. Humans require other species for self-actualization: they are not the great exception but also require air, water, and food.

Consequences for society were discussed earlier. Duty involves consequences for others, and the effect and affect of an act on others is a moral consideration. The most obvious duty qua duty is to others, that is, to society. Duty primarily involves social relations as duty. The basis of duty lies in consideration of others as morally equal to oneself—as an equal self qua self—and entitled to consideration.[12] Consequences for others, justice, are one of the main considerations in moral evaluation of acting. The formulation of duty in universal terms is a technicality: a normative requirement. The "other" universalized includes all intrinsically valuable agents, that is self-actualizing value bearers, including other species. The universalization of duty includes all self-actualizing agents, including oneself. Duty is reciprocal, each agent having the same duties to others as the others. The self has a place in the society in which social norms apply to each agent, including oneself. Each agent is a participant in society, not totally independent. Society is the genus for cultural varieties and some moral consideration should be given to distinct cultures, so long as cultural practices do not violate universal moral principles.

Another principle that we are required to consider in moral evaluation is improvement of the world, meliorism. Since our actions have consequences, the results should constitute an improvement, not only the negative consequence of avoiding wrong. To treat others well is an improvement over treating them badly. Meliorism means improving the world as a whole by our actions if we can accomplish this goal at the same time. More, if a plan is evaluated best we are pragmatically required to creatively actualize the best as the best. Since the best always includes the morally best in any evaluation of a plan, moral improvement is a principle basic to moral evaluation. The examination of moral principles is a requirement of any evaluation of a plan of action. Moral consequences rule out immoral plans in evaluation, the first principle of evaluation. Thus not only universalizability but consequences must be considered. Since we aim at betterment in some sense, moral progress is implied by evaluation of possible courses of action. Meliorism is a slightly weaker obligation than reciprocity, since we cannot always improve any situation by our action, although we are obligated to do so when we can. But reciprocity itself is melioristic, and meliorism should be reciprocal. Meliorism is also measured axiologically to some degree, not in terms of moral worth. An economic improvement within moral limits is not strictly a moral improvement, but may well be melioristic.

Another consequence of reciprocity will be plural self-actualizing agents living together. Indeed, reciprocity is creatively actualized as a value in compossible self-actualizing agents.[13] Evaluation of individual acts should consider consequences for pluralism that arise from reciprocity toward the other. The compossibility of plural intrinsic value bearers is a consequence whose condition is reciprocity. Plurality is also melioristic as an increase in the kinds of goods in the world.

The environmental ethicists have distinguished moral considerability from moral obligation. Moral obligation is duty, but moral considerability is weaker and only creates a norm that in moral evaluation the good of the other should be considered, but that duty toward this other is not required. Moral considerability may end by creating obligations, but may not. Moral obligation does not necessarily follow from moral consideration. In environmental ethics, the reasoning is that animals (or other non-traditional bearers such as the environment as a whole) have intrinsic value, that what has intrinsic value is morally considerable and what is morally considerable generates obligations. Therefore, we have obligations to whatever has intrinsic value. The premises in this reasoning have been criticized from within environmental ethics, however (McDonald, 2003, ch. 1) Each of these premises requires justification; none is either self-evident or indisputable. Upon evaluation of a moral problematic, moral consideration may result in actions or forbearance from action, but is only a check on action, not an obligation. Duty is confined to those to whom an obligation is due. Obligation is independent of moral considerability.

That some good is better or more worthy in the sense of having greater value than some other good does not create an imperative to creatively actualize it. No duty to produce it arises both because obligation is not grounded in non-moral value and because consideration of the worth of creative actualization involves other considerations than inherent or comparative value. Producing an atomic bomb that is superior at destroying cities, people, and other living things and making an area of the earth uninhabitable for centuries is hardly a duty. The relation of duty and good, of obligation and value is to value in general with respect to overall consequences, moral and non-moral. Our duty is to goodness as such, including the goodness of the whole biosphere, to different kinds within it, and so on, not to the inherent value of an instance. On the contrary, moral obligation is the condition of inherent value, not the reverse. Obligation is grounded by conditions of actualization of all value whose deactualization would have negative consequences for habitats, other species, and our own. Creative actualization of moral obligations in acts of duty affirms this relation, since moral principles are enacted, not nonmoral inherent values. Our treatment of inherently valuable goods must be melioristic, an improvement, but there is no obligation to such goods per se.

Moral consideration follows from value, since destruction is destruction of value and we should consider the value of the normative principles and elements I outlined above in any situation. Value is a consideration as value: since the bearer contains value, its value should be evaluated, and evaluation always includes moral principles as the ultimate test. Many forms of creation involve destruction in some sense, including creative actualization, but some may be justified while some other creative actualization is not. Destruction of a rock to make a statue may do a habitat little or no harm, whereas destruction of a river may alter a huge habitat for the worse. Creative actualization of inherently valuable goods should be within moral limits.

Intrinsic value alone, then, does not necessarily entail moral obligations to individuals, although it does entail moral consideration. This norm applies to humans as well, given war, famine and the death penalty. Whether war or the death penalty are moral is another issue. But self-defense is a perquisite of all life. Intrinsic value is morally considerable as intrinsic, as good in itself. Intrinsic value implies that the good of the bearer is intrinsic to that bearer, not extrinsic, and thus creatively actualized by itself, self-actualized. Since creative actualization is a condition of obligation, each agent of creative actualization, self-actualizing living value bearers, is a condition of creative actualization of obligation. Obligation thereby requires self actualization as a condition. Each self is an other to all others and thus also morally considerable in that respect.

In sum, moral considerability does not necessarily imply or entail moral obligation and moral obligation extends only to necessary conditions of actualization whose deactualization would have negative consequences for life, the bearer of intrinsic value. Such conditions are required for all actualization and thus practically necessary or obligatory. Generically, obligations extend to the biosphere as a whole, as well as essential elements within it like air, water, and soil. These are conditions for life that are required and thereby obligatory. Obligation also extends to preservation of species and their habitats, including the special norms of human habitats, civil and rural, that is, duties to individuals in these habitats. In the wild, preservation of the minimal number of individuals of a species required for its preservation and genetic viability is an obligation, but there is no duty to preserve or increase more common species. Rarity increases value, just as in economics. An individual of a rare species has more comparative value, if not intrinsic value, than one of a common species (Callicott, 1989).

Moral Evaluation: I argued above that the actual cannot control our destiny, what we should creatively actualize. When transcendence was the central problematic, in ancient and medieval times, what is "real" or has "existence" could dominate thinking. Despite attempts by Heidegger, materialists, and some positivists to revive this problematic in our age, the overwhelming changes brought about with technology mean that the world has surpassed this problematic. A new problematic has emerged, namely, what "ought" to be creatively actualized in relation to the future shape of the world. What will be made, and should it be? Moral evaluations are at the center of these problems.

One potential choice lies between a return to "nature," and increased technoligization without any limits. Various intermediates are possible. Will the entire world be subordinated under bureaucratically implemented political ideologies, the managerial ethic, rationality, and technologization? The problematic of being was justified hierarchically, higher beings controlling the destiny of lower. Hierarchy is not absent from the problematic of change, however, since creative actualization of norms reflects evaluation and evaluation is assessment of the ostensibly best choice, that is an evaluative hierarchy of potential values, which have not yet been implemented. The choice creatively actualized can represent a theoretical hierarchy brought about in practice. An immanent value hierarchy is then realized as a goal.

I distinguished evaluation from value in the first section. Creative actualization as such is neutral on the question of whether any possibility should be actualized. The question of possible creations is one for evaluation; the decision includes moral evaluation of different possibilities both because the question is posed in moral terms and since a good decision is within moral limits. Since morals are the condition of values, which we are required to creatively actualize, they are a critical factor in or aspect of evaluation. Any potential good or act—or for that matter actual ones—can be critically evaluated and moral evaluation is an essential component of a critical evaluation.

Reciprocity of obligation implies mutually limiting relations of inherent values, and this goal must be considered in moral evaluation. Working out the reciprocal relations of individuals is moral evaluation, not value theory. New goals must be compatible with older goods. The worth of what may be creatively actualized is not limited in terms of some absolute *summum bonum* but mutual limitations in relations, moral relations. While comparison of values for worth is required in evaluation, only the comparison between the values as such is requisite, no reference to a *summum bonum.* The distinctive value of the novelty, its inherent value, is in a relation to other inherent values and must be compossible with them. An instance of a kind may creatively actualize in a distinctive way. This moral consideration applies especially with regard to unique instances of a species, different in some prominent respect from the rest of its species, for example, seedless oranges. Another example is plants that grow outside of the normal habitat and thus have adapted to unusual conditions. Here the value of the individual is distinctive enough to warrant special moral consideration.

Critical evaluation also raises the issue of whether a possibility would be moral if an agent creatively actualizes it. Justification is the moral element in evaluation. Justification is required because evaluation and choices exclude some alternatives and result in changes in the world. Through evaluation of different, competing ends one is chosen as the superior alternative. Creative actualization of that end will have results in the world different than if the alternative is chosen. This decision must be justified, both to do justice to different possibilities and with respect to consequences in the world and society. If society pursues some values, others will be excluded from creative actualization, perhaps forever. Prosperity may exclude theft; peace, rape and murder, and so on. Pluralistic meliorism is not relativism and there will always be individuals at the fringes and margins of society who are not reconciled to its rules: the rapists, murderers, and other anti-social minorities against whom the laws are directed. Social choices mean that not all values are equal, and choices, individual or social, require justification. Are they right? Just to others and oneself? Is their environmental impact justified?

Critical moral evaluation includes doing justice to competing moral considerations in a conflict or problematic. Our moral evaluation of the situation should join what is unique in the situation to principles while doing justice to each. Justice should be done to each aspect of a situation, including the particularity of the situation as such: the unique time and circumstances of the situation. Our decision should involve choosing among the rules for the best to fit the situation. Thus the rules governing the situation should also be evaluated, that is, determining which rules to apply in the situation. Moral evaluations end in choosing which rule, norm or principle will best improve the world.[14] Rules aim at achieving specific values and are chosen in relation to such values. Distinctive rules for a situation would have no function without their distinctive values: values differentiate rules. In other words, the consequences agents will creatively actualize should be another factor in critical moral evaluation, as I argued above, including consequences to others, and to the environment, as a correlate of obligation or duty. Intent is not enough if duty is to others. However, critical moral evaluation is melioristic in tendency, always aiming at improving the world. Our actions will collectively make our world and so ought to aim at improving it. The function of the rules is improvement, including moral improvement in the world. Critical moral evaluation of consequences should aim at improvement, not merely avoiding harm. Agents should creatively actualize the best achievable possibility, the excellent, as a virtue.

2. Morals as the Ultimate Justification of Grounding

Why be Moral? Why are moral values practically necessary? Moral values are obligatory, since in some sense we are required to creatively actualize them: we cannot do without them. In one sense, this requirement is tied to moral value, as opposed to moral principles, since moral principles that have not been creatively actualized are without effect. The imperative is practical. The question we are raising in this section is different: not why creative actualization is required for moral principles, but why we are required to creatively actualize moral principles at all.

First, we cannot do without the environment as living creatures, and therefore have obligations toward it. Society is also necessary for the survival of individuals, thus social obligations are axiologically necessary: we are required to creatively actualize obligations to others. However, moral values are not necessary in terms of consequences alone, the effects of our actions on others. Justice, equality, and other moral values do not refer beyond themselves to some other values, whether utility, wealth, self-interest, convenience, or consequences, for justification. They are valuable for their own sake and thus worth actualization for their own sake. Subordination of such moral values as ends also would have the result that they are no longer actualized as distorted: injustice, inequality, etc. Justice or equality in proportion to wealth or at the whim of the "greatest number" is injustice. However, moral values are also the condition of all social values.

Reductionistic ethics, which for systematic reasons attempts to deny the reality of relations, uses the model of individuals or their psychological states as the basis for morality. But moral issues arise precisely due to our social membership, which is required if only by virtue of the long period of development that babies, especially human babies need. To deny this would be so preposterous it would need no refutation: society and social relations constitute individuals. Individuals are made under social conditions and instilled with social values. An ethic based on the individual's self-interest or ego is an oxymoron.[15] Moreover, relations still remain in such an ostensive "ethic": relations between individuals, between them and their psychological state, between the psychological state and its value, between psychological states themselves, between the values attached to good and bad psychological states, and so on.

Good in the biological sense is what aids survival, although this is too narrow to define value in general. However, no values would be possible without life: life is a condition of all values. In this distinction lies also the distinction of good and right, since what aids life is good in some sense and its good is conditional on life. This relation is reciprocal, not hierarchical, since both elements are required. Moreover, life can be taken in both the individual and collective senses, and both of these have moral bearing. What is in the interest of one individual life may be good for that individual without being at the same time good for all of life. The latter notion brings in social considerations that may mark the distinction between what is good and what is right. The right is the obligation to treat other human beings like oneself, that is, to treat others consistently with our own values: "as you would like to be treated."

What is good may not be right, for example, in conflicts between economic values and social norms. Economic goods are often life supportive, but social norms also protect life. Social norms reflect what is right, not always what is good, either as a good, or what is good for the individual. Survival of individual life requires the biological necessities of life and these are therefore required or obligatory. But social norms are also required for the survival of life: that individuals do not murder each other, *inter alia*. To live as individuals we must survive; to survive we must depend on others in a society; and society requires and is constituted through moral norms. Morality is not a set of ideal principles here but requires creative actualization, moral value, in a society. Creative actualization of moral norms is required for individual survival, individual life. We require others to refrain from murder and violence and they require it from us. The reciprocity of such duties constitutes society and creates right. Human life is social and requires creative actualization of specific moral norms as moral values, the *sine qua non* of social life.

Should society receive priority over the individual in the form of duties to others? The answer to this quandary is necessarily complex, since a number of factors and circumstances may require, determine, or allow for different answers for different situations. When there is little or no consequence for society, there can be no duty to others; few as such situations may be, they give prima facie priority to individual, personal values. Only when another is involved, or many others, can duties to them arise. Other circumstances involve the distinction between moral acts and acts within moral limits; these will be treated subsequently.

Why should we recognize the value of the other? Firstly, we are already members of society and also different communities. There is no absolute "opposition" of self and society given the facts of reproduction, acculturation, participation and the like. Society is not an abstraction here; we are all as much members of society as individuals. No one is "free" of society. Every individual is social if only because they have parents; life proceeds from other life. Parenthood, as Held has rightly noted, is a social relation that precedes individuality and thus is the condition of any "ethic" based on the individual: self-interest, egoism, individual relativism, and all the purely subjective theories (Held in Cahn, Markie, 2002, ch. 39). One's individuality is within a particular society: as individuals we are in and members of a society.

A society's values are expressed in its total culture, and creatively actualized in individuals. Individuals are raised in a particular society, a culture, and this cannot fail to affect their moral views and values. Indeed, their sense of morals derives from the culture. Many of our values are obtained from our society, including cultural stress on individual conscience in moral matters and on calculations of "self-interest." Development of character, achievement of potentialities in youth is socialization. To have values we must learn them from our society: their origin does not lie in the individual subject considered in abstraction from any society. Another consideration in moral evaluation of a situation, then, is our membership in society.

As speakers of a cultural language, humans are already social, in and from a specific culture. Language ability is the ability to communicate to others, involving a community. Each person who has learned speech is already socially acculturated. What is wrong is defined by cultural taboos, which may differ from place to place, in response to habitats. The Inuit practice of geronticide and occasional infanticide is consistent with British Utilitarianism and reflects the harsh conditions of life in the polar regions, the necessity of putting the majority's good over that of every single individual. The lesson of cultural ethics is the influence of the environment upon cultural ethics, the necessity of establishing rules for living that are consistent with survival in a particular habitat. Another lesson is the priority of society to the individual: the individual is a product of a culture as much as the bearer of culture.[16]

Social life is valuable also in that it is life-supportive, and not just in infancy. "Society," in the sense of civilization, a community removed from wild conditions, has transcended jungle life and needs actual norms of behavior to function and continue to function. As members of society outside the wild, we need social relations to survive: our self-actualization requires social others. Indeed, each person is valuable precisely as a member of specific societies, which can protect the rights of individuals. No such rights would exist in the "state of nature," as Thomas Hobbes argued. "Others" are collective members of the same society, with the same rights in principle. Treating the other inconsistently with oneself violates principles identical for all as well as disrupting the social bond. Association with others is also for the good, whether for intimacy or just to improve the world for both. Finally, wronging another is intrinsically bad; causing wrong cannot be a good reason to act.

The goods available through society, the kinds or categories of social goods, require such a functioning society. Such goods include most of the kinds in our lives, since social transmission has made such goods available to individuals. They include institutions of knowledge, health services, an economic infrastructure, a political order, family life, and many other goods. Individuals who partake of such goods require social norms. The categories of goods are reciprocal with social norms: both come in a package: a community of goods available only to individuals in a society. Social categories require a functional society; society requires moral principles, values, and limits. Both of these factors are manifest and both must receive consideration in moral evaluation. The goods of society require social norms.

Can social responsibility be reduced to value for life in general? While moral considerations must reflect human life, what is good for it and right for it, moral values in particular may reflect social concerns instead of value for life. Values are not reducible to value for life, but values must be consistent with life as a whole. Life is the bearer of value while it endures; values that conflict with life as a whole are evil in practice. Social norms protect the lives of others and in this sense are coextensive; but other norms may go beyond the protection of the lives of others (rape). Moreover, morality is an alternative to coercion, that is, violence, and as Hobbes pointed out, generalized violence is an inferior way of life to a non-violent, moral one. An amoral world is one possibility, but not a superior one, for randomized and general force and fraud would soon put an end to all values. Society and social relations are an alternative to civil war and generalized destruction of goods, intrinsic and instrumental.

Such terms as "ought" and "duty" indicate the relation of superior moral values to practice. We ought to tell the truth because it is right, because of the bad consequences of not doing so, and so on. Superior values should be practiced as superior, and they are superior both for themselves and in terms of consequences. While we cannot avoid consideration of consequences, the ancient arguments for a higher, purer morals must also receive their due. Socrates arguments that we should do our duty because it is right and virtuous; and Aristotle's caveat that virtuous acts are done for their own sake, and not for ulterior ends are the spirit of ethics. Realists may scoff that humans act only for self-interested motives, but it is more difficult to argue against the superior value of dutiful and virtuous acts. Acts done on virtuous grounds are good in themselves, that is, inherently valuable. Acts done for the sake of others, in consideration of consequences for society, or reciprocity, are socially necessary. Acts done out of consistency are logically necessary, whether from habit, to avoid hypocrisy, or in conformity with moral principles and norms.

A more difficult problem is the justification of what is right. Just as we may ask, why be moral? so the question of why do what is right may be raised. The answer of moral idealists is because it is right, that is, we do it for its own sake. This point addresses the axiological question, but not the more logical one of the identity of right and "for its own sake." What is right belongs to ethics but for its own sake to values. Are they identical, or at least coextensive? Another problem is that we might supplement "right" with further reasons, namely, that what is right is just and we should be just. We should be just because justice is required for society. The argument may hinge upon conditions of society or consequences for society if we do not act rightly. Alternatively, rightness is justified by the value of justice.

What cannot be contested is that moral values require creative actualization, that is, moral acts are valuable for their own sake. The contrary view would be that there could be exceptions, where immoral acts ought to have priority over moral. Unless we are amoral, such a view is axiologically untenable, since moral acts require creative actualization. This requirement means that such acts have inherent value, and thus that moral values must be creatively actualized for their own sake. Consideration of alternative actions belongs more to moral evaluation than to practice: evaluating what we ought to do prior to practice, what we should creatively actualize. Evaluation of moral consequences also belongs properly to moral evaluation, prior to creative actualization itself. As I noted above, the distinction of value and norms becomes attenuated in practice, since we should pick the best choice. But the best choice cannot be immoral. On the contrary, we may have to creatively actualize moral values to avoid immoral acts. Creatively actualizing the moral norm, standard or principle is an act of moral value. If such acts are distinctly moral, they are done for their own sake and have inherent value.

Since the goods of society also depend on such choices, moral values are the condition of all such goods: the goods made possible by cooperation. Morals are the condition of all values, necessary for all goods of society. Mutual trust is the ground of all possible social relations, but especially of production for market, since a market without trust would soon want for buyers and consumers. If "let the buyer beware" were universalized, markets would soon collapse. Simpler forms of economic life, hunting and gathering also require social cooperation and thereby social norms, as do alternatives such as communal organization. Marriage and friendship also require mutually recognized norms, that is, social values. Above all, family relations, raising the young, require moral norms. Since we all come from such families, the individual has family life, that is, social relations as a condition (Held, ibid.). Cultures that give more autonomy to individuals still require such moral norms: self-interest requires moral values. Values are reciprocal with morals, since they stand and fall together and emerge together. This condition holds for all goods, whether social or not. Even many personal goods require the moral forbearance of others. Values, then, are actualized moral possibilities: the differentiations of creative actualization possible given moral limits. Values are compossible due to and together with morals. Moral norms are the condition of all possible associations and of compossible goods.

Since all social categories require ethics as their condition, ethics is universal. Such universality transcends particular cultures as their own condition, the condition of their own creative actualization as a distinct society with different categories of goods. As James Rachels has convincingly argued, all societies require some rules if they are to continue as societies, for example, caring for the young.[17] Because moral norms are universal, every action should be morally evaluated; every foreseeable consequence of action, including creative actualization of goods, is morally considerable. We must consider the consequences for all relevant others in creative actualization. However, moral obligations do not follow directly from moral considerability. We must consider whether the consequences of our creative actualization are an improvement, but are not required to creatively actualize non-moral values. This norm should be evident from the distinction of moral limits from moral values. Moral values are required, but non-moral values must only be within moral limits (I will examine moral limits below). Their creative actualization cannot be required if non-moral values are ruled out by creative actualization of moral values where they conflict.

The normative character of moral obligations and duties is a further argument for universality. An appeal to duty is an appeal to a moral principle requisite to a case, for example, the duty not to kill applied to a situation such as a jealous rage. The principle would apply to all such situations, for the principle is a rule, identical for all like instances. As such it is universal, although it does not follow that it is absolute.

Ultimate Justification: justification is not equivalent to creative actualization in the world, bringing about a goal in acts of value, or achievement of a good. Agents creatively actualize the goal evaluated worth pursuing and chosen for action as a consequence or result in the world. Justification consists in a critical evaluation of whether we should creatively actualize a goal; the supreme standard for evaluation is ultimately moral. Justification involves moral evaluation of whether we should creatively actualize in terms of moral values and standards such as reciprocity, compossibility with previous goods, consequences for the world and society, and meliorism. Generally, justification is prior to creative actualization and part of evaluation. Justification is not a grounding relation, but consists in critical evaluation of norms and possibilities. Values can consist in creatively actualized norms, standards, ideals, virtues, actions, tangible goods, and so on. But the problematic question for evaluation is whether any of these are of value.

However, duties must be achievable, be capable of grounding as a creative actualization of duty, or they are not duties. Thus evaluation of competing possibilities, justified by critical moral evaluation, requires grounding, while grounding is preceded by evaluation. Grounding and justification have a reciprocal relation, not that of cause and effect, or substance and attribute: any *pros hen*,[18] substantialist model is rejected. In one sense, justification may consist in having good grounds for action, so attenuate the distinction. But the grounds for justification are distinct from the grounds for creative actualization.

Grounding is teleological in structure for creative actualization. The relation of ultimate justification and causality is that of value to telos. Teleological grounding uses cause in a double sense, namely, the goal as preceding the act and "causing" it and also as "ending" it by its achievement. Thus a goal is initially a projected ground, since reference is made to an imagined future consequence. Different "ends" are the cause of acts in one sense, as goals, since they precede the act as such. In another sense, the ends are the effects or result that end the action as successful accomplishment: creative actualization.[19]

Justification is more ultimate than creative actualization, involving a relation of means to ends justified in terms of its value: the most valuable hypothetical goals among competing ends are creatively actualized. Values justify our grounds for acting, as the relation of competing goals. Their justification is their value—personal, categorial, cultural or moral, including their value for life. A goal is chosen and put into practice because we evaluate it more worthwhile than another, a more valuable goal. Agents make the evaluation that its creative actualization will exceed alternatives, competing practical hypotheses, in worth. Creative actualization or value will prove or disprove this hypothetical judgment; achievement of the goal reveals the value of undertaking it. If it proves deficient, it will be revalued and more valuable hypotheses may emerge, and be creatively actualized. "Justification" and "cause" or "ground" are distinct, then: justification is not causal, instead it justifies causes as goals. Thus cause is of secondary importance, not primary. Justification is more ultimate as the creation of new relations between discrete pairs can be justified only in terms of value: the worth it reveals. Evaluations of values justify practice, or what agents creatively actualize, and thereby establish relations of competing ends. Agents are justified in creatively actualizing a more valuable or worthwhile practice. The future is ameliorated: the world is better than before.

The issue of justification does not consist in actuality as such but whether any actuality or possibility should be part of any world. Of any goal, activity, action or other practical affair, it should be asked, is it worthwhile? What is the worth of such potentialities? What is their overall value for life or their impact on life? Are they melioristic, that is, will they truly improve the world as projected? What is its good for society? Grounding is not enough. Potential values must first pass a critical test, of whether a potential value is worth creative actualization, that is, worth acting on. Standards, improvement, meliorism, and other norms are all tests of action, that is, whether action to creatively actualize a difference in the world is justified. Evaluation of such consequences is not grounding as such, which is the result of action and tied to practicality, feasibility in the world. Justification raises the issue of whether the action is worthwhile.

The question of the justification of creatively actualization is ultimately moral: should a possible value or good be grounded in the world, actualized? What of nuclear war, or the clear cutting of a forest? As justification is ultimately moral, a choice or decision should be just and equitable in consideration of all moral factors. The connection of justification and justice is evident from their etymology, but justification has several meanings, some of which have strayed from this connotation. Justified means treating possibilities and ideals justly, since they have been considered morally and evaluated in terms of actual grounding. Justice as a moral value constitutes a reciprocal relation; each inherent value is given moral consideration. Justice must be done to values subordinated in evaluations by moral consideration. For justice to be creatively actualized requires both regulation by norms and a reciprocal relation of inherent values.

Since moral norms like justice are the ultimate appeal, hierarchy is involved in moral evaluation, as morality may forbid some creative actualizations, for example, murder. Moral worth is higher in evaluation than other potential goods. But since this hierarchy aims at justice, equality and other moral norms, it allows reciprocity of compossible inherent values, dissolving the tension of hierarchy and reciprocity. Values of similar kinds may be reciprocal even in evaluation, for example, different moral values. Absolute reciprocity would dissolve differences of value and thus reciprocity between values. Justice should be done to these differences. Only morals can justify hierarchical evaluations that exclude alternatives, since the standard and goal is justice.

One standard in evaluation is compossibility with actual values, involving any future relation of values that may become creatively actualized. New creative actualizations must be compossible with previous goods. Critical moral evaluation regulates creative actualization with an eye to compossibility, reciprocal actuality with previous goods. Moral evaluation thereby does justice to the worth of previous goods, justifying creative actualization as just to other goods and satisfying purely moral requirements. Destruction and violence are minimized.

The justification for all action is ultimately moral, since all evaluation includes a moral test and no new value escapes moral requirements for justification. No action or activity is value-free, but is actualized with or without moral sanction. As I argued above, moral values are required values, or values that ought to be creatively actualized. Moral evaluation regulates the relation of values, actual and potential. Potential values are considered in evaluation and creatively actualized if they pass a moral test. Moral norms and standards form the supreme test within evaluation. (Other elements are also analyzed and evaluated in terms of a hidden or overt value hierarchy). Kinds and categories of value are creatively actualized through action, which critical moral principles limit. All ends must be within moral limits to be worthwhile, justified grounds for action: they must pass the moral test even if they are not moral goals. In such cases, morals limit but do not determine all actions. If an action is taken to prevent creative actualization of an immoral goal, it has moral value.

Each creative actualization is a differentiation, as I argued in the previous chapter. Even creative actualization of an instance of an old kind is a novelty. Since creative actualization is justified in evaluation, which justifies such differentiation in terms of worth, the instances are inherently worth differentiation. In moral terms such instances are compossible with other inherent goods and a creative actualization results in a just relation of reciprocity. Hopefully, creative actualizations of entirely new kinds for the world are melioristic. As instances creatively actualize their kind, a normative identity, norms are also justified by evaluation of the creative actualization of the instance. Differences of kind that are compossible are reciprocal. Principles and standards are also normative and must be justified;[20] to justify is normative in the sense of critical evaluation of moral standards, and evaluation includes establishing a proper ranking for principles. The principles and standards are also creatively actualized as kinds of action; value justifies the pursuit of ends according to rules, while ends actualize and prove norms that direct ends. While evaluation is prior to action, and although moral principles are the ultimate justification, the relation of potential worth in evaluation to value is reciprocal, since norms provide direction for ends but are embodied in the world as results. The practical justification of ends is completion of the goal, achievement of value. Value, creative actualization, justifies the activity retroactively as success or completion in a context or specific situation.

In sum: when is creative actualization justified? Creative actualization is grounding of a goal, and is of practical value when success is achieved: we have grounds for the act as successful. But prior to creative actualization the question arises for evaluation: should some good be actualized in the world? This question raises the issue of the value of a value, of justification distinct from actualization, since creative actualization can be critically evaluated. The evaluation itself is in part practical, raising questions of feasibility, means, timing, situation, and the like. Ultimate justification is ethical, however, since creative actualization is morally justified if it lies within moral limits, including reciprocity, compossibility with other goods, and meliorism.

Otherwise, the project should not be creatively actualized, as it would be a violation of morals. A goal is grounded if it can be creatively actualized. But the end is justified only if it is reciprocal with other values and melioristic, both as superior to alternatives and an improvement of the world. Agents who limit their activities in accord with moral principles and norms creatively actualize moral values. Creative actualization of duty is the conjunction of ought and the world. Their ultimate warrant is ethical, wherein all is as it should be through creative actualization of morals.

In evaluating justification, one axiom is that an appeal to lower principles would not provide ultimate justification, when more important ones are at stake. While other factors of evaluation can provide justification, only morals can provide ultimate justification. Ultimate justification requires evaluation of ultimate worth, of the most important values, and thereby evaluation of importance and worth. Morals are first in worth, since they are the condition of other values, in several senses. Creative actualization of moral principles has reciprocal relations of inherent values as a result. Each value is given its just due, its inherent worth. Morals are thereby compatible with each value and allow categories of value to self-ground, as autonomous and distinctly worthwhile within limits. The limits provided by creative actualization of moral principles thereby avoids the subordination of inherent values into instrumentality as well as the overextension of any category. Morals are not a cause of the inherent value of an instance: from morals we cannot deduce the content of any inherent value or even the scope of value. Morals, unlike metaphysics, do not delimit the content of value and thus allow for compossible novelty. By limiting values, their reciprocal autonomy is made possible. Inherent value requires moral principles.

Moral relations are ultimate as the final justification of differentiated values. Such relations are the condition of inherent values; they are ultimate, as no further axiological justification is needed for inherent values: they are valuable in themselves. Beauty does not aim at knowledge and vice versa. Values are self-justifying as valuable in themselves. A further proof is the normative opposite of such values, namely, that beauty is "proof" of the inherent value of beauty—who would prefer an ugly world? Since we generally pursue beauty rather than ugliness, a pragmaticist justification of the inherent value of differentiated values is evident, that is, the practical consequences of such values. However, practicality should be subjected to critical moral evaluation, as I already noted.

Not all ideals should be creatively actualized: only worthy ideals, not evil ones, should be brought into the world. The Nazi "ideals" of a pure "race," and the like are axiologically perverted. The evaluation of ideals must precede their actualization. Such evaluation presumes some standard or principle of right and wrong, good and evil. Bad in the practical sense is what is not feasible, either a defective instance, in which a goal has not been achieved, or as a whole type (time machines: but never say impossible. Time machines are unlikely, but that is what was originally thought of flying machines.). Bad types cannot be creatively actualized. In the axiological sense, what is bad is what is not worth actualizing, for example, many of the trashy consumer goods of our society. What is bad may also share several features with evil, in cases where creative actualization is not compossible with other inherent goods. Nuclear war could be actualized but would destroy most goods, reducing the total quantity of good: the net effect would be bad as well as evil.

Evil is trickier to characterize, but in one sense is a deliberately bad choice: where bad consequences are chosen over good ones, immoral over moral ones. Evil is destroying goods, good acts or good people. Thus genocide, the deliberate destruction of a kind is the paradigm of evil (cf. Rwanda, Bosnia and other genocides with species extinction). There is no other word for destruction of species than evil and an age that refuses to draw such judgments is morally soft. Random acts of destruction of goods, vandalism, are of this type.

Subverting value is evil, as is a choice that is deliberately destructive of intrinsic value bearers, their society, culture, or habitat. Evil can also attach to subverting another person (Iago) or turning them toward evil. Evil also consists in subordination of intrinsic value, using self-actualizing agents only as a means. Nuclear war would also be immoral, since it subordinates the intrinsic value of persons and living things to some single end, in a hierarchy. Pollution and overpopulation are slower forms of the same result.

Evil can consist in hating goodness, good people or character, or good acts and wanting their opposite. Anti-Semitism has been described as "madness," but whether hatred and envy of so large a group of people is a form of insanity is at best ambiguous, since there are no symptoms. Again, poor rednecks who hate competent blacks that have or are capable of doing well in the larger society hate the good in people who are capable of good. Evil can consist of deliberate exclusion: of individuals from society, of groups from participation or moral consideration (the Nazis).

Power exercised for its own sake without moral ends, but done in the name of another, is evil, for such power is destructive of justice, and other bonds of society. Since power of this sort is based on violence and coercion, evil also may be described as generalized violence, whether in war, especially civil war and total war, or terrorism or destruction of habitats. In general, deliberately creating strife and disunity for its own sake is evil. Violence is anti-pragmatic, despite judgments to the contrary. Violence is used in attempts to work out some problem by forcing the issue, instead of allowing the intrinsic merits of some idea, proposal or program the time to allow its value to manifest itself, or to persuade others. If something is worthwhile, it does not require violent imposition precisely because its merits are inherent: its worth manifest. Misinformation is a form of evil: spreading lies.

Moral Limits as the Condition of Social Goods: if creative actualization of moral norms as moral values is a possible action, what is the relation of such norms to creative actualization of other, non-moral values? In some circumstances, moral principles are creatively actualized directly, for example in acts of duty or virtue. However, not all situations directly involve duty or virtue. Simple exchange of goods, trade or purchases, are neither duties nor virtues, at least directly. However, moral principles also regulate such activities in different forms. One is the honesty of both of the parties, for example that the seller is not selling shoddy goods or the purchaser exchanging counterfeit bills. The issue of creative actualization of moral norms as limits then arises.

Such principles are regulative not only because of what is "present," or actual, but also what is *not* present. Theft has *not* been creatively actualized in this situation, but instead, exchange. Armed robbery has not been creatively actualized. Murder to obtain the goods or the cash has not been creatively actualized. What *is not* "present" may be as important as what is present. The choice not to steal, use a weapon or murder one of the parties is not present in the choice to conduct an exchange of goods. Such a choice may be made on prudential grounds: fear of arrest, imprisonment, or other penalties of the law. However, it may also reflect moral principles as limits upon actions. By limiting actions to the moral choice, or choices, moral limits may constitute a moral order, a civilization. Since moral limits may not be directly "present," but limit actions by forbidding, excluding, or proscribing them, they cannot be grounded in what "is," what is "real," "experience," or any other metaphysic or ontology of "presence." But it does not follow that moral principles are somehow without a ground. On the contrary, they are grounded by creative actualization of moral actions. Moral actions may include acts of exclusion, acts limited by moral principles. Excluding murder because it is wrong has the consequence that murder is *not* "present" in a situation: not grounded.

Moreover, moral principles are the condition of other values, by providing a moral order, by excluding *bellum omnium contra omnes*. By limiting theft and murder, exchange is possible and ultimately an increase in overall prosperity. Moral principles in this respect provide justification, but not grounding for non-moral values, such as wealth or pleasure. Such moral values are not "ontic," not "things" or goods of the world. Moral principles may not be manifest as such, as creative actualization of purely moral values, but are the condition of non-moral values. As values they "are not": the limit is constituted in what choice is creatively actualized by contrast with what is possible. Morals are regulative principles that constitute actions that ground non-moral values.[21] Some possibilities are excluded by such norms. "Limits" may be equivocal in actions, since they may be prohibitions of determinate means and ends that vary applied to different forms of immorality, whether theft or murder. Not acting at all may in some cases be within moral limits: the non-actualization of immorality, its non-presence. However, moral principles may also constitute non-moral values that are within moral limits. Creative actualization of the morally justified non-moral good also lies within the limits.

Morality is a limit on actions done in a social context or with social consequences, and thus limits all acts in social situations. Slave labor camps that maintain the power of an elite or child labor that is profitable is immoral, since evils are creatively actualized that are not justified. Moral values are the supreme condition in justification of non-moral values. Moral limits should be considered in evaluation of all actions and are thereby a universal condition. The ultimate warrant is ethical, as I argued above, and all actions should take moral principles into consideration. Moral values are universal for evaluation and should not be violated, a limit on all actions. Actions that creatively actualize non-moral values that have been evaluated as within moral limits creatively actualize moral limits as a principle in all justified actions. Thus creative actualization of non-moral goods can also creatively actualize moral principles in what has been excluded from creative actualization, even though the action is not directly moral. Creative actualization of moral principles is not limited to direct acts of duty or virtue, but may consist also in acts of forbearance, exclusion, or choices limited by moral principles. Moral considerations limit action, but not by bringing them to an end; they exclude actions from creative actualization or limit them to morally justified actions. The frequent use of negative expressions in duties that forbid actions: "ought not to break promises to others" or "shall not kill," indicates this limiting aspect of morality.

Morality frequently limits but does not direct our actions. Within moral limits we are free to creatively actualize alternatives. Since moral limits are only a limit on non-moral values, and not their ground, the goal of non-moral acts cannot be evaluated on a moral basis for its inherent merits. Economic, religious or medical decisions are made on their own grounds. The decision whether to perform a medical operation is based on the condition of the patient, the need for the operation, the probability of success and other medical factors. However, moral limits exclude medical experimentation of the Tuskegee sort, or the types conducted by the Nazis or the Japanese military during World War II on unwilling victims. Ambiguous value situations are not, then, decided on moral grounds, apart from those that are morally ambiguous, or "hard cases." Marginal returns on investments may be within moral limits but of little economic value. It should be noted that some acts are only limited under specific circumstances, but can be postponed to other times or situations. In such cases conflicts of values are attenuated.

As a universal consideration in evaluation, morality is also a limit on all kinds and categories of value. Morals norms are not the source of authority for the different categories of social goods, in a hierarchy of value. Instead, moral norms provide limits within which distinct values are possible: a moral locus. Actions that creatively actualize a category should also be within moral limits, if they are ultimately justified. Since this norm applies to all of the categories, they have equal moral worth as categories of value. The categories of goods within moral limits are external to one another, that is, they have the same reciprocal relation as social agents and require the same mutually restrictive norms. Limits to each good allow the others their due. To "exceed bounds" would indeed "tempt the fates"—it would indeed jeopardize the distinctive worth of any category of good, whether economic, esthetic or other, in a community of compossible goods. The value delimited by each category would no longer be distinctive if it were submerged in that of the other, as an instrument, based on a hierarchy of value.

Since creative actualization of moral limits is voluntary, freedom is a moral value: morality is a substitute for compulsion. Moral principles constitute a non-violent requirement, as an ought, to perform or refrain from specific acts. They exclude violent compulsion with self-regulation. While we can commit immoral acts, they are not justified, and tend to subvert other values. Lies subvert truth, the condition of knowledge. Morals are not self-enforcing, although societies may set up formal enforcement mechanisms; limits are observed freely and voluntarily. Freedom and duty thereby have an internal, reciprocal relation, since an involuntary act cannot be a duty. This internal relation is distinct from the external relation of distinct kinds of value, since inherent value is not conditional.

The categories of value are externally related, if the limits of each are not violated, but as the relation is external, they are not conditional upon one another. For example, art and religion as such are not politics or politicized. Supreme moral values as a limit, not a ground, have the consequence that the categories, as distinct kinds, stand in a reciprocal, external relation and thereby derive their value inherently, not from a relation to another value. There can be no categorial ends in themselves without moral limits. Moral limits as limits do justice to the inherent worth of each kind or category of value. Only morality can limit itself and thereby be just to non-moral values. Moral limits as limits allow categories their due, the full development of their potential inherent good. Creative actualization of different kinds of values in numerous collective actions over historical time constitutes the manifestation of reciprocity.

Recognition of the worth in other values, the good of the kind, does justice to them. Morality as limits in relation to other values is self-limiting; justice is a moral value, and a supreme requirement of moral evaluation. Justice gives relative autonomy to each category such that its inherent worth can be creatively actualized. Justice is the supreme standard for morals and should regulate our evaluations. Since evaluation requires morals and morals follow justice, justice is basic to all creative actualization. The structure of morals should be modeled on that of justice, in the form of reciprocity. Such reciprocity is not confined to relations of individuals, but of all values, including categorial, cultural, and personal values, since these are inherently valuable. Justice maximizes inherent value, and justifies differentiation of kinds of values. Justice is the moral value in which the reciprocal relation is creatively actualized as a relation.

Morality justifies pursuit of the categories of value, but does not provide their goals. The goals and the worth are distinct, for we can pursue bad goals, or good goals in a bad way. The worth of these is compromised in the latter cases. Their value must then be delimited by morals, not categorial goals as such, for the goals may prove to be unjustified and thereby without worth. Small lies compromise the worth of a new experiment in science, even if much of the information has worth. Similarly, injustice detracts from the worth of economic life. The value of goals, not their content is subject to moral evaluation. To prove worthy, then, the goals must lie within the limits of moral norms.

Morals do not ground categorial goals, nor can anyone evince such goals from moral premises. Moral truth is not the ground of intellectual truth, for it only limits pursuit of truth and does not provide its content. Determining whether 200 distinct species of Iris occur in the world is not a moral matter but a biological one, although a deliberately inaccurate count would violate moral limits. Insofar as intellectual life is social, we should be moral. Truth telling is an obligation in a society and we should not deceive others. Indeed, moral and intellectual truth are closely related, for there is no intellectual truth without moral and no moral truth without a category of truth independent of other standards.[22] Further, moral limits may be variegated in terms of the kind. Truth in the moral sense has to do with prevention of lies, not of murder. However, covering up a murder may constitute a lie.[23] Such duties and virtues are valuable for all activities, but may apply particularly to one category.

The idea that some actions "make no moral difference," then, must be evaluated in the light of such distinctions. The phrase does not rule out moral evaluation of acts, nor the universality of moral norms. It may be a recognition of the autonomy of non-moral values. But it is hard to imagine that any act could make no moral difference whatsoever; the phrase as it stands is overly general, since the degree of moral worth or harm is not considered. While private acts of individuals may not affect others, they may be deleterious to character. A gambling vacation to Las Vegas may not have consequences for others, but may be substituted for other activities that are ultimately more worthy of an individual's talents or time. The degree of disvalue may be minor, but even such minor moral considerations argue against the implications of this phrase.

Moral principles and standards, as supreme in evaluation over other standards, regulate choice in action, not the content of the categories of value. Morality rules out the bad and evil without limiting actual improvement. Since moral principles are always supreme, the question of conflicts between other goods and morals cannot arise, but only conflicts between non-moral goods. The limits on goods and values are provided by supreme moral principles as limits, and creatively actualized in acts that also creatively actualize categories of value. Morals and the categories are distinct modes of value, with a mutual relation, for categories must observe moral limits for justification, while their reciprocal relation creatively actualizes norms of justice. They are mutually supportive, since creative actualization of the categories also actualizes a moral relation. Indeed, the better the worth of the categories, the less likely moral norms will be transgressed. The categories and moral limits are creatively actualized together as a community of values.

Values are also creatively actualized in a culturally determinate form. Different cuisines serve the same ends, nutrition or health, pleasure, and sometimes intimacy, although none of these can be reduced to the others; they are distinctive kinds of good, or categories of value. Different varieties of valued ends result in different cultures; we speak of Greco-Roman values, Judeo-Christian values, modern European values, and so on for other cultures. Morality is the condition of possibility of all cultures as the limit that provides categorially distinct kinds of goods, culturally mediated. Cultures also stand in reciprocal relations.

The categories connote kinds of values that are inherently valuable as distinct kinds, although they may not be regarded as equally valuable by a particular culture. Cultures that place a premium on economic values may have little patience for intellectual, spiritual or aesthetic goods, and may regard the world as a giant resource. Ascetic cultures may reverse this valuation and look suspiciously on wealth, even though economic prosperity is also creatively actualized, a valid kind of good in itself if it lies within moral limits. Ascetic cultures only tolerate wealth if wealth is subordinated to a higher authority: wealth is not then an autonomous category, not a value for its own sake. Only so much wealth can be allowed, and the limit is formulated in a way that will allow for its subordination to a religious or other authority.

Only moral limits can be expressed, not some sort of "essence" of morals. In the form of limits, a series of rules constitutes a philosophy reciprocal with the categories of value that cultures creatively actualize. Morality is an ongoing, historical affair, not a static one. Regard for the environment is a new norm that human predation has forced us to recognize at this stage of history. Institutional racism is slowly disappearing, if not personal prejudices. Thus there is no "nature" of morals in the sense of "essence," only moral principles, creatively actualized. Moral progress implies an increased range for morals over time, however. Furthermore, different nations, cultures, and other social groups follow different norms, an argument against some forms of "essence," although not all. The problem of determining an "essence" of morals is a legitimate problem, but is addressed by the model of morality as limits. An essence defines the logical core of a concept, principle or other norm. A definition, "concerning bounds," is a limit, the boundary of the norm. Since morality as a limit bounds, a definition in term of limits is feasible. As for "nature" as a model, a reciprocal relation of distinctly valuable kinds of goods is an improvement over and departure from a uniform nature for a species, and the uniformity of acts all aiming at survival. Our norms must now be derived from the model of justice, since nature provides contrary norms. For example, shall humans follow predator or prey species as models? (Warriors often identify with predators.) But natural law and natural right models provide so many contradictory standards that they are not helpful. This is to speak nothing of the escape from natural habitat by humans: civilization and adaptation to cold. The environmental ethicists have made the case, however, for thinking of humans within an environmental whole. George E. Moore's arguments that identification of good with anything natural is a fallacy is itself a fallacy—goods are in the same world as anything else—but nature cannot be a source of moral norms for a creature that has subjected "nature." (There are also theoretical problems with nature (McDonald, 2000). Humans and domesticated animals are not jungle creatures and, indeed, both domestic and wild creatures can inhabit the same world, an argument against the strict identity required for a "nature" for all life. Different creatures live different lives. By experiment we discover limits, not natures, since many goods of the world have more than one function.

Morals are supreme as an epochal principle, not as a fixed hierarchy of goods. The differentiation of cultural goods from the wild, the achievement of civilization, marks a new epoch in the history of the world. Moral relations of cooperation among individuals are the condition of these novel kinds of goods, the practical or social categories. The end of morals would be the end of culture and civilization, since the categorial goods are creatively actualized within moral limits. Although this effort is human driven, creative actualization has as a condition an environment that allows punishing treatment, notably, meristematic growth in plants, which permits grazing, and thereby domestication. We are conditioned as much as we condition: a reciprocal relation. We channel water for storage and irrigation of prized plants, but cannot get by without water. We heat the air in order to survive cold climates that are far removed from our original habitat, but still require air, water, and nutrition to survive. Our relation to the environment we alter is reciprocal; we control it but it controls us. We are co-conditional in relation to the new goods of culture, and the sooner we learn this in relation to the total environment, animal and plant life, the better.

I am not going to discuss cultural relativism in any detail, since that has been competently done by others (Ladd, Rachels, ibid.; McDonald ined.). What I will note is that cultural relativists argue that the conclusion we should draw from the actuality of other cultural practices is tolerance: we should be tolerant of other cultures. However, tolerance as a norm for treating dissimilar cultural practices constitutes a universal, not a cultural norm, which violates the thesis of culturally relative norms. Cultural relativists also have not argued for or indicated why we should be tolerant of other cultures if our norms are culturally relative. Why be tolerant of other cultures, if our norms come from our own? Finally, they have not discussed another corollary of cultural relativism as a source of moral norms, namely, that we *should* follow our culture's norms, since these norms are the ostensive source of norms. Few would argue in favor of such a corollary, since it both constitutes a universal norm, contradicting the thesis of cultural relativism; and is hardly a valid moral principle, given cultures such as that of Nazi Germany or extreme fundamentalists like the Taliban.

Cultural relativists have stressed the differences between cultures without noting the equally striking similarities, notably, the taboo against murder, theft, and lies in all societies, at least or especially within the "in-group" (Kluckholn, and Redfield, Ladd, 1973). Since such taboos are not different they are universally forbidden by all cultures. If all members of a society murdered, the culture would not survive; thus murder is wrong from a cultural perspective. A culture with ex-centric values does not constitute an exception to this universal moral norm. Nor does the cultural relativity of some rules make them any the less morally compelling. Equality for women is not a rule for all societies, but that does not make it any less valid a principle. Western science emerged from one culture, (although there were elements derived from non-Western culture, notably, the invention of algebra) but the law of gravity applies to all. If tolerance is required in our dealings with other cultures, reciprocity is also required from them, regardless of their cultural norms.

Finally, and as others have noted, moral progress can only be measured as a universal standard, whether of rights, values or moral principles (Rachels, ibid.). Would anyone seriously argue that the liberation of slaves is not an improvement? Or the belated recognition of the equality of women? Anyone who seriously defends inequality should consider basic ethical principles, like the Golden Rule, a similar principle in Confucius, Kant's Categorical Imperative and other variants. What if you yourself were a slave involuntarily? Few would defend inequality or absolute cultural autonomy if the issue were posed in these terms. Moral progress is the condition of other forms, including intellectual progress, and is measured in both universality of moral consideration and Gandhi's standard of the reduction of violence. By both standards, humans have a long way to go, given their treatment of other species.

Although inherent goods and values are indeed good, their good is finite, if only because of the limits of other goods and values. Goods of the world are good but not "absolute good." Since their good is not absolute, it requires further justification. Moral limits provide such a justification through reciprocity, compossibility, moral progress, pluralism, and meliorism. By limiting the categories of value, supreme moral values allow, and in some cases give them their due: the difference in worth that is theirs. Limits allow for the value of the kind by differentiating one from another. Differences require finitude, separation from the infinite and the other: finitude implies limits that separate kinds. Ends and values that have no limits would be all-inclusive: a monistic one or absolute identity. Kinds imply differences and plurality. The categories of value achieve different ends. For these to be compossible, they must be limited, and this is a function of supreme moral values in evaluation. The distinct value of each category is differentiated by moral limits and thus a function of the extent to which morals limit pursuit of ends.

We should not conclude from a discussion of moral limits that direct moral acts are not important. Duties to others may be creatively actualized that are essential to social functioning. Trust, the key to the virtues, is a requirement of any healthy social relationship, the condition not only of intimacy, but of medical practice, economic life, and other types of goods. Creative actualization of other acts of virtue may be valuable in themselves, whether helping someone in need or admitting unpleasant truths. Again, what is *not* present, murder and rape, may also constitute moral choices and make what is present, a good society, possible. The non-"presencing" of evil, its "non-being," is all to the good. The act of avoiding the temptation to do evil, of not creatively actualizing evil and bringing it about in the world is a moral act. Moral values are not ontic and their presence may be hidden.

3. Principles of an Ethic: Reciprocity, the Other, Compossibility, Pluralism, Meliorism and Moral Progress

Reciprocity: the structure of a radically axiological ethic is neither foundational nor hierarchical, but reciprocal. The "golden rule" involves a reciprocal relation of self and other, and similarly, creatively actualizing some social category reciprocally involves creatively actualizing morals. Morals are the condition of the categories of value, to be sure. But we make a moral choice by pursuing the categories: the two are interdependent. Morals and other goods both represent an improvement of the world that is a conjoint creative actualization. Similarly, the categories of value serve individual lives. Life belongs to individual agents. The relation between personal and categorial values is reciprocal. The person chooses and creatively actualizes different kinds of value that in turn benefit persons as means. While the personal values of an agent can be deferred for categorial values, the latter are limited by moral values that serve such agents.

The reciprocal relation of moral obligation and creative actualization may be taken as the first moral principle in an ethic. Such an ethic presumes principles and relations, that is, both that norms can be creatively actualized and inherent values. Relations have already been justified: to creatively actualize norms is to act on a principle, which in turn involves a relation of instances in some respect. If a norm can be creatively actualized, then principles and relations are grounded in the world. However, the relation of different kinds of goods may involve plural principles. What is their relation? Each creative actualization results in an inherent value or good. Inherent goods do not derive their value from another good, or they would not be inherently good. Plural inherent goods thereby require a reciprocal relation, in which the value of each is given its due. A moral relation of inherent goods is reciprocal. This principle is valid for both goods of the same kind and of different kinds.

Reciprocity constitutes a relation of equals, in which neither self nor other are subordinated by hierarchies, social, political or axiological—and is also the relation of value and obligation, since both are required for moral acts. Duty involves the recognition of others as equal, not subordinate in value. Duty arises primarily with respect to others, but does not put the other ahead of or above oneself but as equal in value. Since universal norms and principles imply equality, equally valid for all instances, equality is consistent with them as a moral value creatively actualized in moral acts of value. Moral values are required also in this sense: we ought to recognize the value of the other as a reciprocal moral agent who is morally equal. Intrinsic value belongs to such agents as self-actualizing actors who creatively actualize moral and other norms. Since they have creatively actualized their own value, for themselves, they are valuable in themselves: intrinsically valuable. Since duty toward self by others is required for my self-actualization, creative actualization of a reciprocal duty is required toward others as an end. The consequence or outcome is the reciprocity of self-actualizing agents, with their own intrinsic value. The value of reciprocity as a required norm or principle is also the condition of plural intrinsically valuable agents, forming a common society. Creative actualization of moral norms is a condition of such a society and thereby required to achieve such consequences. Reciprocity, the condition of society, is a moral requirement whose creative actualization in relationships bears moral value. Indeed, reciprocity is the condition of all moral values, whose creative actualization is involved in all social values. Reciprocity is the normative element in all acts with moral worth, even those constrained by moral evaluations.

Treating moral agents as ends in themselves is also an axiological view. This view is consistent with Kant's treatment of persons in the second statement of the Categorical Imperative. Norms in Kant are not for the sake of rules but for the sake of morals.

Reciprocity is the first principle of an axiological ethic. I noted above that moral principles require creative actualization, but creative actualization has moral conditions. Value, then, is a condition of morals, but morals are a condition of values: co-conditions. Ethics justifies value by reciprocally limiting goods for their compossibility. The relation of value and obligation is reciprocal, bringing differences under one principle. Justification, in the sense of moral evaluation, precedes creative actualization, but justification is ultimately grounded by creative actualization, or it has no efficacy as a standard. Value and evaluation also have a reciprocal structure; inherent value is a standard for future worth, but such standards are justified comparatively in evaluation. Evaluation requires creative actualization but creative actualization is justified in evaluation.

This principle extends to the relation of other kinds as well, for example, to the relation of categories of value. Knowledge, power, wealth, health, and other kinds should not subordinate other goods but constitute a reciprocal relation with other goods. The relation of the basic value to intermediate ends, values, and norms is also reciprocal. The ultimate end is the basic value, but such basic kinds are actualized in the same act as the intermediate ends, norms, and values. For example, a particular act of taking medicine may actualize health, creatively actualizing a norm and a valued end simultaneously. The ultimate value is actualized with particular creative actualizations; since its value is actualized with them, its relation is reciprocal with them. The projected end would not attain value without particular creative actions. The value of ultimate ends is actualized as particular cases of principled actions. Ultimate values are often actualized as values through particular actions and would not otherwise be actualized. Actualizing particular actions is reciprocally actualizing ultimate values.

Reciprocity also characterizes the relation of evaluation and action. I argued this point in the context of obligations; the norm of reciprocity can be extended to the relation of evaluation and action in general. Action is required to creatively actualize the norm, but is guided by the norm. Both are required for creative actualization, the value of the norm. Action without norms is "blind"; norms without creative actualization through action are Utopian and impotent. While they involve distinct moments, their value requires reciprocity. There is a reciprocity of "tasks" or moments in creative actualization required for values. Prior to action is justification of activity in terms of projected value, moral and critical evaluation. Ends must be evaluated, a choice made between competing goals and adaptation of one or more as goals. Means should also be evaluated. Action is then taken to creatively actualize the goal. The action results in changes in the world that are for the better, if the action achieves value. Each plays a necessary part, is thereby required and so justifies a reciprocal relation. Similarly, creative actualization justifies means, but means are required for creative actualization.

Unlike the various dualist value structures—hierarchy, foundationalism, reductionism, and so on—reciprocity allows for inherent value. The value of an instance does not refer back for its value to a justifying first, such as happiness, but is valuable for its own sake. No "referring back" of inherent value can result in inherent value. Nor does its value require a metaphysic of value, whether the "being" of art or the "experience" of art, but the value of art of its own. Art is not instrumental in ultimate value, as "revealing being" or other extraneous purposes, but of inherent value. However, again unlike the dualist structures, the relations are not fixed, but indeterminate: there is room for creativity within limits. Being cannot thereby fix relations for all eternity; the subject does not give value in a reductionist hierarchy. There can be no "ruling" principle of value for inherently valuable goods.

Reciprocity is the principle of justice: justice, the most important and basic principle of social life, is creatively actualized as reciprocal in structure. The reciprocity required for moral value is justified as just. The relation of intrinsically valuable agents is just, not hierarchical. Their value does not derive from an extrinsic source, which would devalue their intrinsic value. Since it is not extrinsic, it cannot be hierarchical. Each agent has intrinsic value; each is equally valuable. The equal value of each cannot be subordinate to other agents and thereby requires a reciprocal relation of equals. Equality is a moral value as the *condition* of ethics—that each individual is capable of acting morally and thus equally capable ethically.

Equality does not require identity, or uniformity, contra Nietzsche. Differences can be as reciprocal in structure as identity; reciprocity is required if different kinds of goods are possible. Since differences in kinds are one measure of good, increasing kinds is for the good. More, actualization of inherent value requires justice to differences, to the distinct value of the inherent good or value. Justice to each demands that its distinct value not be conflated. The condition of such distinctive worth is a supreme value in evaluation that involves and allows for differences. Justice to the worth of each is required, for distinct kinds and levels of value require a ground justifying the worth of their distinctiveness. Similarly, different life-enhancing values require relations of justice to constitute distinct values. Justice is reciprocal with such life-enhancing values, since creative actualization of different kinds is creative actualization of their reciprocal external relation: justice for each. Justice is not only an ideal but is actualized in the equal axiological relation of different values and goods. The result is an inseparable community of values in the world, which can accommodate new differentiations and relations that better the world, not a mere "ought." Where justice prevails, a vertical, hierarchical, sovereign authority is superfluous.

Justice is required if different categories of value are equal in their reciprocal difference. The inherent worth of the kind is distinct as a kind from other kinds; this worth would not be actualized without justice as a condition, that is, that each category be treated as valuable in itself. The worth of the kind as different should be independent from other values, or its value is not in itself. Thus no value "underlies" such independent values as the essence or "substance" of value. A just relation to other kinds of good enables each kind to receive its due and thereby reveal its distinct, autonomous worth.

Moral equality involves reciprocal differences—indeed, moral equality could be defined as reciprocal difference, thus equal autonomy. Autonomy is required for moral agents, or indeed, any intrinsically valuable agents, self-actualizing agents. A "higher" identity consisting in a universal identity, in which intrinsically valuable agents are instances of a single norm, is an absolute equality, incompatible with intrinsic value. For the relation is not reciprocal, but derived from an extrinsic, supreme principle. Such a principle would subordinate intrinsic value to extrinsic value, and subvert the value of the bearer, as well as the difference that inherent value actualizes. The only truly moral relationship involves equality, in which others are treated like oneself ("doing unto others..."). Duty to others can never be destructive of their inherent value, only supportive.

Reciprocity is thereby a duty, a moral principle. Since all values require reciprocity, reciprocal relations should be a limit even on creative actualization of non-moral values. Several norms can be derived from reciprocity that provide a basis for ethics. These are justice, compossibility, meliorism, and participation. Justice is duty to others; participation, to self. Compossibility includes duties to plural inherent values. Meliorism is the duty to improve the world, including moral progress. Moral value includes compossibility along with maximization of good and thereby rules out atomic war, destruction of the environment or any other threat to previously actualized goods.

The Other: Society. Social relations mean that one's self and others stand in a relation. This relation can be one of subordination, or of equality. If the subordination is based on force, as in slavery, the relationship is not a moral one, since it is not free or equal. Some within the society are unequal, slaves, and their slavery is forced upon them, not voluntarily accepted. What then of social relations that are freely accepted but unequal, such as those in bureaucracies, governmental or business? In such hierarchies, some members of a society are subordinate, and thereby unequal. However, such positions are ultimately accepted by all who are not in rebellion. Moral equality and freedom are the conditions of any society not based on violence.

Since other agents also self-actualize, they also have intrinsic value.[24] The others have intrinsic value as self-actualizing agents within society. For multiple intrinsically valuable agents to associate, the relation must be reciprocal. The condition of such intrinsic value is reciprocity, that is, mutual restraint, common norms, and their creative actualization. Each agent must accept mutually binding norms that do not unduly affect or restrict the activity of the other. As Peter Caws put it, "For the individual, the mark of moral responsibility is a sensitivity to the probable consequences of action as they affect the freedom of others" (Caws, 1967, 162). They must practice such norms or principles in the form of creative actualization of moral values, actualized moral ends. Their own intrinsic value as a self-actualizing agent is made possible by such norms, since others then recognize the agent as a moral agent, subject to the same binding norms, and may trust them to creatively actualize moral values. They will respect their own intrinsic value. Those who except themselves from such norms are mere freeloaders on the morality of others, depending on the moral goodness of others. The question, "why should we be moral?" is the freeloader problem of moral hypocrisy: freeloaders live off the morality and character of others, but do not practice it themselves. Morality is axiologically necessary even if some get away with not practicing it. We can be immoral exceptionally, but not collectively. (Note to elitists: superiority to and from norms is an evaluation).

Reciprocity is not a principle of difference any more than of identity, however, although it involves elements of both. Recognizing a different other as nevertheless reciprocally having intrinsic value is to place both under an identical principle, whether justice, equality or intrinsic value. An identical ethic for both is creatively actualized. The other may be socially related (kin) or different (alien) but still have intrinsic value and due reciprocal rights and obligations. Similarly, the value of one intrinsically valuable agent is as great as another; their conjoint intrinsic value requires reciprocity. While two cultures may each have different religions, both have a religion. This is not, then, an argument for a single "true" religion, a "higher" identity, although both cultures may share some religious values. A just structure does not consist in atomic individuals, out of all relation to others, but reciprocal differentiation, with recognition of the intrinsic value of the other.

Moral values may not be realized as a distinct good, but generically or as a principle. They defy analysis in terms of "things," that is, material objects, since they are then normative identities. Rights and duties are neither embodied as such as a material object nor transcendent, but a principle of intrinsically valuable agents. They arise from reciprocal and mutual recognition. They are normative, not "ontic"—mutually recognized aspects of reciprocal association, chosen as superior to the potential alternative of non-recognition. They are actualized as a distinct kind of relation, not as "things"; "present" only in the form of non-violent association. Since a world with such rights and duties is manifestly different than the ontic world of being and matter, the value of ontology is at least problematic. Moral behavior is often actualized in non-violence. A choice not to rape is a choice not to use violence for personal gratification. The "non-presence" of rape in a situation, its "not-being," is a creative actualization of a moral principle. But morality is not distinct from society, for non-hierarchical social relations creatively actualize moral relations. Relations are one norm that moralists influenced by nominalism try to avoid. Such a view is impossible for morals, since morals in the first place have to do with social relations. Morality is primarily social; moral values are fundamentally social values. To deny the actuality of relations is to deny society, social transactions, culture, and the entire plethora of social goods. Such a universe "exists" only in the imagination of nominalists, who, in any case, are themselves members of a distinct society. Indeed, "relation" may originally have been derived from social relations; just as truth was originally a moral imperative in society, relations are primarily social in origin. "Logical" relations are derivative from social by analogy; "ontological" relations are doubly derivative and analogical.[25] A hierarchy of relations and quantitative relations are thus also derivative. Similarly, the "balance of nature" in predator-prey relations, appeals to justice, although the most dangerous predator of all has upset this balance.

Although the interests of individuals are served by any functional society, there are duties to society as such, apart from individual interests. Soldiers, who put the duty to preserve their society above their own survival, exhibit responsibilities to society apart from interests. Courage, the virtue of the soldier, is a social virtue. Other virtues may also serve society instead of individuals, such as generosity, care, courtesy, and the like. Love of the other is not required for society, unlike justice. However, love, benevolence, sympathy and compassion are certainly virtues, and improve society.

The argument has been made that by serving the society, individuals advance their self-interest, but then virtue is neither egoistic nor altruistic but transcends the distinction. Self-interested motives are also irrelevant to the moral worth of the act, since the issue is whether the act conforms with moral principles, not if the motive does. If we refrain from murder, our motive is irrelevant. It would be hard to make a case for the self-interest of a soldier who is killed in war.[26] In any case, social good is served. The world is frequently a better place if virtues are creatively actualized.

Justice, as reciprocal, is the basis for the principle of obligation, a duty toward the other person. The other is granted equal moral status with the agent as another agent. The supreme social value, justice, creatively actualizes a reciprocal structure in the actual world. As supreme the principle is a universal; as universal it is required, since anything less than full universality would entail injustice. This supreme social norm is creatively actualized as obligations to others. Duties are primarily to other agents and derived from the reciprocal structure of social norms. Duties are not undertaken or exercised for their own sake, apart from socially regarding norms, but for the sake of society, that is collective others. Duties in this sense are reciprocal, as my duty to you is reciprocal with your duty to me. Non-reciprocated duties belong to a hierarchical society, incompatible with moral equality. The structure of moral values, that is, their relation, is not hierarchical, not authoritarian, with one value determining the rest from on high as it were; but just and reciprocal. Justice as reciprocal implies both equality and freedom, in the sense of rights. Duty is reciprocal with rights, as my right is your duty and your right is my duty.

While the other is morally compelling, our duty to the other does not exceed the duty of the other to ourselves; Levinas' ethic is not equalitarian, for it errs on the side of the other: subordination of self as sub-ject. At the root of justice is the principle: "What if what happened to others happened to me?" (Reciprocity also encompasses the Golden Rule and its variants). Some self-interest lies in justice, if only an abstract self-interest. Society benefits each individual equally, despite the abstract language of the principle. All of us are participants in our society and involved in its history. The subject is not in solipsistic isolation from society but within it. Moral solipsism is impossible in principle and unwise in practice. Thus "*ethike*" and "*mores*" are the conditions of speaking freely: universal and trans-cultural. Injustice to others affects oneself. Another moral requirement, then, is consideration of consequences for oneself: the principle of participation. This third ethical norm further answers the question posed above, namely, "why be moral?" Why not freeload off the moral actions of others? The answer is the question posed above, that is, what if injustice happened to you? How would you like it? Would you want to be gassed (Auschwitz), kidnapped and sold into slavery (Africans), have your land stolen (Native Americans)? Would you want your life sacrificed for the greed of others (the environment)? The questions are rhetorical, since the answers are obvious. The principle of participation rules out freeloading of all kinds, including any claim that "the rules don't apply to me." Yet moral freeloading is widespread in our age, since some lack a sense of evil. However, although justice is slow, it is inevitable. For justice is ironic: if we do not observe it, and practice it, we shall receive it.

Compossibility is also required for reciprocity: plural goods. Different goods mean other kinds of good; the condition of such differences in the same world is that they are possible together. Compossibility means that plural goods are possible together in the world in a relation of reciprocal difference, that is, an external relation to other goods as inherently valuable. Value does not refer back to a justifying supreme value in a hierarchy, since compossibility, a reciprocal relation, preserves the value of different inherently valuable goods. Thus our relation to other species and the world should include compossibility with other species and with the good of the world as a whole. (I will examine such duties and the place of humans in the environment in chapter four).

Compossibility of values is a moral restriction that limits the collective good. Values whose creative actualization would destroy other goods are not compossible with the others. Nuclear war and environmental habitat destruction would destroy all intrinsic value, the one quickly, the other in a long drawn out agony. An atomic war is not compossible with intrinsic value. Destruction of the environment is not compossible with life. Creative actualization requires compossibility with other required values, especially the basics of life: air, water, and food. Living things require each other—the food chain and symbiosis—and therefore air, water, and soil. Creative actualization of goods that are destructive of such necessities is destructive, not constructive: bad not valuable. Grounding a project in the world requires reciprocity with survival, or such a project would be bad or even evil. Creative acts should be compossible with each other and with the world. Moral limits require compossibility with inherent good as a norm; the novelty should add to the good in the world without net loss of good.

Creative actualization requires evaluation of compossibility with other goods already in the world. A good may be inherently good but destructive of previous goods, for example, destruction of the wild for "suburban development." Economic goods may indeed be creatively actualized by suburban "dream homes," but at the price of previous intrinsic and inherent goods. No net good accrues, and in most cases, there is a net loss of good. Compossibility as a condition of creative actualization means that no value accrues without reciprocity with other values. This requirement excludes nuclear war, destruction of the environment, genocide, and other bad and evil tendencies.

Compossibility does not rule out a net increase in good and indeed should result in a net increase as the number of goods is increased. Multiple values are compossible, including different levels of value. Thus the norm of compossibility, which was an aspect of reciprocity, is melioristic. Compossibility involves an improvement of the world in terms of a greater number and kind of goods: it is pluralist but also meliorist. Thereby, the norm of compossibility means a project should not only be feasible, actualizable, that is, valuable as such, or as an instance, but should increase the good in the world. Such an increase is an ethical imperative by contrast with destruction, although ethics only provides a limit. Proof of excellence lies not only in grounding possibilities in the world, the possible creatively actualized; but doing so within the limits provided by reciprocity, compossibility with other inherent goods. I will discuss meliorism in more detail below.

The compossibility of intrinsically valuable moral agents requires duty as its condition and as a limit on the actions of such agents. Evaluation of any impact upon collective others brings consideration of consequences into moral purview. Moral evaluation should include projected consequences for others, whether members of one's society, other cultures, the effect upon other species, other inherent goods, and the biosphere or the world. Consequences for ourselves should also be evaluated.

What is the relation of values as a general ground of differentiation, and value categories as limited in scope? Compossibility of inherent values and goods is a moral requirement that preserves and allows for different goods. "Values" also imply differentiation, that is, pluralism in values. Different goods imply plural goods: creative actualization means pluralism. The different kinds of value pursued by humans are evidence for pluralism in values, for example, the pursuit of beauty, of spirituality, of pleasure, of knowledge, prosperity, and other kinds of goods. Creative actualization is pluralistic, and pluralism in value is also incompatible with a *summum bonum*, a hierarchy of values, or a referring back or reduction to a single kind of intrinsic value.[27]

Moral limits, by differentiating but not constituting categorial values, are consistent with and allow them their differences. Values justify differentiation as creative actualization of goods valuable in themselves. As valuable in themselves, they are both distinct, different from other goods, and of value. Differentiation of inherently valuable goods, good in themselves, in accord with reciprocal relations results in pluralism of values. This value relation is consistent, I believe, with the "Pluralistic Universe" of William James. The distinct good of one culture does not derive from another and the value of wilderness cannot be reduced to instrumentality for humans. True difference is irreducible, as without a common ground; relations of difference retain autonomy for differences. The categories of human value pursuits are already plural and do not have a common element like rationality to identify them as essentially alike. They constitute plural grounds in the world. Pluralism requires reciprocity in relations of worth. Differences must be insubordinate of rank or values are not truly plural, but derivative: inherent value reduced to instrumentality.

Justice should be done to all values, through insubordinating norms. The result is a plural structure of different inherent values in reciprocal relation, with morals constituting the limits to categorial and personal values. Though the structure is reciprocal, the relation is not a form, as a limit. Moral values justify differentiation by limiting the scope of any value to its own worth, preventing subordination and thereby allowing creative actualization of plural sources of value whose own value justifies their differentiation. The value of what is different is not composite in structure, as the "unique." Values as reciprocal are grounded in this difference, and the condition of this difference. Such value cannot be an abstraction from what is common, an identity, as good in itself: not a composite of identity and difference, although constituting creative actualization of an instantiated norm. Thus ultimate justification is not a foundational ground, for values within moral limits are justified in themselves.

Pluralism, as differentiation, is not the "leveling" that Nietzsche claimed. On the contrary, genuine differences are compossible only with reciprocity of plural values. Differences in rank constitute hierarchy, not plurality: the value of the lower is subordinated to the higher, an ultimate reduction in the kinds of value. Hierarchy is reduction of value to a *summum bonum*, a singular value. Pluralism of value is not identity of value, since identity belongs to norms. Distinct values, including intrinsically valuable agents, are not instances of a general identity, but good *in themselves*. They are not tokens of a universal norm but reciprocally different agents, or other inherent goods, independent as autonomic. The conceptual identity of differences as instances of "difference," which is difficult to avoid, requires a careful statement of the uniqueness of inherent value. Values are conceptually or normatively related as "values," but distinct in content as inherently valuable. (Nietzsche was aware of this difficulty). Pluralism in the qualitative sense means that "being" is not a universal concept. A community of different values of inherent worth is justified, not a leveling.

A pluralistic community is more just, as relations are reciprocal, not hierarchical. Nietzsche was concerned about justice for the exception, the "higher" values. But a world with pluralistic sources of inherent value is just to the differences that are exceptional. The inherent worth of a good or value has a place in the world, and is not subordinated, sacrificed to a "higher" good, whether of rank or number. Diversity or pluralism of values within a life is also the condition of a good life, for we would not pursue a life containing *only* one kind of value: only wealth without the goods wealth can buy; only work and no health, esthetic life or entertainment. The general judgment is for a diversity of goods in a good life, judging by people's actions: intimacy and family, prosperity, health, pleasure, spiritual life, intellectual development, and so on.

Meliorism: moral values consist in creative actualization of both duties and virtues, constituting an improvement over a previous situation in which moral values have not been creatively actualized. Moral values, like other values, improve the future as a consequence of present action. Doing one's duty or creative actualization of a virtue will have better results than not doing so: moral improvement, even if the measure is purely moral. The principle that the world can be improved is meliorism; meliorism is consequentialist, since the improvements resulting from creative actualization are consequences of acting.[28] Meliorism also involves evaluation, both preliminary evaluations of competing goals and means, and of results. Since we must evaluate potential consequences in and for the future, including compossibility with other creative actualizations, meliorism is the basis for an ethics of future times. Meliorism suggests an improvement in which we are obligated to try to improve the world by our creative actualizations. We have a responsibility to act such that the future is improved for the better. Since such creative actualizations are both active and constitute value, the passivity of value theory in the modern age is ended.

Action aims at creative actualization in the world, an improved world, although not a perfect one. Perfection cannot arise from *a posteriori* relations. New ends may arise; new means of greater value. There were no personal computers 30 years ago. If the action succeeds and the result is achieved, it has value, but has not necessarily made for a better world. Creative actualization of norms is of value, but not necessarily better than before the action, not necessarily a better world. The norms are grounded but not justified, not necessarily an improvement or compossible with other values. Thus both action and the results of action require critical evaluation and improvement is one aspect of moral appraisal.

As our present creative actualizations will make the world of tomorrow—our acts will become manifest in the world—our responsibility must be to act for a better world. We are thereby blamable for a worse one. The statement of meliorism in evaluation is "what good would that do?" The criterion or test is improvement: would creative actualization constitute an improvement? Of the many activities and actions available, I should evaluate and choose what I imagine will have the best results. Whichever potentialities I choose will have repercussions in the future. Since ethical considerations are supreme in evaluation, consideration of the future always comes under moral consideration, as all actions, including ethical actions, aim at a future state: an ethics of the future constitutes ethics in general. As I argued above, evaluation constitutes the logic of freedom. However, ethical considerations are supreme and ethical limits a constraint upon freedom as its condition. The logic of freedom is manifest over time as creative actualizations and therefore must be evaluative, not interpretive. A corollary of an ethic of future times is to concentrate on improving the future, not regretting the past. Instead of accepting injustices, for example the destruction of Native American cultures and the appropriation of their land, we ought to work for justice for the Native Americans now and in the future. We also ought to work for continual improvement of the world.

What is the relation of meliorism to the other ethical norms required for creative actualization, especially reciprocity? Moral considerability arises as an issue for evaluation, the critical appraisal of different possibilities. Since we are obligated to take what we decide is the best course, moral evaluation is inherently melioristic. Evaluation of competing possibilities, courses of action, norms and principles, and even hypotheses aim at improvement, since evaluation consists in the activity of ranking alternatives and choosing the superior one. Reciprocity of inherent value does not preclude evaluation of superior alternatives in a ranking, judgments based on hierarchical evaluations. Competitive evaluations in any category of value can be justified, as well as personal evaluations of quality. Evaluation must consider, however, the moral conditions of reciprocity, compossibility with other values, participation and meliorism in the world. Subordination of distinct intrinsic values is both non-reciprocal and non-meliorist and therefore ruled out.

Since moral norms are supreme in evaluation, evaluation is both moral and meliorist. Further, if a creative actualization is not reciprocal, it would not be meliorist. An act worthwhile in some sense by itself but which subordinates or destroys other autonomous values is not an improvement. This norm also applies to means, to instrumental values. Pollution as a by-product of combustion requires improvement, since the build up of pollution is a threat to health, a distinct kind of value for all intrinsically valuable agents. It might be argued that all agents believe that they aim at improving the world in acting and that meliorism is thereby either platitudinous, tautological or superfluous. As I argued above, however, creative actualization can result in a good that is not an improvement in the world. Aiming at as much good as possible, the ultimate denouement of meliorism, includes as much moral good as possible, in the form of a reciprocal relation of non-moral goods, which are thereby insubordinate, that is, of intrinsic and inherent, not extrinsic value. Moral values are the universal condition of the inherent value of goods, including their increase, and the means used to creatively actualize them, instrumental values. Moral values are always operative as a limit: universal over time.

Justice and the ethic of future times, meliorism, are also in harmony; if more justice is creatively actualized, the future will be improved.

Creative actualization is undertaken to improve the world in some respect, to increase value. Both novelties and qualitative and quantitative increases in goods aim at improving the world. A project is worthy of creative actualization because it is an improvement, a bettering. Meliorism justifies the project, and thus meets another standard for moral principles. Improvements in quality, however slight, and in quantity of good in the world is nevertheless improvement. Novelties and new kinds of goods are also of value so long as they are compossible with other goods.

The implication of melioristic creative actualization is that the world can be improved, that it is improvable. The world is to some degree malleable if we can improve it, that is, change can be actualized. We are not prisoners in a fixed world of "being" or a world determined by "matter." The world changes through amelioration, so there is no "substantial" basis for improvement. Change means improvement is possible; improvement is proof of change. Meliorism is the disproof of an ontic world. The world is a pro-ject. Even our principles have been discovered and practical ones invented: not all principles are fixed, since some are novelties tied to inventions. Indeed, our characterization of the world can improve over historical time as our theories of the universe and its contents improve. Even scientific principles get better and better: they are melioristic, not absolute, fixed or final. They are hypothetical norms subject to evaluation, not final truths. Similarly, our previous creative actualizations can always be improved in some respect.

Improvement also implies a historical change from the past to the future. Time is not a static flow but creatively actualized as historical, as deliberate change. Change is creatively actualized in ameliorations altering the world in a historically determinate sequence that is fluid and pragmatic. Changes are evaluated as improvements in relation to the complex of values in the world, that is, a historical situation. Good is in relation to this point in time, whether because life is made easier or any other amelioration. Such changes are not absolutely good; consider the automobile, which altered transportation but creatively actualized traffic jams and increased pollution. The automobile is an improvement in some respects given the historical time in which it was creatively actualized; it could now use additional improvements, or better, elimination in favor of some less polluting, more public means of transport.

Since value in the world and novelty are part of the character of creative actualization, there can be no absolute standard of values for evaluation. Fixed standards are superceded by amelioration. Improvement in the world has historical reference, and although history is not the sole point of reference for value, improvement is in an evaluative historical relation. Such improvement is proven by practice: if the value of some practice has been adopted, it has proven its historical superiority in some sense. While we cannot know whether an historical improvement will have absolute value, we can evaluate if the change is better or worse. We can list the merits and demerits of a novelty or any other creative actualization, with reasons on both sides. There is not "absolute best" given historical changes and improvements. The best auto of 1935 is not as good as any below average car today, since the standard technology in contemporary vehicles had not been invented in 1935. Change is the enemy of perfection.

Undreamed of technologies and novelties of other kinds cannot be included in a present day ethic, anymore than a right to health care could be retroactively assigned to pre-medical humans of the Stone Age. Nor actions unimaginable to us. Future ethicists must critically reevaluate duties and other moral considerations for future conditions. However, they are bound in the loosest sense by norms of meliorism; their efforts must also be an improvement, just as civil rights and the inclusion of the environment in moral evaluation are moral progress for our age. "Absolute" value is thereby excluded by a meliorist theory. However, creative actualization is not relativist, since a better and worse are possible. The rejection of fixed standards of value is not relativism, since values are melioristic. Improvements are grounded and measurable. Moreover, there is a "best under the circumstances," that is, evaluations that determine and creatively actualize the best solution for the historical situation. Consider, for example, contracts and labor-employer negotiations, which neither side thinks are perfect, but both can live with. Finally, inherent goods are good in themselves, even if they are not absolutely good. A car may be good without being perfect. Absolutist theories of value, tied to ontology as eternally of value, enduring, are actually worse, since they are timeless, that is a-historical and decontextualized. Although we may evaluate improvements historically, in relation to a determinate situation, evaluation should not be confused with value.

The ethics of creative actualization, then, involves a teleological dynamic. Relations are across time, without reference to a timeless absolute that gives value in a hierarchy. Differentiation over time, historical change, is based on the evaluation that such differentiation in quality or other values is worthwhile. Evaluation is not from higher to lower but in relation to a historical situation. Morals are supreme in evaluation and thereby first in value, but as a limit are not the ground of non-moral creative actualizations. Temporal relations are melioristic and improvement is a requirement of evaluation and moral consideration. Meliorism is an obligation for an ethic of future times, since ethical acts are an improvement, a novel good of some sort. Improvement is projected, a goal of creative actualization and thereby teleological. However, the *logos* of the tele*ology* is the logic of freedom, of evaluation. Improvement is by choice and cannot be determined, since the future cannot determine the present, although visions of the future can influence present action. Evaluation of the future in creative actualization makes a new world, not a copy of the old.

Moral Progress: a moral rationale does not seem to practically require reciprocal relations with aliens, but only the small circle of those within our own society. Why then extend obligations to those outside our "tribe," however the last term is acknowledged or defined? (A "tribe" may consist of kin, either extended family or also including clans, and larger blood relations). Is there a duty to aliens? Here the idea of moral progress is requisite: moral progress consists in the progressive expansion of moral considerability, rights, and corresponding duties to larger and larger numbers and groups.[29] Roughly, the "rights of man," which were largely confined to white men, were extended during the nineteenth century to non-white men, in principle. Later, women were included. The thrust of the environmental "liberation" movements in the twentieth century was to extend such moral consideration beyond adult humans to future generations (children), animals, and the entire biosphere. Since this would be an expansion of good, it is consistent with meliorism, that is, improving the world. Recognition of the other as different includes not only toleration for different cultures, but different species. In part, species preservation is required, since an environment is the condition of any rights and thereby morally required or obligatory. The air, water, soil, plants, and animals in the food chain must be cared for humans themselves to survive.[30] Species preservation is a condition of all values, and a moral obligation.

Our relation to the world is also reciprocal, that is, we should recognize that we are autonomous agents only within the world. The world is other than us but not in the same respect as other agents. The world is part of us even as we act within it. By our actions we may discover new values, but we need the world for a sphere of action. We bring out the value, but create it in the world as a ground. We belong to the world and are not subjects outside of the world. Even the inherent value of goods as instruments requires an agent to recognize the value of the good, (although in another sense, such recognition differentiates self from the other as well). The good we have actualized is not imaginary, for such differentiations must be grounded in the world as a test: the good must prove worthwhile. The moral justification for moral progress is that the expansion of moral consideration is an expansion of morals, resulting in a larger morality. The world is creatively actualized as more moral.

Moral progress is parallel to technical progress, for both involve an increase in quality and quantity of good. Moral progress includes expansion of moral considerability to greater numbers of self-actualizing agents, and should also include refinement of moral principles. Since such progress can be measured, we thereby have a measure for morals. Thus moral nihilism is unjustified; moral progress can easily be demonstrated, if only in the elimination of slavery. The moral nihilist has the burden of proof.

Morality has always been characterized as a way of life in respect of creative actualization of duties and virtues for their own sake, because it is right. Moral agents, however, are in the world and require the world for creative actualization, including creative actualization of moral norms and virtues. Self-sufficiency is a false ideal since it would be impossible to practice for creatures that require air, water, food, nurturing, and other necessities of life. Self-sufficiency is also axiologically false, since life within society is superior to life without any social or communal norms and goods, whether language, family, techniques of survival (economy), virtues, culture, and many other goods.

Justice as reciprocity must be life-enhancing also, not only in the relation of social justice, regulating and supporting multiple lives in association, but also the relation of personal values. Justice must be done to all the values in life for the part they play in making it worthwhile and meaningful. Wisdom is the excellent evaluation of possible goals with an eye to their place in our lives and thereby includes moral limits. Justice should be due to the inherent value of each of the goods in life. Doing justice to the different ends in our lives requires their mutual compatibility, a type of compossibility but over time.

A commitment to morality is a commitment to moral limits in creative actualization and thereby creative actualization of compossibility, inherent differences and mutual reciprocity among the diverse values and goods. Moral limits should be observed and creatively actualized in all activities. Morality is of wide scope and moral limits should regulate improvement in all our values and institutions.

The ethics of creative actualization requires moral limits. Moral limits are sometimes creatively actualized directly in acts of duty or virtue valuable in themselves. We should act within moral limits, as an immoral world can never be a good world, a world of value. Such limits confine us to reciprocity, compossibility with other inherent goods, participation in the moral life of our culture and community, and meliorism, leading to moral progress. We should act so as to improve the world in the future; improvements include quality as well as novel kinds. The consequences of our actions on others and on the environment that is a condition of all goods should be considered morally. Justice as reciprocal must also be life-enhancing: we cannot reciprocally deal death, in a nuclear exchange. Moral limits should be supreme in evaluation over other non-moral ends, categorial, cultural, and personal. Moral limits include not only duties to others and virtues but entail a way of life, with justice done to different values in a life. Creative actualization involves moral limits and reciprocity, not just feasibility, functioning well or inherent worth. The relation of plural inherent values is reciprocal, and moral principles require creative actualization for value at the same time as they limit creative actualization to compossible ends.

Four

CREATIVE ACTUALIZATION AND THE WORLD

In this chapter I will examine the relation of the moral principles that we analyzed in the last chapter to the theory of creative actualization. I will begin with creative actualization in the world, the condition of all values. I will then take up creative actualization by living things. Finally, I will briefly discuss philosophical anthropology and its relation to values, especially creative actualization.

1. The Value of the Whole: the World

In the first chapter I argued that creative actualization by agents takes place in the actual world, not some ideal realm beyond the world. Moral values are also creatively actualized in the world. This value world is not distinct from the world described by physics, but the same world, contrary to the view of some naturalists. Indeed, the world of the naturalists is, at best, an abstraction from the value world, the world that was creatively actualized by past evaluations and thereby made possible the science of physics. Goods of the world are not in our minds but in the world. Value is the ground of the changes that have improved our world—or at least transformed it, for better or worse. Houses are not in some ideal realm of "ought" but in the same world as the abstraction "nature." That same world prompted hairless apes to build shelters, with the goal of keeping out unwelcome elements.

Martin Heidegger made a similar point against subjective phenomenology: that we are "beings in the world." Heidegger accused Immanuel Kant of this same sort of subjectivity (Heiddeger, 1927, 1990). However, he missed Kant's larger point. Heidegger is indeed correct that we are in the world. But we live in the noumenal world of moral association, not the phenomenal world of imagination (Kant 1785, 1788). Heidegger's world is a figment of imagination—the "phenomenal" world is a representation of our minds. Our minds meet in moral association, not perception. And the "problem of other minds" dissolves in moral association: a pseudo-problem based on perceptual-subjective foundations. The phenomenal-subjective world is imaginary; the world must be acted upon to establish value, including truth or knowledge. Heidegger never completely escaped the problematic of phenomenology: he took phenomena as ultimate, and ignored the genesis of many phenomena in creative actualization.

Values are the ground of the actual world that includes goods, while at the same time grounded in that world. However, unlike phenomenology and naturalism, value as creative actualization does not reconcile us to "this" world or a passively received world, but to a world of our making: we create our world. Values also create connections to the world: they move us to action in the world in the attraction of a goal, a good, a result.[1] But such values are not confined to humans. A change in species is a change in how the world looks. Yet the world is differentiated from novelties *in* the world even as they transform it. It has a value of its own. I will examine these points below as well as the place of humans in the world.

It was J. Baird Callicott who first challenged the nominalist model of value, in which value inheres in individuals, in our age (Callicott, 1989, McDonald, 2003). Value nominalism was largely a product of Protestant culture, and the immeasurable value it put upon the individual soul.[2] Since this view excluded the value of non-human agents, "wild value" (Rolston, 1989) was not considered a value at all. Callicott argued that value could inhere in a whole, the world, with ordered relations. The whole is inseparable from its parts. Similarly, it can inhere in a habitat inseparable from its many distinctive elements: animal and plant life, and their functions, soil, water, and climate, air, and niches, a micro world. The locus of value is not in any one individual in this wild condition, but in the whole. An individual may perish but the habitat survives, providing a conducive environment to many forms of life.

What of the notion that the relation of parts to a whole is that of instruments, since each part of a habitat has a function that it contributes to the whole?[3] Is functioning in an environment or habitat a form of instrumental value? For example, is the use of plants by predators as an instrument of their own survival an argument for the instrumental value of the parts in the whole? I would respond that there is no actual whole without the parts, although a whole cannot be reduced to a collection of parts. The whole is qualitatively but not quantitatively distinct from its parts. The parts are, then, an actual condition of the whole and thereby essential to its continuation, its continued functioning as a whole. If such conditions are also instruments, they are not mere instruments, but morally considerable as irreplaceable.

The holistic model has other problems, however. One is that "functioning well" is not a justification of species preservation, since some species may have little or no function in the whole. We may also ask, Is "functioning well" self-actualization? That is, does it meet the criterion for an intrinsically valuable agent? If a tool functions well this does not mean it has intrinsic value. But intrinsic value is the condition of moral considerability. Holists might argue that we may not know the value of niche species, but such species may be in truth crucial in ways that we do not understand. More importantly, the parts that groups of species play in the whole--microbial, plants, insects, herbivores, and carnivores—may be so interrelated that the groups are crucial.

For example, insects are required to pollinate the plants that are eaten by the herbivores. Carnivores are required to keep the herbivores in check. However, even this model can be questioned, since the function of the herbivores is unclear. Plants may require insects but they hardly require predator species; competition would serve as well as herbivores in selection.

The relation of parts to the whole may be a useful theory for analyzing environmental relations, but it will not do as a complete theory of value. Functional relations are not value and should not be confused with value, for if individuals and species are to bear inherent or intrinsic value the value cannot be relational.[4]

A theory of relations should be based on the inherent value of different species and individuals. This would constitute a different approach to the value of the whole, since the whole would be a collection of plural kinds and sources of value, but with a distinct good of its own, holistic value. This whole is the world, which contains numerous different habitats, climates, soils, species, and other goods. The world is a positive good in itself, better than nothing. If there were no world there would be no value: nihilism in actuality and in value. Since there is a world, nihilism is excluded. If nihilism is arguing for the value of no value, it presumes a world of value that could be devalued. The world is the value-ground, that is, the locus of actualization, the value world, as the source of all goods and thereby their condition.

The different goods in the world are a part of this world and thus a part of a good whole. Anything in the world is either a previous creative actualization or a potential or actual instrumental good, even if it is an undifferentiated part of the world, for example, one grain of sand. Thus any part of the world can be value-laden. "Matter" or "minerals" are good as part of the world. Since the same world contains different goods, different goods are possible in and can be part of the same world: compossible. Compossibility is one value of the world, in which different goods are compossible. The value of the actual is tied to compossibility, although not reducible to it. Different rock strata are not valueless, for they are part of a valuable world and compossible with other goods in the world. Similarly, other stars, planets, and the like are compossible with our own. The value of each good remains inherent, but collectively the goods are compossible with each other in relation.

The world is necessary for all theories of value, even relativism: thus the world is the condition of all values, and practically necessary or required. Since the world itself is a necessary condition of good on any theory, preservation of the world is morally required, a moral value. We are thereby obligated to preserve it above and beyond self-interest. Blowing up the world would be the ultimate evil. But destruction of habitats or the larger environment, in the form of air pollution, soil depletion, species extinction, water contamination, and other human shortsightedness is a slow form of destruction of the world. In short environmental destruction is morally evil. The world is required for all goods and its preservation is morally required.

The inherent value of the world is a condition of life and thus a moral value. The parts of the world compossible with intrinsic value are also conditions of values and thus morally required.

Is the world an *a priori* good? It can be considered such as the condition of any theory of values. The ultimate moral value of the world does not mean or imply that the world is fixed in value for all time. Were this the case, life could never have arisen, nor new species, and new kinds of inherent goods. On the contrary, the world can change and also be transformed for the better, so its value is not fixed. Indeed, moral development in children implies changes in character and behavior. Children are intrinsic value bearers but can improve. The world can also change for the better.

Goods of the World: If value is creative actualization, the question arises, what of "natural" goods, that is, goods in the world that preceded human agency? What of water, essential to human as to all life? What of air? What of soil, in which we grow our crops? What of other species, on which we depend for food? None of these have been creatively actualized by human agency, but they are surely goods. Water is good for quenching thirst, soil for growing crops, and so on. I will discuss this issue in two parts, since I will argue that of these goods, only water preceded life. Some goods that are not creatively actualized by humans were creatively actualized by other species or represent self-actualization of other species: their own survival.

Water is certainly a good, necessary for all life, but was not creatively actualized by life. The instrumental value of water is not a matter of dispute between theories of value. The issue is how we are to treat of non-artificial goods given our theory of creative actualization? How can we argue that water has value? I have distinguished value from good in this work. To have value is to have been creatively actualized in acts of value. Value is tied to action and only derivatively to the products of action as their actualized ground. The success of the actions of the Wright Brothers in building an airship successfully is distinct from the plane they built. The plane is a good. Some may speculate that water was created *ex nihilo* by an infinite power, but this must remain speculative. In any case, water is a good, but not a good that was creatively actualized by living things. A few goods of the world, therefore, have not been creatively actualized, or at least not by biological agents.

However, this judgment requires qualification. Water as distinct from the world as a whole involves an abstraction from the actual world of which water forms a part. In other words, water as a good may be considered part of the world as a whole as a good. Water as distinct may be considered in two ways, in intellectual abstraction or in practical abstraction. Both forms of abstraction require creative actualization. Water in intellectual abstraction from the world—as a chemical, a geologic formation, or a resource—requires intellectual activity and the creative actualization of knowledge under the logical and moral norm of truth. That water is so much H_2O is the end product of much research, chemical experimentation and classification, and other intellectual activity that has resulted in the creative actualization of knowledge. Water as a practical matter requires creative actualization of it for our use. If we are parched in the desert we must act so as to obtain water. To obtain its good, we must still drink the water, and this activity itself is a creative actualization. Going to a river is an activity that is successful or not. Water must first be discovered to quench thirst or a supply must be creatively actualized in the form of a stored supply, a reservoir. That water is good for quenching thirst also requires an evaluation that water is good for quenching thirst, or good enough.

Another example is provided by scenic landscapes. The Grand Canyon is a beautiful sight, but was not creatively actualized by living agents. We may even say that the Colorado River "carved" the Canyon, but this would be metaphorical; the river did not do so deliberately or purposively. The Canyon is the accidental or chance result of the flow of the river through the geologic formation in the area. (Even if it we accepted the premise that it was the hand of God that created this magnificent art, the Canyon would still form an exception to human, or living agency.) Separation of the Canyon from the surrounding areas is an abstraction of the imagination. The Canyon is not physically distinct from the rest of the world, its own world, but is a distinctive formation for the imagination in perception. As for human agents who wish to view the beauty of the Canyon, they must successfully travel there: they must creatively actualize a journey to the Canyon in order to view it. They must creatively actualize scenes of beauty, whether in the world or in the museum. Again, the discovery of the Canyon was also a creative actualization and the judgment that its beauty is worth a trip an evaluation.

Goods of the world that were not creatively actualized by living agents still have inherent value: they are inherent goods. Water is good for all life in many respects, as a constituent of life; scenery adds to the quality of life. Thus I would argue for the inherent value of instrumental goods, since the characteristics of water that are good for life belong to the water, not the life. A living thing that does not have water dies; the characteristics of the water belong to the water not the living thing that lacks the water (except insofar as the living thing is composed of water) or the living thing would not require them. These characteristics are extrinsic to the living thing. Since such instrumental goods are also conditions of life, instruments may have more value than *summum bonum* theories recognize. *Summum bonum* theories are structured such that means obtain their value from the end. But water has inherent value apart from the living thing that needs it and moreover is a condition of the self-actualization of the intrinsically valuable agent. Instrumental value, then, is revalued by creative actualization. Instrumental value that is a condition of intrinsic value is practically necessary, and thus morally considerable.

Goods of this type are not values, but inherent goods. Good is inherent in actual instances; value attaches to actions as the agency of changes in the world. Value is novelty changing the world for the better. Water is still a good, still a component of life. The Grand Canyon is still beautiful. But not all such goods are the result of the actions of living agents. Esthetic good in the world—the Grand Canyon—is also distinct from instrumental value. Both esthetic and instrumental values require action to actualize, but one is practical: drinking water from a river, for example, while the other is contemplated for its own sake. Water and beauty are actualized but not creatively actualized, although they are creatively actualized in acts of and for agents. They are part of the inventory of the world's goods. Value is creative actualization but some goods are already actual in the world. Goods are not novelties without creation, they are not an expansion of the kinds of goods, but they are part of the good of the world.

Creative Actualization of the World: Inherent good is an internal relation or characteristic of a bearer and is tied to actualities—value has been achieved or not already and persists in the bearer. Inherent good can belong to means (airplanes) or ends (agents) but in either case the bearer is *good* in itself. Thus means and ends are distinct from goods. Ends are goals, but goods are inherent. Aristotle ran the two together in a reductive move, in which the goodness of good was subordinated in a hierarchy of ends, with human happiness (*eudaimonia*) at the apex, the *summum bonum*. In order to bring out this difference, I have distinguished value from good above. The problematic of "the Good," which dates back to Plato, is distinct from the problematic of plural inherent goods that have been creatively actualized. Value in this philosophy is not a transcendent one in many, but a normative identity of distinct inherent goods.

Value does not derive from an extrinsic source; instrumental value is the use of inherent good by self-actualizing agents, that is, living bearers of value, who subordinate goods to themselves. The reduction of inherent good to a means, becoming a tool, is instrumentalization. Value becomes or is actively made into an extrinsic relation to self-actualizing agents. Plants turn minerals and decomposed organic matter into soil. Animals use plants to survive; carnivores use other animals. Evaluation of the instrumental value of an inherent good bearer is distinct from *reduction* of an inherent value bearer to instrumental value—to extrinsic value. Inherent good as a character of a bearer is in the world.

I am also arguing against any attempt to make an absolute separation of fact and value, or the so-called "naturalistic fallacy," the idea of Moore's that since fact is one thing and value another, good is a "non-natural" predicate that can never coincide with nature. This "fallacy" is itself fallacious. Firstly, Moore contradicted himself, since he argued both that nature includes everything and that it does not include good, which he judged is a "non-natural predicate." If "nature" includes everything it must include good, especially "natural" goods like water that were not creatively actualized by humans, or even living things. Moreover, water both quenches thirst and is good for quenching thirst. Since our bodies are composed mostly of water, and we lose water during normal biological processes, we require water to survive, as do all living things. Does Moore propose, then, that drinking water is not "natural?" That an animal sipping water by a stream is doing something artificial? Or that water is not good for quenching thirst?

Creative actualization of "artificial" goods is in the same world as rocks, water, sunlight, animal and plant life. There are not two worlds, a value world and a non-value world, but one.[5] If good were "unnatural" in one sense, it would be a violation of the so-called "laws of nature" and no good would be possible. The same point applies to creative actualization. Growing crops is both in the world and a kind of creative actualization. It uses the "natural" tendencies of plants in a non-wild method of growth utilizing cultivation, creative actualization of preferred foods in the world.

On the contrary, the environment and the world are the source of all goods, even "artificial" ones. Automobiles are artificial goods, but the components of automobiles—iron, glass, plastic, rubber, and so on—are all derived from the world. Iron ore is both an actuality and a good. I am not arguing that "is" and "ought" coincide, for some "oughts" have not been creatively actualized. However, values and goods are in the world, the same world as "facts" about the world. "Is" and "ought" can also coincide: when affairs are as they should be, for example when standards are met in creative actualization. If water quenches thirst, it is as it should be.

Water both quenches thirst and is good for quenching thirst. In this case, the "facts" of water quenching thirst, and of being good for quenching thirst coincide, although they are not identical. Benzene is not good for quenching thirst, since it will kill. The deadliness of benzene rules out its value in quenching thirst. A "fact" coincides with the disvalue. Similarly, soda is not as good for quenching thirst as water since it contains sugar, which can make a person thirsty all over again. A fact can explain comparative value. Facts and values or goods are not identical, since, among other things, facts consist in propositions about actualities while goods are actualities.[6] However, actualities and goods can be identical: water.

Naturalism as a philosophy ignores the changes in the world made through creative actualization. Creative actualization has transformed the world with inventions, institutions, culture, and other goods. But such transformations are not confined to human goods. Animals, plants, and other life forms continually transform the world. The world is not static: not a "being" in any of its senses, including "states of affairs." The world changes and life contributes to such changes. The world before the creation of the biosphere by the slow action of plants was quite different from the world after this creative actualization. Similarly, the world of the hunter-gatherer was different than the twenty-first century world of technical transformations. The world is both a condition and is conditioned.

2. Life as the Bearer of Value

A life can be analyzed as a biological life. As such it has finite value, since life is of limited duration. However, life can be treated in terms of its values, either a way of life, an ethos (Socrates); or the value of life itself, a question on which Arthur Schopenhauer dwelt. Ethics is a way of life in the widest sense, considering what is a good life. In an ethical life, the biological life is only one factor, since the question of the quality of life is raised. In this view, an ethical life is lived within or located within a biological life but cannot be reduced to it. Life in the moral or ethical sense has higher value than a mere biological life, for example, that of Socrates is higher than that of a clam (Mill 1863, Singer 1975). A closely related issue is value in relation to life, namely, whether value is grounded in life and also whether quality of life can be evaluated as a basis for the value of life.

A life may also be treated in terms of social categories such as an economic life, a political life, and so on, that is, other ways of life as the predominant value pursued in such a life. This includes life considered as sacred, the religious view in which each individual life has infinite worth or absolute value. This view unites Christian agapism, love for each soul, with Jainist and some Buddhist views of the absolute value of all life, although agapism has a smaller scope, limited to humans. Religion may be the source of specific cultural value scales in its function as articulating what is sacred or inviolably valuable as the source of value. However, there are complications with this view, since it is unclear whether life or the soul has value. Philosophers and others who argue for the infinite value of each individual soul are not arguing for the infinite value of life. They would agree that life itself ends, even if the soul is immortal. But this complicates their efforts to justify life as such, in arguments against abortion, euthanasia and even the use of medicine.

The vitalist view of value is that all value is grounded in life, and that all instrumental values are values for life.[7] In this view, life is the absolute basis of value. Even most psychological and subjective theories depend upon it in some sense, for example, hedonism argues that pleasure is the supreme good, but that only a life which is pleasant in the balance or preponderantly is a good life. However, this view has difficulties. Since some life is used instrumentally by other life in predation, *inter alia*, which life is to be accorded value? If only some life is accorded value, the view fails, for example, if only human life is accorded intrinsic value. For then value does not reside in life, but in a specific quality of life, a worthwhile life; or a species. Science is in the service of human life while art is for the sake of its value to (human) life. Vitalism is only feasible as speciesism, in which some ends are more equal than others.

However, the view of value as located in life and thus life as the basis of value has more plausibility in another sense. In this view, life is the condition of all value, since value for its own sake or good in itself is located in life. Life is also the locus of species value, the value that attaches to distinct kinds of life as unique goods, apart from individual bearers.[8] Thus life may still constitute a moral value, despite its biological limitations. Life as a moral value is distinct from life as the ultimate ground of value. What then is the relation of life as a good to value as creative actualization? Does life become a mere instrument? Are all lives worth living according to this theory? Do even a short or an unhappy life have meaning which makes it worth living? Is there a standard implied in quality of life? Do lives that do not meet this standard lose intrinsic value?

Intrinsic value is self-actualization, that is, creative actualization of an agent that creatively actualizes itself. Such an agent is intrinsically valuable since the actions of the agent themselves have intrinsic value—and at the same time creatively actualize their own conditions, the agent. Self-actualization means an agent creates itself through action and thereby is *valuable* in itself. The actions succeed, or fail, in actualizing the agent, the source of the actions. Agents achieve value or not at every moment by themselves. An animal must with each step decide if this step aids its survival. A prey trying to avoid capture may create a new pathway through the undergrowth. This pathway may or may not prove to be the correct strategy for the animal to survive, to creatively actualize its own life. If the strategy is defective, it will lose its life. Perhaps it might have succeeded if it remained on the older pathway. But what is now considered old was at one time new and creates a means for new acts of motion by different creatures. Both pathways are creative actualizations, however.

Since all living things share the actions of an agent, all of life is the bearer of intrinsic value. Animals and plants actualize themselves through action and activity as much as humans do and must do so to continue living: to actualize themselves. Sometimes they fail, and thus intrinsic value is not absolute.[9] However, value is not relative either. Instead, life is finite in value, since value is borne by life and life is finite in duration even if predation does not end a life prematurely. Life is not of infinite value, as it requires self-actualization and this cannot, apparently, continue indefinitely. Life can lose its intrinsic value through death by a predator, what Rolston has called "capturing natural value" (Rolston 1988) Other forms are death from organ decline, destruction or depreciation. Agents of life retain their intrinsic value so long as they continue self-actualization, but this is temporary, finite. Thus life is the bearer of intrinsic value as self-actualizing, but intrinsic value is limited. A finite life cannot have unconditional value, since it is finite, that is, its value is contingent upon staying alive.

Anaerobic microbes, and perhaps green plants, liberated air, or more precisely, creatively actualized free oxygen. Since free oxygen is a necessary condition for all other life, we practically require this element, making it morally significant. Its condition is also morally significant, that is, green plants and other living things that create, maintain and filter our air supply. Soil was also built up over the millennia by plants since plants contribute to the breakdown and coagulation of mineral rock into soil particles begun by erosion. (The dung of animals is another element although microbes, earthworms, and other agents also break it down. It took 100 million years of dinosaur dung to make enough soil for human crops.) Thus plants are agents that creatively actualize themselves, create their own food, and are the source of air, food, and soil for other living agents. They are self-actualizing in a way that humans, dependent upon them in the food chain cannot be, and also conditions of all other forms of life. Moreover, the creative actualizations of plants: themselves, soil, air, and food for other species, mark actual changes in the world. New plants are obviously novelties in the world, but the slow accumulation of soil also changes the world. The self-actualization of plants is evident in the world, and in no way dependent upon human subjects.

With all this intrinsic value, plants should receive the highest moral consideration. But of course "anthropocentric," that is, greedy, speciesist "theories of value" ignore this irreplaceable and incredible value of plants in blind self-interest. If no other argument refutes subjective, self-interest or other anthropocentric theories, the horrendous treatment of plants alone by such theories should refute them. Plants are the condition of subjective happiness, subjective feeling, subjective attitudes, subjective desires, subjective "wills," and all the other rationalizations of human destruction of the environment. Other species are the condition of the "subject" with all its selfishness. Such theories are morally reactionary.

Plant life is not only self-actualizing, intrinsically valuable, but self-actualizing in a higher degree than humans, since humans depend on plants, but a world of plants could survive quite nicely without humans. Indeed, life self-actualized for hundreds of thousands of years without humans. Consciousness, then, is not required for creative actualization, but only for deliberation. To argue that all of this good was for the ultimate goal of humans is so preposterous it is laughable. Without moral consideration of plant life, to speak nothing of bees, cows and other livestock, and soil microbes, humans are doomed. So much also for "*summum bonum*" theories that place humans, human capacities, or human feelings and desires at the apex of value. Biologically, humans are parasites on plant life.

Animals also creatively actualize themselves as self-actualizing agents: intrinsically valuable living things. Indeed, animals too deliberate in their way. Each moment of an animal's life is a struggle to survive, to escape predators, and continue self-actualization through nutrition and other bodily processes. The higher animals make deliberate choices that will affect their survival. Animals and other organisms also act for an end. Even plants bend toward the sunlight, grow, and extend their roots even in scabby soils to obtain needed nutrients. Flies flee if a human hand is coming at them: they must be able to see their own death as a future event if they act accordingly. The goal of survival is in the future, so cannot be the result of a stimulus alone; a stimulus is in the past. Many stimuli do not result in any change in "behavior," for example, the wind, brushing up against other organisms, and so on. Non-humans act as selectively as humans do, that is, for the good, not for the hell of it.

Some have argued that animals and other non-human life operate by "stimulus-response," that is, mechanically, and thus do not value in the same sense as humans. In this view we can only compare animal "behavior" to human valuing by analogy. However, this view is indefensible, even if the validity of behaviorism, itself a dubious approach, is granted. In a famous experiment, a monkey entered a room from whose ceiling some bananas had been suspended. There were also some boxes lying around. The monkey piled up the boxes to obtain the fruit, clear evidence of teleological behavior without a previous response, that is, creative actualization of means to an end. Animals do exhibit practical reasoning of the ends-means type, without language or concepts, evidence against the dependence of logic on language or the exaggerated extent to which practice is viewed as mediated by concepts.

Since air, soil, and most habitats were creatively actualized by other species, other species can creatively actualize. In this respect they do not differ significantly from humans and are thereby morally significant. Other living species are alive, and thereby bearers of intrinsic value. They have intrinsic value as self-actualizing agents; if they cease to actualize themselves, they die, and their intrinsic value ceases. Since air, soil, and other species are necessary conditions of human survival, they are practically necessary, or required and become moral values. Maintaining clean air, preservation of soil and of other species is a moral obligation. Culture is also a condition of the subject and culture reflects the environment to some degree, as I will argue below. In sum, the subject, subjective states, interests, attitudes, feelings, and other subjective "values" require culture. In turn, culture requires a habitat, including such elements as clean air, pure water, and good soil. These elements are part of a world that is the a priori condition of trivial subjective whims. The conditions of the subject are moral values that we are obligated and required to creatively actualize. Humans in the world are deliberative creative agents; subjectivity is only a part of a much larger whole.

3. Wild Value

The other issue raised by consideration of these background conditions is whether they are necessary in the same sense as life, society, and even other species. Are nonliving factors necessary in the same sense, that is, morally considerable in the same respect as one's child? Is poisoning the soil as morally reprehensible as killing one's offspring? Do we have moral obligations to the soil? Or only to the beneficiaries of the soil? This issue raises problems for any environmentally conscious ethic, since the ethics of reciprocity does not fit the wild ecology. Preservation of species requires predator-prey relations, both because some species feed off others and for the health of the whole. Carnivores must keep herbivores in check in habitats to avoid overgrazing, which would be detrimental to the herbivores themselves but especially in its devastation of the habitat. Humans exempt themselves from such relationships, although they live in a habitat altered by their own forms of creative actualization and are primarily a predator species. Domestication, both plant and animal, is intermediate between wild and civil conditions. Since the food chain on which all three depend is based on air, water, plants, and soils, these elements of the environment are generic for all habitats, wild or artificial. The question is how we can have obligations to wild species, given predator-prey relations: how can obligation be extended to a totally different kind of habitat, where civil rules and laws do not apply?

What of intrinsic value: do we have an obligation to self-actualizing agents? Due in terms of their intrinsic value? Here the problem of distinct generic habitats or "zones" (Rolston, 1988)—wild, domesticated, and civil—is most acute, for wild creatures are as much self-actualizing agents as humans, but live in different conditions. We should consider these conditions in moral evaluation, not necessarily individuals within the habitats. Civil society includes individual rights among its conditions, but this would be "ludicrous," (Callicott) in the wild. Gazelles do not lecture lions on individual rights or take them to an animal court; they flee. The wild is "other" than the civil, distinct in kind, and as other may carry obligations as a whole, its preservation as wild. This requires that it not be domesticated or civilized and thus that its norms of predator and prey be preserved. Indeed, domestication would end the self-actualization of individuals in the wild and thereby their own intrinsic worth: they would become dependent. Self-actualization is not absolute independence since all individuals are dependent on their environment, civil or wild. Humans are as dependent on habitats and other species as wild creatures, although humans have domesticated their food supply. Nevertheless, independence includes relative autonomy within a habitat and destruction or deactualization of the wild would end this.

Preservation of the wild includes essential elements within it, including air, soil, water, and wild species. But it does not extend to individuals in the wild, since they are self-actualizing in a way other than that in human, civil society, independent of human norms. Such wild individuals are morally considerable: preservation of their conditions is required, but not preservation of each individual. The latter would be impossible, and creative actualization as a condition of ethical obligation rules out impossible tasks. (As I argued previously (McDonald, 2001) individuals of rare species, as the bearer of species value, might form an exception (whooping cranes), since each individual is practically required. An individual of a species going extinct, but not due to human agency, is a borderline case.)[10]

Species preservation, however, is consistent with compossibility, since each of the species is irreplaceable and part of the incredible variety of life in the world. The relation of species is a type of compossibility, especially given predator-prey and niche relations. Animals live in and partly shape a distinctive environment or habitat that is conducive to their life. Polar bears live in the arctic; parakeets in the tropics. Carnivores serve a function in keeping the herbivores in check, resulting in a "balance" of species in the environment. Since too many grazers and herbivores might destroy the plant life in the habitat and end all life, carnivores have a place in the whole that is necessary for the health and continued functioning of the habitat. If neither predators nor prey are driven to extinction, they have actualized a form of compossibility, since predator species require prey and prey species, and herbivores would probably overgraze without predators to keep their numbers in check. Their relation keeps each alive and thus preserves each. They are possible together, and also together with their habitat and the larger world. Since humans live in and have adapted to such habitats, they should respect the elements of the habitat.

Compossibility among wild species and their interdependence is referred to as symbiosis by biologists (Leopold, 1949). As co-conditional value, symbiosis is not a denial of inherent value but recognition of interdependent parts of a whole. Most flowering plants require pollinators while the pollinators require the plants for food. Their relation is a type of reciprocity, in which functions are exchanged cooperatively. Thus it is a larger, emergent form of value, in which the cooperation of intrinsically valuable agents benefits all. Human domestication and agriculture may be derivative forms of this value. In both the wild and in domestication, symbiosis is the creative actualization of a close type of compossibility.

The Value of Life: Life creatively actualizes itself, that is, each organism is a center of value, or good in itself, since by nutrition and avoiding predators the organism survives as a good by itself. Each organism is responsible for its own survival and thus is a good by itself, since its survival is not instrumental on another. By reproducing successfully the organism creatively actualizes its own species as a continuing good; unsuccessful mutations and unhealthy offspring are defective at actualizing themselves, and thus lack biological adaptability, survival value. Life is where such value is located and thus its condition in that sense. However, mere survival does not include quality of life except at the most basic level: of superiority to fatal illness in the health of the organism.

All of life has value collectively as self-actualizing. The value is intrinsic to the self as the agency of its own creative actualization. Life is the condition of agency, that is, of self-actualization. Living organisms are the bearers of intrinsic value, since they self-actualize through action and activity. The goal is continuation of self-actualization. Life is connected with agency, action, and activity. Life consists of self-actualizing agents, not of feelings, desires, or other psychological factors in life; indeed, it may include agents without much of an inner life. On the contrary, life is the condition of such feelings and desires. If the feelings and desires have any value it is as part of life, and because they contribute to life or enhance life. Without life, we could never actualize such feelings. Without life, creative actualization of objects of desire would not be possible. Goods of the world have inherent value but may also have instrumental value for life. Other values regulate acts aiming at some good in life, but require life to enact such goods. Life is the bearer of intrinsic value, although, as I argued above, not absolute. For life alone aims at itself as the goal of its creative actualization.

Life is not a mere condition, since life is its own condition, and thus alone has intrinsic value. While individuals' lives are finite in value, the continuation of life is not, at least potentially. Life, then, is of a different order of value than other values. These others have inherent value, but are also instrumental to a good life. However, life is also dependent upon an environment. The values proper to life itself in the biological sense are that minimum necessary to sustain life, including ecological goods such as water, air, food, and the like. Shelter is only necessary out of habitat, for creatures that have left their original habitat and require creative actualization of an artificial habitat, one mimicking the original, to survive.

Some values are directly life supporting, while others are life enhancing. What enhances life must go beyond merely aiding survival. Goods that enhance life must be creatively actualized: discovered and reduplicated. We do only not pursue goods that improve life for the sake of life but for their own sake. Life is a condition of such values, but not the ground of such discovery and duplication. Life as a condition of an enhancement or improvement is distinct from life as a value in its own right, as self-actualizing. This distinction lends some plausibility to the distinction of needs and non-needs, necessities and luxuries.

We may in some circumstances have to rule out life-enhancing "goods" on the basis that they are a threat to the necessities of life, life supportive goods such as air, water, and soil. Our moral evaluation ought to rank life as supreme over such "goods" as cars that pollute or increase global warming through excessive use of fossil fuels. We should not forget that economic prosperity only has value for living agents. Our biological roots are of far greater moral significance than our standard of living, since without our lives we could not increase our standard of living. Other values are subordinate in moral evaluation to the value of life as a limit. Such limits do not rule out a range of permitted goods, only abuses of life. Goods that enhance or fulfill life within moral limits are permitted, whether kinds and categories of goods or personal goods. Revelations of the worth of goods in enhancing life should be within this framework, that is, the goods should result in an improvement in the quality of life within the limits prescribed by life and its necessities.

The kinds of value also have value for life. Prosperity is a value that has meaning in the context of living organisms. Prosperity is not a necessity of life, not a condition, but may enhance the quality of life. Thus the inherent value of prosperity is not only as an instrument of life but has a value of its own, which enhances life. Life is also the condition of creative actualization of such goals, although not of their inherent worth. The different sorts of value add good to the world and thereby enhance life: amelioration. Wealth, for example, is creatively actualized where before there was less or none. Such creative actualization of inherent goods is not necessary, but freely enacted for its projected inherent worth. Economics is not biology, although it works within the parameters of what is life supportive or good for life in some respect. Life supportive goods add to our lives within the limits of life.

Life is enhanced by a community of values, which are distinct from one another but collectively support life. This community is more than just a plurality of different values since they may be mutually supportive, for example, economic life of artistic creation. Artists may earn their living through their art and thereby economic institutions that enable purchasers of art to come into contact with producers. Thus different categories of goods may enhance life, that is, categorially distinct activities aiming at different ends which cannot be reduced to one another, for example, beauty and prosperity. Such categories of activity describe different ends and goods that are pursued to enhance life, and are general or generic to different cultures, that is, they are social categories. These categories describe values in different cultures, for example, religion as a category includes pagan, Eastern, and Western religions. As descriptive they are not value judgments although they involve the evaluation that a cultural institution is best described as under a specific social category. Thus anthropologists describe the art of other cultures under the normative category "art" without any other value judgment being advanced. Values are being scientifically assessed in such a normative judgment.[11] Value categories are the species of activities common to different cultures taken at a general, normative level.

Are there conditions under which life is not worth living, as Aristotle and some others have believed? Or is the belief of some Stoics that no one can take virtue away and that it alone is worth possessing more valid? A life cut short is considered a tragedy in almost all cases, apart from evaluations of the quality of life. Also, the taking of a life in murder is considered a crime, sin, immoral, and so on, apart from whether the life is evaluated worth living by others, since the agent's own life, self-actualization, is at issue, that is, its intrinsic value. (Euthanasia is not a serious counterexample, since it is only raises ethical issues, if at all, as a species of suicide.)

The evolution of life may provide a type of argument here, or at least evidence for one, since although the fittest or best survive, all life attempts to survive. Animals act for an end, their individual survival, although they may sacrifice themselves for their offspring. They react to predation by fighting, fleeing, or other goal-directed activities that will preserve their self-actualization. Not only humans creatively actualize, since animals use elementary forms of technique (birds' nests, beaver dams, ant colonies, beehives, and so on), although humans may be more successful at amelioration. More, humans evolved out of animal ancestors and share an "instinct" of survival with them, a fear of losing one's life and an urge to preserve it. All of life is self-actualizing and this constitutes evidence for the value of life, and an argument for the intrinsic value of life, despite predation (Callicott 1989, McDonald 2001, 2003, Rolston, 1988, 1989).

Can pain, angst or other causes overwhelm the "survival instinct," the urge to life? Or on the contrary, is life worth living in spite of pain, for example, cancer victims or others with incurable illnesses whose pain cannot be alleviated? Is the evaluation of most humans that life is worth living even in illness or other adverse circumstances, and thus of health as valuable apart from the distinct value of life valid? If health were not a value in itself, life could not be of value but only as "perfected" by health. Although health is of value, so is life, even in illness.

Is life only livable under an illusion, as Nietzsche argued?[12] Is it "solitary, poor, nasty, brutish and short," in Hobbes famous words and thus in need of illusions to sustain it? Is pessimism the only realistic choice, as Schopenhauer believed? Whether life by itself in the form of mere survival has value may not be answerable. Life always exacts its price: death for a good life; life itself for a bad one. Is life still worth living—is it better to live when we are oppressed and there is no hope of alleviation? Consider slaves who would rather live than die, or extreme conditions of life such as those of Arctic peoples or workers in the nineteenth century, who chose to live despite day-long labor for subsistence wages: is this only a personal choice? I would answer that life can often if not always be improved: that creative actualization as melioristic is the attempt of agents to improve their conditions of life and thus make life at any rate more worth living than previously.

Is quality of life more important than life itself? Is a way of life more important than a long and healthy life? Some people persist in smoking even though smoking and cancer have been definitively linked. Is the quality of life contained in smoking so strong that it supercedes life? Of course tobacco is addictive and so the issue may not be something without which life is not worth living. But for a few it may be a life-style choice. One smoker I knew proclaimed that he knew he was going to die anyway, so "it might as well be from cancer." Consider those who jumped instead of facing a life of reduced means after the 1929 Stock Market crash: is life on reduced means ultimately intolerable to those who have become "accustomed" to a higher standard of living? Some forms of Buddhism argue that life is worth living on any level: that living in any mode of activity is worthwhile. Is suicide by this standard a mistake? Or is this a matter of different temperaments having a different perspective about life?

A bad life is better than no life, as generations of slaves can attest. But just as surely, we can make the evaluation that a good life is better than a bad life. Given any value apart from life itself, if life is added, there is more value; if a good life is added, the total of good is always greater. This evaluation creates grounds for meliorism, improvement in the quality of life and thereby of the future. Other values of worth cannot be appreciated in a very bad life: the basis for nihilism. Life itself must be of value for other values in life to have value. However, if life is not judged worthwhile, there is something wrong with the circumstances in such a life,[13] another argument or basis for meliorism. The problem is the circumstances of the life and not life itself, and the circumstances can be improved.

The meaning of life is not self-evident; it cannot be determined a priori. Life is given meaning as a teleological project in which each life creates meaning through activity. A meaningful life is a good life full of significant purposes and goals that have been creatively actualized as projects. Life must be given meaning as a project: active not passive meaning. Value is discovered in the world: investigated, revealed, and uncovered. Each life derives its own meaning from such discoveries and their assessment. Giving meaning or value to life is a creative, imaginative, active project. Such a life is full of small, incremental acts of value, which collectively make for a good life. Detached judgments as to the abstract value of life or contemplation of its meaning are replaced with incremental improvements of limited action. Value provides the worthwhile element of purposes that give life a meaning. By differentiating actions, values differentiate the moments in a life and thus give them significance as well. Value as creative actualization ends a futile, meaningless life.

Such values as give meaning enhance a life and improve its quality; value is not only connected with survival value, supportive of life, since goods may enhance a life. Mere survival is a static or cyclical process. As I noted above, this would tie surviving on a subsistence level to being or becoming. Whether such a life is worth living is another question. Life enhancement is a betterment of the quality of life, an amelioration or improvement of life in various ways, whether as healthier, wiser, or still other respects. Life enhancement through discovered value differentiates our era from life in the jungle, where mere survival is at stake. Life enhancement involves improvement of the quality of life. However, the speculative or critical question remains whether true improvement is involved in living some kinds of lives in our era, such as a life spent in prison, or boring assembly line routines.

The relation of grounds of action as values in distinct categories to ways of life is that an actor, a self-actualizing agent, evaluates these by prioritizing them into a hierarchy. An agent may consider an activity only a means to some larger end, for example, a "day-job" to an artist. Many earn a living but only to pursue their "vocation," dream, or life-style as ultimately valuable. Thus the ultimate importance of the employment is minor to the actor. Individual actions in a life are evaluated under one category or another, whether a means or end, by agents. The unity of the categories of activity lies in the actor, not in life. Action is what unites such categories, and the actor who pursues such ends.[14] The actor acts for self-actualization as a good in itself. To deny the worth of such categories of value, as valuable in themselves, worth pursuing for their own sake apart from life, would be to deny that life is worth living in any case.

No single kind of value, however, exhausts value in any life. The life of the artist is also filled with other values, including social and economic relationships. The different categories of value are within a life and part of life as a whole. They are not parts in the sense of ongoing attributes of life as a substance, but activities that occupy our time for a period and then cease, while we take up another activity. They do not endure, but come to an end, only to be taken up again in the future. Valued activities are a moment of our lives to which we give our efforts and attention. We creatively actualize some goal for a time and when the goal is reached, can move on to other values. What we value adds something with meaning to our lives, which had been missing.

"Value for life" should be understood in these respects: not as instruments of life but as constituents. While any categorial values that fulfill our life have value for life, the value is inherent in the activity that we are living through. Valued activities in life are valuable in themselves and collectively make up a whole life. A good life, then, consists in creative actualization of diverse activities. Most agents pursue different ends at different times of the day, different days of the week, different seasons, and different times of our life. What is appropriate for childhood may be unappealing to the senior citizen; summer activities end with the return of the cold. No single piece of music, no political movement, no institution, no activity is enough of itself alone to make life worthwhile. Instead, several institutions, activities, and goals collectively constitute a good life. But what of people who "give their life" to art or to a political cause? Such individuals dedicate their lives to a category, but rarely give up all other activities. Artists have family connections, and so on. The categories of value are, as plural, finite in value and each should be given its due within a whole life. Since all these activities are valued in some respect, they form a community of inherently valuable ends in a life, not a hierarchy of ends under a ruling type; even the single-minded pursue some other goals. Indeed some individuals do not let one end interfere with another, work with pleasure, realizing the inherent value of each.

Morals constitute a limit to life-enhancing categories. Categories of value that are not explicitly moral still require moral limits, that is, justice to their distinct good, a reciprocal relation with other kinds. The alternative is a hierarchy, with one or more distinct values becoming an instrument of an external value without its own value, without value in itself. This subordination is unjust to different goals in a life and thereby constitutes a less full, less valuable life. Indeed, modern careers, with their demands of time, detract from family life, intellectual and spiritual development, creativity, health, and many other goods that are potentially part of a life. Despite our ritually invoked admiration for the well-developed, polymath figures of the Renaissance, we sacrifice all to economics. An increased "standard of living" has meant a decline in quality of life.

4. Anthropology

The Origin of Values: values are learned by the individual from the culture, which has adapted to a place and learned to survive in that place. In other words, "subjective" or personal value is also culturally instilled, tolerated or taboo. The values we would learn in Inuit culture are quite different from those in desert or jungle cultures. These factors of culture and place must influence what values we learn and their origin. All of these are factors in anthropology, a field in which philosophers tend to overestimate learning and underestimate place. Cultures are marked by differentiation of values that constitute the individuals of the culture in terms of their habitat. Humans cannot be given a "logos" without considerations of their environment, since the culture generically reflects the habitat.

I say "generically" since different cultures with different languages may develop in similar habitats. Both may share a primitive "economy," that is a hunter gatherer subsistence or an agricultural one. But they may have different ways of life in several other respects. However, these are specific, and not generic cultures. The culture of the Inuit is not at all like that of the Yanomano, an Amazon tribe: they are generically different based on the extremes of a tropical rain forest and the polar tundra. The Yanomano practice slash and burn agriculture, a practice impossible in the Arctic. Seal hunting, with its attendant cultural requirements is totally foreign to the tropics. I am not arguing for an environmental variant of economic determinism here, since I am arguing that economics itself is shaped by habitat and also that there are other factors of culture that are more independent of habitat, for example, language. However, many words in a language may reflect habitat. I doubt if the Yanomano have a word for snow or the Inuit one for cashew. Moreover, entreaties to the supernatural world are for survival in a particular habitat. Cat gods are only thinkable in a habitat containing cats.

Several authors have argued for a close relationship between the environment and culture, and indeed that human culture is heavily dependent upon the environment. Jared Diamond, in several impressive studies, emphasizes this point in his comparison of different culture areas and the ascendance of some over others:

> I'll argue that continental differences in level of civilization arose from geography's effect on the development of our cultural hallmarks, not from human genetics. Continents differed in the resources on which civilization depended—especially in the wild animal and plant species that proved useful for domestication. Continents also differed in the ease with which domesticated species could spread from one area to another.
> Storable food surpluses grown by some individuals permitted other individuals to devote themselves to metallurgy, manufacturing, writing—and to serving in full-time professional armies. Domestic animals provided not only meat and milk to feed people, but also wool and hides to clothe people, and power to transport people and goods. Animals also provided power to pull plows and carts, and thus to increase agricultural productivity greatly over that previously attainable by human muscle power alone (Diamond, 1992, 36).

Diamond notes that Europeans had sheep, goats, pigs cows, dogs, and horses by 4000 B.C. Humans on other continents could not domesticate as many animals, with as many functions, either because there were not as many available (New Guinea, South America); or they were harder to domesticate (Sub-Saharan Africa, North America) (Diamond 2005, 2006). The resources available in a habitat were decisive in allowing some cultures to conquer others. Even earlier, John Dewey and James Tufts pointed to social cooperation as a condition required for hunting in a habitat having such game, (also, hunting developed as a reaction to environmental changes, namely, the drying out of north Africa). Hunting must take the environment into account. Similarly, agriculture is impractical in Polar Regions, deserts, and wetlands. In agriculture, harvest and planting rituals include festivals, ceremonies, and rituals.[15]

Humans, like other animals, are biological agents, and as biological have limits on what will allow them to survive, to continue their life. As agents they must creatively actualize their survival, by creatively actualizing their food and other biological requirements. Creative actualization is not restricted to humans, since animals and plants also creatively actualize their survival conditions. Thus the primitive origin of values is in the context of life, since all of life must self-actualize to survive. In turn, life takes place in a habitat that is the origin of culture and cultural values. By this perspective, the Cartesian separation of the subjectivity of humans from the objectivity of the world, including the body, is at best an abstraction. The environment is the origin of our values, not a "valuer," since it delimits what can be valued. The environment, as Dewey realized, is not "out there," an object of the contemplative gaze. It is what we are "in and of": inseparable from us. But a limit to choice does not necessarily directly determine choice. (Before Columbus, no one in Europe desired chocolate.) Moreover, a "valuer" emerges out of culture and social relations, as Virginia Held (2002), and before her George H. Mead argued.

Human values are culturally derived and mediated actions, goods, and principles by which we choose and act. An "*archai*," a principle is a beginning and the identity or rule of an actuality. The principle is the beginning as the differentiating identity that normatively identifies a kind, or even an individual, and thus marks it off as distinct. The historical "event" of value is the creative actualization of a novel good, including new self-actualizing agents, living plants and animals. Value marks the origin as the principle of the good as different, including intrinsically valuable agents as goods in themselves. In this sense also, value is the most general term of differentiation, since the unique good of the intrinsically valuable agent is distinct from the value of any other. This principle is normative in that it endures through changes, up until death, as the principle of individuation. In this sense of principle, values are relatively constant in a dynamic universe.

Cultural values are a requirement for survival of any species not dominated by its instincts, that is, any species that learns from its parents. Cats learn that prey is food from their mother, and humans learn to cope with their habitat from their culture. Such cultural values enable us to confront our world with proven methods and the goods that can be normatively actualized, that is, can be remade. Culture is also our frame of reference, not subjectivity. Culture is our first teacher regarding our place in the world, other cultural practices, what is natural or real, and what is good or bad in individual behavior. Culture pervades all that is human. Humans as "creatures who value" have developed values as social creatures. We do not emerge from the womb as Cartesian subjects but as helpless infants who must be taught almost everything beyond nursing.[16] We are biological lives first, not Cartesian minds. We must learn such cultural values prior to becoming individuated, presuming we live in an individuated culture, like those influenced by Protestantism. The values that are the ground of cultural norms in turn regulate the limits that a culture prescribes in every area of our life. Even something as simple as norms governing urination are learned from the culture. Why do we go to a specific place to relieve ourselves? Due to sanction by the culture. Indeed, the hygiene of it requires the individual to control their behavior at all times, even while sleeping, to avoid bedwetting and sopping clothes, which are not culturally acceptable. Whether a person individually, personally or subjectively values such hygiene, the culture does, which is why individuals are taught it. Taboos on murder and incest are learned in a culture, not by rational reflection on human nature or empirical experience.

Individualists might argue that cultures ultimately serve their members, so acculturation can be fit within a self-interest model. But this thesis is extremely dubious, since survival of the culture is almost always put above individual survival. Cultures are usually willing to sacrifice individuals to preserve themselves, both in war and in internal conflicts, that is, inflicting death for the violation of cultural taboos. The death penalty has been banned in Western culture only recently; theft was punished by hanging as recently as the seventeenth century. Self-interest is only thinkable in a post-Protestant culture in which individual conscience in moral and spiritual matters has been secularized. On the contrary, moral values are the ground of possible individual values, since the nurturing relation is moral and cultural. I am not arguing against the idea of personal values, or that a culture does serve its members. However, these should be seen in context.

Values in one sense are the operating ideas of the culture, both its norms for survival and its judgments of what is worthwhile, important, and significant. As such, values compose the culture as constitutive norms of the institutions and language of the culture. Humans cannot be separated from cultural values. A society develops such cultural values over its history and preserves them as its unique culture. Values thereby differentiate a culture as unique, as distinct from other cultures. Are values then connected only with culture and not with pre-cultural life? But animals and other living things also creatively actualize their own survival; recent indications are that they have a culture in the sense of a localized "economy" (Yoon 2003). As I argued in the last section, culture must be redefined as local, as belonging to place, since a culture responds in part to local conditions: tundra, grassland, wetland, jungle, and desert. The values culture exhibits are environmental, and habits reflect habitat.

While some animals exhibit culture, adaptations to different places by a species, culture has flourished most completely with humans, who are displaced from their original habitat, likely due to hunter culture. Humans probably followed the wandering of the herds, seeking out new prey, and thus moving from the originary habitat of humans in Africa to all types of new habitats. We are African in origin and thus our epistemological categories and capacities are universal – they evolved in Africa – but our metaphysics is diverse, based on local habitats. For example, the Inuit as Africans facing a frigid environment or "reality." Culture is a response of Africans to different habitats. Culture is the adaptation of a displaced species to new environments and the creative actualization of the means to preserve life in new places. The context of culture is biological survival in a specific environment. Culture is not cut off from biology in that culture consists in norms and values that have been creatively actualized to preserve life in a specific habitat. Included are animals and plants radiating outward to new territories. Subjective values are the superficial level of such cultural values.

The origin of values is historical, that is, values consist in creative actualization at a time and a place. Novel values begin a new segment of history in which the principle of the value-production commences. The principle is learned through action and is copied in the actions of others. Learning a value is learning creative actualization: values are neither innate nor self-evident. Achieving a goal by action, creative actualization, involves the whole person acting. Psychological states are not the origin of value, since value is the point of creative actualization in the world. Psychological states must also show their value or a culture discards or alters them.

Viewed in this way, humans as historical, biological, and cultural agents acting within a habitat, exposes the artificiality and error of connecting value to an abstract subject by the modern age. Supposedly "rational" norms are the product of a culture that values rationality, even if it forms a subculture. Rationality derives from its practical value in coping with a specific habitat, it does not simply "spring to mind" like Athena from Zeus' head. The context of this treatment of rationality is the epistemological project, but value is not strictly a form of knowledge, at least in origin. The agent seeking new goals does not usually know the path to successful creative actualization of the project. The normative principles involved are discovered, not "given." Once such principles are grasped, they are generally repeatable, and thereby can be learned by others through repetition of practice. There may even come a point at which they can be taught by language alone. Cultures preserve such principles and pass them on to novices.

The condition of such culturally shared values is "recognition of the other" (Levinas) as alike. The moral integrity of each fellow must be recognized for shared values and association. Such association is not limited to humans, but belongs to all gregarious species. What arises with humans is cultural differentiation, that is, morals specific to the tribe, reflecting the recognition of some others who share norms, by contrast with aliens others who do not. Such tribal mores in turn reflect place: survival of life in a habitat. Recognition of the other is limited in tribal ethics to identity of norms and values, excluding those with different norms and values. Cultures in contact do not deal with detached subjective substances but personalities of living cultural members. Feelings and individual psychology are regulated by such mores far more than they reflect them. While individuals may be the source of changes in a culture, group acceptance is a prerequisite. Individual deviances are evaluated by the culture, that is, morally evaluated, to see if they are compatible with cultural survival.

Culture is also pre-economic: contrary to Heidegger, economic value is not the source of value. Reading prehistoric cultures as having an economy is using contemporary categories to interpret non-contemporary activities. While tools were used by ancient hunter-gatherer societies, the activities of hunting and gathering are consistent with those of animals. They are biological not economic, instruments of life, not luxury or personal whim. There was no "market" in the modern sense and minimal, if any, economic exchange. There was no "political" or public, organized economy. Hunting and gathering were biological, that is, conditions of the survival of life in a habitat. Economic value grew out of such life-sustaining value in historical time. Indeed, the overwhelming balance of contemporary economic goods are not needs, or necessities of life, but valued goods, which enhance life. Moreover, such goods receive economic worth in the form of demand because people value them. Mere creative actualization of a good will not assure its economic value. Value is pre-economic and grew out of the interaction of culture, habitat, and biological survival, that is, value for life.

Cultural values are environmental values. Society, environment, and biology cannot be separated, as they are in the artificial "distinction of 'is' and 'ought'," and of culture and nature. What ought to be is creatively actualized in the world, grounded, and becomes part of the world. Dwellings are not in some detached normative realm of the "ought," separated from "fact," but quite actual parts of the world and reflect habitat, that is, the need of a hairless ape for shelter in temperate and arctic climates. A home is not "given" but made in a cold climate. Its value is not "subjective" but vital for biological survival in such an environment. Subjective values are only possible at a late stage of culture, when more basic, biological values for life have been taken care of. Goods that enhance life in the form of personal whims are epi-axiological. Subjectivists stranded in the wild in winter would not worry about pleasures or attitudes but survival.

Cultural values cannot be derived from "human nature" either, whether the intellectualist value of reason as the specific difference of humans; or any other reading of "human nature," for example, that it is "human nature" to feather our own nests, to loot the treasury at others' expense if we think we can get away with it. Culture is the adaptation of humans to a place, not of a place to humans. Humans modify places through creative actualization of goods, dwellings, agriculture, and so on. Such modifications reflect the place, however, since they are creative actualizations that are responses to the habitat.[17] But does this reveal something of human nature, namely, that humans are plastic animals that learn to cope with new environments? If so, then it argues as much against any "human nature" as for one, since humans do not have the same responses, but vary according to place. As for rationality, cultures allowed it mainly insofar as it allowed humans to cope, that is, as an instrument for human survival in a specific environment.

Any strong distinction of nature and culture is fallacious, since cultures are in an environment and remain so even when they become highly technical. Culture is not one "substance" and the world another. Cultural goods are creative actualizations in the world, that is, in the environment or habitat in which the culture dwells. Those that cannot be creatively actualized in the environment are of no value to the culture, for example, agriculture in the Polar Regions. I would not deny for a moment that human culture has grown and added categories of value that are supra-biological; indeed I argued this same point above with regard to economics. However, economic activities and goods are in a specific environment and the last must be taken account of in creation and preservation of the good. Culture has expanded but cannot escape environmental limits. I also would not deny either that some valuations can originate as subjective or that some anthropocentric sources of values do not only reflect the environment. However, these are not the origin or source of value and still have limits consistent with the originary biological, environmental, and cultural limits of value. Humans are not totally determined by their environment, since creative actualization of novelties can transform an environment. But as biological creatures, they must respect the environment requisite to their own survival in what they creatively actualize.

Heidegger's view that early Greek philosophy dwelt in being and the cosmos is valid up to a point: there were figures who argued for becoming and thus against being but in reaction to the Parmenidean philosophy of being. However, he missed the shift in later Hellenic philosophy to the problem of humans in their relation to nature, which up till then was conceived in terms of the logic of the dichotomy being and becoming. His view of Greek philosophy was based on an exaggeration of this early period. The problem in the later period was the notion of change, especially change in human fortune, first made problematic, perhaps, by the Greek tragedians. By the time of the Sophists, the problematic has shifted from nature as being and becoming to the distinction of nature and culture: that culture cannot be explained in terms of nature, since nature is "everywhere the same," while culture, or convention, is marked by differences. Culture cannot be explained in terms of the dichotomy of being and becoming. A new problematic had arisen, in which the problem of being was, at least by some figures, explicitly rejected.[18] Similarly, making or *techne* was problematic for the philosophy of being and becoming since something novel arises out of "not-being" with new inventions.[19] Technology is not entirely confined to humans but has achieved its fullest flowering in human hands. Heidegger too was human; the attempt to articulate the problem of being without reference to humans, the problematic of the later philosophy, is folly. Heidegger could not escape human values in his thinking, the values instilled by culture. Thinking "being" apart from the human is similar to the attempted project of articulating scientific "objectivity" by some analytic philosophers. Both presume normative distinctions made by humans, whether human being and non-human being, or subject and object. Science is as much a product of culture as the philosophy of being was a production of pagan Greek culture.

The problematic of human identity, culture, and technology was articulated within the framework of value, that is, the realism of Sophistic relativism and the Socratic-Platonic idealist response (*Protagoras, Gorgias, Republic* I). In this later phase of Hellenic philosophy, many important figures recognized that the problematic of the good was central, replacing being and becoming, since the diversity of cultural practices could be problematized as different values; the emergence of the novel in technology as genuine change. Since humans were enculturated and creators of technology, the identity of the human with value was evident: humans are creatures who value and create values. Identification with a culture is identification with its values and norms. Invention of new technologies is one form of creative actualization; production of knowledge, another. New technology goes beyond the naïve philosophy of being and becoming by revealing the potentialities of the world of "not-being." More, since human life is enframed by the philosophy of value, nothing human is value free—indeed, nothing at all is value free (McDonald, 2004 esp. ch. 4). The value of all of human life can be problematized and evaluated, the spirit of the dialogue between the Sophists and Idealists over nature as a standard for convention.[20]

The Judeo-Christian tradition added the notion of the intrinsic value of each individual.[21] Each soul is precious, irreplaceable, and intrinsically valuable. Humans are defined axiologically in terms of the extrinsic-intrinsic distinction. Love is the recognition of the intrinsic value of the other, of the individuality of the individual. Augustine also raised the problematic of free will. Humans have liberated themselves from the wild and, as Karl-Otto Apel phrased it, from "instinctual barriers," and thus are capable of good and of evil. Freedom means choice and raises the problematic of morals anew. Moreover, we are free within specific limits to creatively actualize our destiny. Our destiny is not fixed by the problematic of being and becoming. Humans consist in specific value commitments but can become others. Our selves and our cultures can change, where change is distinct from "coming to be." Change is not a coming to be but creative actualization of novelties: of what previously "is not." Humans make themselves: their choices determine their own lives. The changes they creatively actualize impact the world, for better or worse, since they are in the world.

Moral limits are requisite for the sake of human survival, but humans are not an ethical animal in essence. Morals are necessary but not sufficient for humans: a condition of non-violence within culture but not of the human species. A limit does not define content: burned as a limit to what is edible. Humans cannot be defined as "the moral creature." Some violence, especially toward those outside the culture, is considered acceptable, indicating a moral defect or at least limitation of tribal ethics. Morality is social not individual: the individual is inculcated with social and cultural values in education and enculturation. Individuals are expected to creatively actualize cultural values in appropriate situations: courtesy, honesty, generosity, and the other virtues. In other situations, moral norms limit what ought to be creatively actualized by individuals, in the form of rules, taboos, and other principles. But in such situations moral values *per se* may not be the goal of acts, for example, in art, economic life, and other categories of social activity. Moral values are creatively actualized where they conflict with other possibilities or where requisite as virtue. Again, character is a limit, which does not delimit the content of acts that do not violate norms of virtue.

There is no life for humans outside of some type of association; each culture is one species of such a type of association. Birth is a product of association. Association is thereby a limiting factor for human life, as a universal condition of such a life. Association is a relation, not an essence, since it does not belong to humans as a characteristic of individuals as such. Each individual is thus a participant in a society in some sense, from family to culture. Each individual is both a distinct person and one with such social relations; these multiple roles of each individual constitute evidence against any strong dichotomy of the individual and society. Since each individual also participates in and is dependent upon a society, the principle of participation has a relation to the universal human condition. Association is proper to human life and thereby a moral limit, as delimiting human life.

Historically, humans have emerged from "the jungle," that is, wild conditions. Civilization and culture are different in kind from the wild, a new development that requires association. Human emergence from the wild argues against any destiny. Culture, language, food cultivation methods, and many other goods characterize humans in a civil or cultural state. But the emergence of culture has also meant a radical change for humans, from biology to culture. Humans are plastic, since they have removed themselves from their habitat, forming diverse cultures, but they also require the discipline of culture. A *prima facie* duty to the culture is implied by this dependence: respect for the values peculiar to our own culture, as a participant in it.

The historical alternative to culture, in a return to the wild is impossible. There is little chance of survival for humans outside of culture, in a utopian return to the wild. Everyone is free to try, if they truly believe the jungle is better, but a legitimate return to the wild would have to leave behind any of the paraphernalia of civilization, those that enhance life, like tools, methods of food preparation, and the like. There is less value for humans in the wild and a stricter delimitation, in which the conditions of survival are more limiting. However, association is also better for human life, since it is life enhancing. Predator-prey relations rule the wild: *bellum omnium, contra omnes*. Humans are protected from such forces in a life characterized by non-violent relations, civil peace, where humans have rights. Rights are proper to human life in a culture. A life of force implies the possibility of victimization, and the end of one's life, the end of the value of life, in the wild.

What I argued above is an apparent contradiction, namely, that culture, unlike nature, is everywhere different, but is a response to a place, a specific environment, that is, a natural setting. The contradiction lies in the ontological understanding of nature as both the whole and specific parts. Nature as the cosmos includes everything. In this understanding, nature must include change, including "artificial" changes. But this collapses the natural-artificial distinction in which all that is artificial is excluded from the part that is "the natural." Similarly, if nature is the whole, it must include the different cultures that variegate humans. Each of these must be within nature, including their peculiar conventions. But if all the different conventions are "natural," the nature-convention distinction that preoccupied the Sophists is problematic. Perhaps this problem is the origin of Protagoras' famous dictum, since human cultures make "being" and the ontological understanding of nature as "essence" problematic. Culture and technology are part of the whole but also something new and quite different. If nature is the whole, it is a very different affair than what its ontological understanding encompasses. Since these are human creations, anthropology makes nature problematic. Another way to see this is that if values are connected with humans, what is their relation to nature as a whole, the cosmos?

What remains is the environment as the dwelling and origin of culture and technology, as well as humans and other animals considered biologically. Humans are in the world, part of a community of plants, animals, and other organisms. Soil, air, and water are essential elements of the environment of all cultures and the origin of all technologies. The development of wheat, rice, and other staples was local. Brass required local supplies of tin and copper; steel required iron and coal. Humans make their world in accordance with values; we can hardly speak of the world apart from humans. Humans live in a world of their own making but one that reflects a habitat and makes humans at home in an alien habitat by dealing with the potentialities of the habitat in creative ways. The alien is no longer alienating since creative actualization of human values humanizes and gives meaning to even forbidding habitats, the desert, and the tundra. Humans live in their world, to which they belong, not in some abstract world of "being" or "mind." What culture and technology reveal are the wonderfully diverse potentialities of the world and its habitats. The world is not fixed in ontic patterns of coming to be, since its hidden potentialities "are not" and may never be limited. The potentialities of the world cannot be encompassed by any philosophy of being: their creative actualization is the bringing forth of what "is not" to actuality, permanently changing what "is."

Freedom: a free act is the criterion of a value. Since a free act is the criterion of a value it follows that values and ethics are not only concerned with limits on freedom but are inherent to freedom. Ethics is not confined, as in the thinking of much of modern philosophy, to the realm of the ought, to the pure potentiality of what should be but "is not." What humans actually do indicates their values: their idea of what ought to "be." Even scientists reveal their values: by the pursuit of truth, the scientist reveals his/her belief that s/he ought to be engaged in such research.

In the light of the environmental, biological, cultural, social, and other limitations on the human condition, what is the place of personal values, the values of the individual? To some degree, these are values of the individual person that are shared by others, social values. However, there is still a place for personal activity and thus for individual freedom. The things with which we are concerned or involved reveal our personal, categorial or cultural values. Decisions and choices also reveal such values. Values, then, are not subjective but in the public space of the world, manifest as action. What then is the place of personal values? These too must be creatively actualized. In another respect, all cultural values are instantiated in individuals. That we as individuals can imagine goals as actualized in the future is an individual affair. That we can actualize the imaginable is not a "subject-object" relation. Value is not metaphysical. The source of such evaluations lies in the person, but people also actualize values and goods in the world. The results are neither subjective nor objective but transcend the distinction. They are "objects" that reflect the creativity of the "subject," like works of art, personal values actualized outside the person in the world, changing the world. Such values are both wondrous and wonderful.

The person can be the origin of novel creative actualizations, that "were not," undetermined by past conditions. Judging which circumstance calls for which action, including classifying acts (for example, is this a situation that is a threat to my life?) is evaluative. Freedom is intrinsic to such evaluations, for were they determined, we could not respond to the new circumstances, but would be physically or genetically determined to act in a determinate way by the past.

If humanism is defined as the view that humans are in some sense different, then radical axiology is not anti-humanistic. For human culture has liberated us from a place of origin. By creating culture, humans have liberated themselves from a single place, although they share this proclivity with some other species. By living in different environments they have been able to discover different goods in the world and share them, for example, grains from different areas of the world. Creative actualizations of so many goods is peculiarly human and follows first from culture, liberation from a specific original place; and secondarily from dealing with varied environments. Thus if humanism is the view that humans have an essence, a fixed destiny, then radical axiology is not humanistic. By rejecting the problematic of being, a fixed destiny for humans is also rejected. Humans do not have an "essence" defined as their species being. There is no "essence" of human, if only because human freedom means humans are virtually unlimited in their capacity to grow and adapt, even in outer space, an airless void filled with lethal radiation!

If humanism is the view that humans are somehow supreme over nature, a view that can be traced back to the Old Testament, then radical axiology is not humanistic, since humans require an environment. If humanism is the view that humans are the central value, the only creature with intrinsic value, then radical axiology is anti-humanistic. All creatures creatively actualize themselves, self-actualization. Human freedom and human well-being are not the only values or even the most important values. The environment is the most important limitation on human freedom, as the condition of human freedom. Better, human freedom dwells within a specific environment, which is part of the free act. The action and its environment, its place, are inseparable. Humans are free within moral and biological limits. Freedom is freedom from some biological limitations, but not all. Yet they have expanded the limits. This freedom of *homo liber* means that "nature" cannot be the model for humans, but only the home for humans. The danger is forgetting the origin of values in a specific environment in the rush to economize the world. Greed is blind to its own conditions.

Five

CRITICAL EVALUATION OF METAPHYSICAL VALUE THEORIES

What is the relation of creative actualization to other prominent value theories? In this chapter I will critically evaluate the major theories in the light of the theory of value as creative actualization. The other theories will be treated under certain rubrics, namely, transcendent theories, ontological theories, and action theories. I will examine subjectively-based and relational theories on their own, in the next two chapters. The theme of these chapters is that creative actualization is the denial of sovereignty to any *summum bonum*, whether human perfection, happiness, self-interest or others, over "subordinate" ends. I will address this point below.

1. Metaphysical Value Theories

I argued in a previous book that value theory has been treated metaphysically in the tradition, that is, value has been treated in relation to a metaphysical first (McDonald, 2004, 3). This approach includes both hierarchical models based on transcendence (Plato, *Rep*. VI), and the subordination of value in the ontological relation of attribute to substance (Aristotle, *Meta*. VII, VIII).[1] I will treat hierarchy first and ontological theories below. The hierarchical model of value is also metaphysical, in that the "Form of the Good" in Plato is also the origin or cause of the being of the other forms, which individuals in the visible realm copy or in which they participate. The Neo-Platonists, including Augustine later use this model to argue that all good derives from God, who is identified with the highest good.

A hierarchy of values is a model in which subordinate values derive all of their value from a relation to one supreme value, a *summum bonum* that justifies the rest. Subsequent theories of value may substitute other values such as pleasure for God as the ultimate value, the *summum bonum*. But incredibly, given their otherwise non-Platonic approach, they retain hierarchy from Platonic Idealism, which is the origin of value hierarchy in the form of *summum bonum* theories, the idea that there is one supreme value which encompasses or grounds the rest. In this model, all other values must accord with the one supreme value, the *summum bonum*, by participation, since their value is conditional upon the supreme value. Subordinate values derive all their value from the highest good.

In the *pros hen* relation, Aristotle argued by extension that instrumental value "referred back" to intrinsic value for its value. For the value of the instrument refers back qua instrument or "value for," to that for which it is valuable, that which is valuable in itself, or intrinsically valuable. As more than one critic has pointed out, the *pros hen* relation may not be known to the average reader, or even to philosophers. Aristotle's *pros hen* relation is the relation of equivocal terms to a first term to which they "refer back", a "referring back to a first." For example, being is "spoken in many ways," but the being of the attributes refers back to the substance. "Grass is green," involves the being of the green which is grounded in the being of the grass which "stands under" (sub-stance) changes in color and thus endures through changes or has being more fully. For Aristotle, the subject-predicate relation corresponded to the substance-attribute relation. Just as the predicate refers back to the subject in propositions, so do attributes refer back to substance for their being. The relation of different properties is reflected in the grammar of language. The relation is also reflected in the relation of essence to accident and ground to consequent. The ground is the cause of the consequent just as the being of the substance is the ground of the being of the accident.

Although the instrumental value has its locus in a different being than the intrinsic value, the value relation as such is a *pros hen* relation. That is, the value of the instrument refers back to intrinsic value. The instrument derives its value from the intrinsic value and would not have any value without the intrinsic value. The value of the instrument is not intrinsic to the instrument, but refers back (*pros hen*) to whatever is assigned intrinsic value. Since the *pros hen* relation subordinates to a legitimating first, it also creates monistic hierarchies, borrowing hierarchy from transcendence, although of a slightly different sort than the relation of participation. Subsidiary values derive their value in relation to a first by referring back to a first (*pros hen*), which creates a more foundational hierarchy based on relations. A monistic, unitary value is the source of value, to which all value ultimately refers. The value of the many is contingent, which is a devaluation of them.

In *Nichomachean Ethics* I and other texts, Aristotle subordinated the good to being in the relation of attribute to substance. The good was not an independent "principle" but subordinated to being. Thus the being of the good is as an attribute of being in a perfected mode. Being still stands under good or value as substance that changes from potentially good to actually good. The *pros hen* relation is of aspects of being, namely, substance and attribute. This relation is the foundation or basis for the relation of good to being as modes of being in reality, since the being of the good depends on or refers back to the being of the substance, being in its primary sense.

This relation was retained by the subject-object metaphysics of Descartes in that the object refers back to the subject as knower and valuer. The value of instruments, then, refers back to the intrinsic value of the subject as its source of value. In turn this creates a basis for a hierarchy of value, with the subject as the one to which instrumental value refers back in a *pros hen* of value. The subject is the *summum bonum* or apex of a value hierarchy. This relation is also retained in the move to language by Moore in *Principia Ethica*, which, like Aristotle, examines "what is said" as evidence for what is (Moore, 1903, 1). As the subject-predicate relation is retained by Moore, and used as the basis for analysis, the *pros hen* relation is surreptitiously used as the basis for "meta-ethics," and is the basis for meta-ethical studies of the grammar of ethical statements or propositions. Thus analysts often begin an analysis by asking "What x term means" or "what is x" as if questions of meaning were equivalent to questions of what something "is."

Finally, some theories atomize value in the constitutive relation. Individuals are composed of atoms in materialism. Similarly, value is constituted by random subjective events over time, whether pleasure, attitudes, emotion or other feelings; or random desires, willing and interests. The good life is atomized into random, fleeting events that come and go but collectively are good. A good life derives its goodness from having more of such episodes of feeling, desire or attitudes than their contraries. Goodness is the collective sum of atomic episodes of subjective events. How such atomic episodes form a whole, that is a good life, is problematic: an example of the fallacy of composition. Atomic theories are generally subjective and since this type is especially and notoriously prominent at present, I will treat such theories separately, in the next chapter.

The ultimate ground of value lies in the *pros hen* relation to an object as a supreme good or to a valuing subject in metaphysical theories, whose scale forms a hierarchy, and not independent of either the subject or the metaphysics of subjects confronting objects. Thus the problem is defined as metaphysical instead of as axiological. Value has been treated metaphysically and issues of value have been confused with issues in metaphysics. In this chapter I will critically evaluate such theories. However, I will concentrate on the particular content of the theories of value more than the metaphysics of value, insofar as these can be separated.[2]

2. Transcendent Theories

Absolute value: absolute is in contrast to relative, whether defined as cultural, personal or temporal relativism. A hierarchy of values involves a *summum bonum* that is generally absolute or unconditional, that is, defined by contrast to "relative." A *summum bonum* is not conditional upon subordinate values in the hierarchy, since it conditions them; nor relative to cultures or persons, as unconditional. However, absolute values may be treated in distinction from unconditioned values as unchanging, that is, not historical or relative to a time. Absolute value is a species of universal value, not contingent upon circumstances, persons, cultures, and so on. Thus it is universally good across cultures, and indifferently to persons (although not all universal values are absolute.). Such values would not be contingent upon the vagaries of fortune or pragmatics. Since they are absolute, they transcend cultures that do not practice them, and would constitute a standard by which to evaluate the such cultures (Kant 1785, 1788, Münsterberg, 1909).

Even subjective *summum bonum* theories may evaluate all goods in relation to the *summum bonum*, whether pleasure or feeling. Subjective theories have retained the *summum bonum* model from idealism. Ultimately there is only one inherent value for such theories, namely, the *summum bonum*. Thus such a subjective *summum bonum* may constitute an absolute for such theories. The hidden premise in reductionistic theories of value is that the *summum bonum*, whether pleasure, feeling, attitudes, will, desire or reason, gives value to the person/human/body/world as a whole. Thus only pleasure is an end for John Stuart Mill: he is explicit in his reduction of all values to means except pleasure. The larger scope of value is devalued since it only has value as derivative from the subjective absolute. Pleasure, or another candidate *summum bonum*, constitutes them as of value: the reductive part, pleasure, gives value to the whole, the human person and thus constitutes them as worthy or of value. Thus the specific difference of humans is not a natural but a value difference qua loci of reduced value. As Foucault realized, the body is devalued and perfected by such absolutes.

However, meliorism constitutes an argument against absolute standards. Value cannot be prescribed *a priori* in a world of change, since this would not allow for improvement. The question then arises of whether absolute values have a place in such a world. Changes tend to undercut fixed standards: what was best at one time is now a historical curiosity which has been surpassed, for example, top of the line automobiles from 75 years ago, or computers from only ten years ago. Historical change is the foe of the best, since the continuing improvements may eventually change the standards. Standards can change for the better, and absolute standards, a species of duration, superceded. Absolute values might freeze history and thus preclude melioristic developments. Moral progress constitutes an argument against absolute fixed standards. The treatment of slaves in the Old Testament may have been considered just at the time, but is now almost the paradigm of injustice. However, some standards may endure, for example, certain kinds of justice. We may see injustice in the present with the same standard with which future betterment is evaluated. Present value may contain betterment in terms of relative improvements over what had been good prior to their creative actualization. Wholesale rejection of them in terms of some absolute standard of reason is a rejection of meliorism (Marcuse, 1964).

Different categories of value imply different values, that is, a plurality of values, and thus constitute an argument against reduction of values to value, that is, any supreme value or one. Pluralism in kinds of values provides evidence against a reductive *summum bonum*. Absolute theories have no principle of difference in values, since all values derive their value from an absolute. In such theories only one ultimate value rules, whether pleasure, love, etc; not many different values, for example, beauty, health, knowledge and the like. Thus these different categories of value cannot be differentiated *as values*: beauty is not valuable only pleasure. Beauty is not a value, but an instrument of pleasure; health not a value but only the pleasure derived from being in good health. However, beauty is not reducible to wealth or vice versa. Nor are personal values social, since personal values are what individuals value apart from socially shared values. Plurality of values and modes of value constitutes an argument against reduction to one *kind* of value (but not against one theory of value). Reductionism does not do justice to the different kinds of values, to distinct personal values, and above all to value in itself: what is uniquely valuable, the individuality of the instance, its inherent value. For all value is derivative from a reductive first.

Differentiation of values is an argument against absolute values, since their inherent value cannot be derivative from an extrinsic value if it is inherent. Intrinsic value and inherent value cannot be subordinate to an absolute as inherent or intrinsic. Inherent value, including intrinsic value, cannot be subordinate to a *summum bonum*, an absolute value, since it would then be extrinsic, not inherent. Its value would not be inherent but derive from an extrinsic relation to an absolute. The intrinsic value of the person would be conditional on the relation to some *summum bonum* and contingent upon its actualization. This evaluation devalues the person, to say the least.

Another argument against absolute value is the distinction of value from evaluation, made previously. As a standard, absolute value is not "objective" since it transcends bearers of value: absolutely valid or of value regardless of whether it is instantiated. Thus absolute value is not necessarily inherent, or at least not confined to inherence in a bearer. It does not require creative actualization in a particular case to remain obligatory. However, this places such values in the realm of evaluation, not value. Their value requires creative actualization. Similarly, reduction of all values to one *summum bonum* that gives an identity to all differences is a logico-normative exercise, as identities are normative. Norms as such are in the realm of evaluation, not value.

Absolute standards that cannot be creatively actualized are Utopian. Thus their value is contingent upon practicality, which is an argument against absolute standards, if not standards. For value to inhere or accrue, a project must be actualized: it cannot be confined to the realm of the normative, evaluative or subjective. Creative actualization is an improvement in the world: there can be melioristic values apart from evaluation.

Absolute value is untenable, since it devalues any but the unconditional *summum bonum*. Thus the value of goods other than the absolute value would be conditional, not inherent. Since intrinsically valuable agents have inherent value and intrinsic value is the source of rights, absolute value provides a justification for violations of individual rights. However, if any actual good is good, its goodness cannot refer beyond it, to an extrinsic good. Absolute value as a standard lies in the realm of evaluation, that is, potential goods. Such standards are norms by which activities to achieve them are measured. Since a standard applies to all potential goods, including future goods, the standard is not actual as such; absolutes are evaluations and must be grounded in world as goods. The absolute is grounded in and through creative actualization, but is not the exclusive ground for creative actualization.

3. Idealist Theories

Idealist value theories, apart from subjective idealisms, hold that value consists in a beyond of value: that the locus of value is not in the world but in an ideal realm. Value idealism includes Platonic, Neo-Platonic, and Schelerian theories. It may also include Kant and the Neo-Kantians such as Heinrich Rickert and Hugo Münsterburg, although Kant and Scheler may better be treated under subjective theories, or absolute theories.[3] However, such ideals as ideas are in the realm of evaluation only, even as standards, since until they are creatively actualized they remain Utopian. An evaluation that creative actualization of some ideal of value would be worthwhile may be valid, but requires creative actualization. The ideal may indeed have potential value but remains untested in the actual world. An ideal can be evaluated as potentially valuable, but not yet an actual good. Idealists confuse value and evaluation and reduce the multiplicity of values to some absolute value.

An idealist might argue that ideas have value as ideas and are actual as archetypes or forms. Ideas do indeed have value as ideas but not as actualized unless they prove feasible. As a teacher of mine once remarked of the Maginot Line, it was a "great idea, the only problem was, it didn't work." However, ideas are actualized as goods, not as ideas. As ideas in the mind they only have personal value. The idea of owning a million dollars may be attractive but not yet creatively actualized. Potential value or worth is not value. However, ideas may motivate us on a personal level to creatively actualize them (inventions). As for formal value (Kant 1788, Hartmann 1974),[4] formal value in one sense is inherent but not intrinsic, namely, in the sense that formal value is the value of repetition, of an identity of instances . Thus formal value is not the value of difference, that which makes an instance unique and of inherent value as distinct. A formal kind is instantiated and so inheres, but formal value is not intrinsic to an instance as such as a novelty, distinct from other instances. As normative, its value is contingent upon creative actualization, both of a distinct kind and of its duplication. In the other sense, formal value is a copy of a standard or idea and so transcendent and normative, not actual and inherent.

Thus values are not ideals. They do not "exist apart" in some objective ideal or intelligible realm (Plato), since they are actualized by successful action in the world. Again, they are not subjective ideals of the mind, for successful action actualizes such ideals and this creative actualization is value.

Without it, they are Utopian. Nor are values norms in the deontological sense of what ought to be creatively actualized, including duties. Apart from moral values, most values are not morally required but instead evaluated as worthwhile. Their practical actualization can hardly be required if it is novel. There is no imperative to build a time machine, if this were considered worthwhile, since it may prove impossible. Values are not normative and value is part of world, not a normative realm of evaluation.

Another idealist theory is the harmonic theory of the Pythagoreans (?), Plato and Leibniz. The good is realized in the "order and harmony of the parts" in interrelation to form a perfected whole. This could be treated as a relational theory, or a holistic theory. Either good is a proper relation of the parts or is a holistic quality of the parts in a proper or optimum relation. George E. Moore may also have held the theory in the latter form.

Moore's Holistic Non-naturalism: G.E. Moore proposed a number of innovations in ethics and in his approach to value theory which make his work unique and significant. From the perspective of this book, perhaps the most significant is his value holism. Value is neither subjective nor wholly objective, nor does it involve a simple relation of a subject to an object. Both subject and object are included in larger wholes as parts in which neither part necessarily has value alone, that is, apart from such wholes of value. Most figures from the history of philosophy locate value in individuals in some sense; only a few are holistic. This affects the character of value as well, that is, value theory, as I will argue below.

Moore argues that the field of ethics, or "the general inquiry into what is good," includes the fundamental question, "what kinds of things ought to exist for their own sakes?" (Moore, 1903, 5). Thus the question of intrinsic value is investigated and intrinsic value is connected to actions. For Moore the good is "the only simple object of thought which is peculiar to ethics," and argues that the definition of good is the "most fundamental question in all of ethics." There is a subtle shift here from questions of psychology to questions of definition. The evidence in ethics is drawn more from spoken language and not facts of psychology; the emphasis is on what is said, not on the relation of psychological states such as feeling and desire. Moore believed his method of asking for the meaning of basic ethical terms was both novel and in accord with "common sense." Moore thereby began the analytic revolution and this in a work on Ethics.

However, (analytic) judgments of good have the peculiarity that they are indefinable: the good "cannot be defined" (1903, 7)[5] The good is indefinable because "good," according to Moore, is a simple, basic term like yellow, in terms of which more complex terms are defined. Analysis must stop at such simple terms since if they are defined any further, they are complex themselves. Simple terms are the basic terms out of which more complex terms are defined. This presumes Moore's theory of definitions, that is, that the meaning of complex terms must be derived by analysis into simple, indefinable terms; that meaning is distinct from reference beyond terms. It is not clear whether the analysands are simple or also the parts of which complex objects are made are also simple. Simple attributes constitute a foundational argument, that analysis can go no further but must stop at such basic terms to avoid infinite regress. More complex terms correspond to complex objects composed of simple parts. Such parts are correspondingly simple and indefinable: they may be "perceived" but "you can never, by any definition, make their nature known" (1903, 7). The good as a "quality" belonging to complexes that can be judged "good things" is indefinable. Synthetic judgments about the good are feasible in this view, for example, that "pleasure is the good," but this is not, according to Moore, a definition. "I believe the good to be definable; and yet I still say that good itself is indefinable" (*ibid.*, 9).[6] The judgment that pleasure is good is not a definition, but a proposition that good attaches to, or is a simple predicate true of pleasure: an attribute of a more complex thing.

Good, however, is not a "natural" property: in this stance Moore argues against Aristotle and most of the tradition. Good is unique in being applicable to natural objects, but not a natural property itself. The term is a non-natural predicate of natural subjects. More strictly, good cannot be defined in terms of natural properties, the "naturalistic fallacy," although it can be a property of natural things. "I do not deny that good is a property of certain natural objects" (*ibid.*, 41). To define good in terms of natural properties would make it a complex term instead of a simple one. The confusion of the meaning of goodness with other natural properties, for example, pleasure is what he calls the "naturalistic fallacy." Thus the naturalistic fallacy is a corollary of the distinction between meaning and properties. The good must be attributable to natural objects, given Moore's definition of nature as including everything that has or ever will exist. But natural existence is "in time" for Moore, and the good cannot exist "by itself in time." "It is not *goodness*, but only the things or qualities which are good, which can exist in time…." (*ibid.*, 110-119). Thus good is not connected with the intrinsic nature of the object.

Good can be attributed to pleasure as a quality of it without being the same as or meaning pleasure.[7] For it can be asked of any natural property in terms of which good might be defined, "but is it good?" Good can still be distinguished as a property of such natural properties, and thus is distinctive. If it is not natural, however, how can it be attributed to natural things? Moore's answer is that good or value is not like other "intrinsic nature predicates," for example, yellow. Although attribution of intrinsic value to something "depends solely on the intrinsic nature of the thing in question," (Moore, 1922, 260) as an intrinsic quality of it, value is not itself an intrinsic property. This excludes natural intrinsic value, according to Moore. The good is unique but depends on the intrinsic nature of the thing to which good is attributed. As an example, Moore uses "better," meaning, "better fitted to survive." The latter quality "does not depend merely on the intrinsic nature of…" the object, (*ibid*., 1922, 257) perhaps because survival also has to do with circumstances, environment, and so on.

Although there can be good objects, goodness is not necessarily a property or character of objects. In its fundamental sense, goodness is a property of wholes. Intrinsically valuable things consist, as I noted above, in "highly complex wholes"; value is a character of such wholes independently of whether it is also a character of the parts. The character of good of a whole is a quality of a whole, which is usually not an object. Nor is it an overall good like happiness. Moore's holism is distinctive in that good can be ascribed to a whole that is not embodied in a particular object, and is more than the collective good of its parts. True, each of the parts is a subject, an object or the character or relation of these. But the good of the whole is not equal to the sum of its parts: "the value of the whole bears no regular proportion to the sum of its parts."[8] The value of the whole can be greater than the sum of its parts. This whole includes subjective consciousness as well as objects: both may be included in the whole, but "consciousness alone does not always confer great value upon the whole of which it forms a part" (1903, 28). The good in good objects is generally instrumental to intrinsic good or a part only of a larger whole. The whole contains the part but not vice versa.

In addition to states of consciousness and objects, the good can be included in a larger whole that includes bodily and material embodiment, actions, cognition and appropriate emotions. For example, appreciation of beauty may include a beautiful object in which the beauty is embodied, belief in the existence of the object, cognition of the beauty, appropriate feelings toward the object of beauty and pleasure in the contemplation of the object. All these, Moore claims, are requisite to beauty as good. If some elements are lacking, the result may be less good than the more complete whole; if the elements include bad parts, for example, inappropriate emotions, then the good of the whole is mixed. Thus parts may also have their own intrinsic value (*ibid*, 130). Intent may form a part of action, and can have value, but does not necessarily have it, according to Moore. Complex wholes are the end of action; instrumental value is defined in terms of its effects upon the whole as the consequence or goal. Thus value can be embodied in means but the value derives from the consequences.

As I noted above, "good" is a unique predicate that Moore connects with intrinsic value. "Whenever he thinks of 'intrinsic value' or 'intrinsic worth,' or says that a thing 'ought to exist,' he has before his mind the unique object—the unique property of things—which I mean by good" (*ibid*, 17). If a thing to which the predicate good attaches is good, when it is "asserted" of something, this means that its "existence or reality is good." Thus Moore does not mean a "supersensible property" by good.

As for what sort of things or experiences exemplify such wholes, that is, the content to which "good" may be attached as a predicate, Moore uses the experience of beauty and relations of personal affection as paradigmatic, although intrinsic value can belong to many different kinds of wholes and things. Beauty as such is indefinable, like good, but as a simple component of a larger whole, which includes beautiful objects, consciousness of them over time, and so on. Beauty is a necessary element or part in an intrinsically valuable whole (1912, 202).

Moore raised the issue of the application of the term "good" in light of such questions. Of any ostensive good we can ask, "but is it good?" Moore thought that this implied that good was a distinctive, "non-natural" predicate, since the goodness of any "natural" good is problematic: we can always raise the question of its value (1903, ch. 1).

In criticism of Moore, we could go him one better and ask: is "good" in the sense of a simple, non-natural predicate good? As he states that one can always ask this question of any value theory, presumably we can ask it of his own. But if that is the case then the infinite regress is not avoided; good is forever indefinable not in Moore's sense, but because the same question can be posed to the tautology: is good good? Further, if the distinction is between the meaning of a term and its attribution as a property, the distinction is untenable. *Intrinsic* value may be inseparable from its attribution as a property of an object as such, in which goodness is inherent to an object. Intrinsic value and the problem of the mode or locus of value may be inseparable, as intrinsic implies intrinsic or inherent to something. If meaning is defined by essential attributes, this essence has its origin in the problem of the locus of value. Meaning refers back to a ground, at least in the case of intrinsic value. Moore might argue that questions of meaning are distinct as logical from questions of mode or locus of value. However, intrinsic value must be defined essentially as a quality, as it is intrinsic to something. Thus its being a property is essentially inseparable from its meaning.

However, his view involves a serious confusion of evaluation and value. Moore confused evaluation of the value of the actual with actual value. A plane that flies is good in some sense, at least as compared with one that does not. It is good if the project has been successfully grounded in the world, creatively actualized, and the plane actually flies, as the project intended. The aircraft is inherently good in that respect. "But is it good?" in the sense that Moore asked is a question for evaluation, not a question of value as such, whether conceived in some ultimate cosmic sense, comparatively, or in terms of inherent value. Moore might argue that goodness is only an evaluative term and thus that raising the question of such good is to ask whether we may regard application of the term as legitimate. But then the problem of intrinsic value cannot be raised, as I argued in chapter two. For no actual good would be good in itself, if good were confined to evaluation, but only good in evaluation with respect to some standard. We can indeed raise the critical question of the value of any actual good. But if moral principles can be put into practice, as Moore also maintains, he cannot seriously claim that nothing actual is good. For moral practice embodies goodness in the world. Moore's question belongs to evaluation, not inherent value.

Moore christened this the "naturalistic fallacy." We may ask, is Moore's naturalistic fallacy truly a fallacy? This question has been raised by Harold Osborne (1933, ch. 9) and rightly so, since Moore accuses practically the entire tradition of this fallacy. In one sense, Moore appears to equate natural with what is perceived. He states that "it is not *goodness*, but only the things or qualities which are good, which can exist in time…can be objects of perception (1903, 110-11)." But Moore also has compared goodness to yellow as a simple non-analyzable property which by analogy with colors cannot be defined but must simply be "seen," or pointed out. Moore is inconsistent in arguing both that goodness can be perceived and good cannot—but also that simple properties cannot be defined, as basic to definitions, so must be seen. True, Moore argues that good is unique in this sense. But then how is the goodness of good things or qualities to be determined? On Moore's account, applying judgments to things is impossible, for example, in "grading" the quality of fruit, cars, diamonds or any other object (Urmson 1950). Good is often applied to objects or actions as a *descriptive* epithet, for example, a good tree, whether for shade, as a specimen, or in terms of its quality of form or by comparison with other trees. How can one validly make the judgment that something is good or better, if one does not know the meaning of good?

If good can be a property of natural objects, how is it unnatural? Since natural objects exist in time and good can be a property of them, then it must "exist" in time as such a property, for as long as the object "exists." If it "exists," an object is natural by Moore's own definition; indeed, the very existence of certain wholes, such as beauty, makes them intrinsically valuable. Their existence is good in itself: "when we assert that a thing is good, what we mean is that its existence or reality is good" (1903, 120). Further, good could not itself "exist" without such bearers in any account in which good is a property. Its "existence" is contingent upon the "existence" of its bearers. On Moore's account of nature, "it may be said to include all that has existed, does exist or will exist in time" (*Ibid.*, 40). Then either good cannot "exist," cannot be a property of natural objects in time, or it must be somehow capable of becoming natural, since natural includes everything which "exists" according to Moore. As for "goodness itself," does "yellow itself" or any other simple, "natural" predicate "exist?"

On the contrary, the naturalistic fallacy is itself a fallacy, since the argument is that values both are "in nature" and are "non-natural." Values creatively actualized are the basis for "facts" whether defined in the etymological sense as "deeds" or in the more contemporary sense as established knowledge.[9] The deeds and their consequences were undertaken due to evaluation; knowledge of goods is thereby possible. Once a goal is creatively actualized an "ought" becomes a good in the world, for example in the invention of planes. Natural or not, it is a "fact" that such goods are in the world, and the attempt to exclude such goods from consideration, with ritual invocations of a fallacious fallacy, is a failure. The fallacy is the attempt to argue that everything is in nature, but good is in a different class than "fact," one natural and one "non-natural." Goods are in the same world as everything else. Creative actualization is in the same world as "nature" and in many cases involves "natural" tendencies, notably in the growing of crops. The growth of fruit on a cultivated tree and a wild one is in the same world, and the picking of the fruit as a good is an event in the same world. Creative actualization of fruit trees in an orchard is a design in the world; the husbandry of increasing domestic herds increases economic goods through "natural" processes and would be impossible without them. The "fallacy of naturalism" implies that the widely accepted norm that "it is wrong to take the life of (murder) an innocent person" is invalid, since life is natural. Hopefully this example is enough to demonstrate the fallacy of calling this a fallacy.

More, "intrinsic" is *defined* as part of the "nature" of the object, so I have difficulty in seeing how intrinsic value can be an unnatural property. If such objects are good in themselves, goodness is intrinsic to their existence, thus natural. Otherwise, good is not indefinable as a property of the intrinsic nature of the object. If intrinsic value is both indefinable and an end, all value bearers or goods as properties must be means, or the naturalistic fallacy has been committed. No synthetic judgment that, for example, "pleasure is good" can be correct, for if intrinsic value does *not* mean pleasure, or any other natural property or combination of natural properties, then natural properties are not intrinsically valuable, but only a means to realization of intrinsic value.[10] It cannot be an intrinsically valuable end or such an end is identified with "non-natural" goodness.

Another consequence of Moore's non-naturalism is that action cannot be a means to effect good ends, as goodness cannot be brought about, or effected by action. If good is separated from either nature or reality, it cannot be brought into nature or reality. But if good can be realized in the future, if it can be made "real," it must be natural, by his definition of nature, as it can be made to exist in time. However, Moore argues precisely this point, that "it is certain that our actions can only affect the future, and if nothing can be real in the future, we can certainly not hope ever to make any good thing real" (*ibid.*, 116). Moore does not distinguish artificial and natural good but the point applies to either, as a seed may grow into a "good" tree, that is, realize goodness in whatever respect—health, beauty, and so on—and thus good may come to exist in time. If action can realize the good in time, this contradicts his point that it cannot exist in time. The point applies *a fortiori* to artificial goods. Morals and virtue are also affected as actualizing an ought, doing what is right in the form of an action, or acting virtuously, can be brought about in time. Instrumental value is also affected; there is a denial of value cognition that follows from the denial of definition. We cannot know if an object may be good for something else, if we cannot know if it is good; the inference follows from indefinability, as I argued above.

Moore argues for a separation of Ethics from reality (in *ibid.*, ch. 4). This may be a point about boundaries, instead of a literal separation. But if reality is more or less coextensive with nature, as it is in many naturalistic philosophies, including Moore's, then to identify good with some aspect of reality is a version of the naturalistic fallacy, as indeed, Moore accuses the idealist tradition of doing in the same text.

Finally, if wholes, which are intrinsically valuable in Moore's view, are composed of parts, and complex objects are defined by reference to simple attributes, then good cannot be attributed to wholes in the same respect as to complex objects. This is because wholes often have such complex objects as parts, and the good of the whole is not definable with respect to that of its parts. Moreover, a part may be inherently valuable apart from the whole. Water may be a part of life, but thereby has inherent value apart from life. Moore confuses intrinsic and inherent value.

Although both Moore and Kant attempted to objectify morals, both wished to exempt morals from nature. Kant based "objectivity" on the subject while Moore defined good non-naturally and considered any attempt to define good naturally as fallacious. Neither, then, offered an alternative to the problematic status of the value of the non-human world. Both committed the fallacy of confusing norms and values, since Kant took the "ought" as the model for morals, morals as the condition of goods, and the normative as in a separate realm than nature, the world of phenomena. In these judgments, he was followed by Moore, who argued that good is a "non-natural" predicate and equated ought and intrinsic good. But values are not norms since the value of a good is not in its identity but is inherent as instance. Creative actualization of one instance is a differentiation in time that has only a logical relation to other instances. On the contrary, norms, in the sense of rules, aim at certain values, that is, at the achievement of values, and thereby require creative actualization. Nor are such inherent goods in another world, the "noumenal." Water is both in the world and good for quenching thirst. Water is good for all living things. Water is natural. Therefore what is good for all living things is natural. QED. A good plane is phenomenal, since we can perceive it flying. Even norms are in the world as instances of an identity.

This framework for treating value was also in the background of the axiology of Raymond Polin, who otherwise anticipated our theory of creative actualization (Polin, 1972).[11] Polin correctly saw the dynamic thrust of value in creating a new world, and thereby overcoming "being." But since he took Kant's separation of noumenal "oughts" and natural "facts" as axiomatic, he was forced to conclude that value is "unreal." Even the product of actions aiming at a value does not result in a world of value, according to Polin, since the result is a thing. Inherent value is thereby denied. Polin attributed value to the creative acts of the subject, without realizing that the subject too is in the world. In other words, he belonged to the era of anthropocentric ethics, before the revolution in values marked by the rise of environmental ethics. The subject indeed transforms the environment in creative acts, but is first shaped by that environment, of which the subject is a part.

Environmental ethics has been treated as a "branch" or part of ethics proper by those who have not realized the revolution it has effected. Environmental ethics is not the part of ethics dealing with human relations to the environment, as if somehow the subject is outside of this environment. The subject cannot be separated from the environment. Environmental ethics marks the end of anthropocentric ethics and its unworldly subject. Anthropocentric ethics has now become a dated historical oddity, like the earth-centered universe. The ethics of the future is environmental ethics

As for order and harmony, it is clear that this theory, which is probably Pythagorean in origin, is esthetic, a metaphor from music, despite the use J. Baird Callicott made of it in application to environmental ethics (Callicott, 1989). If we have learned anything in the twentieth century from our art it is that beauty does not require either order (Pablo Picasso's "Guernica," anything by Jackson Pollack) or harmony (numerous composers, for example, Arnold Schönberg's "Moses and Aron," Dmitri Shostakovitch's "Lady Macbeth of Mtensk," John Coltrane's "Om" and so on). Disorder may be a more fitting artistic description of contemporary life or our contemporary condition; lack of harmony may have its own beauty. The order of a formal garden may be rivaled in its beauty by the disorder of a mountain meadow. The harmony of modern music may have its origins in the random and aharmonic noises in the environment that are without rhyme or reason, from frogs croaking to birds chirping (compare Ludwig van Beethoven's sixth symphony, or Igor Stravinsky's "Rites of Spring"). The order and harmony manifested in Plato's *Republic*, in which each class does its proper part, requires fraud and manipulation by the rulers and censorship of the poets. Its good is moot, if not ambiguous. Order is the virtue of hierarchy.

Harmony as a type of holism is, as I noted in chapter two, one mode of value but not the only mode. Inherent value is not holistic, for example. Harmony also ties in, metaphorically, to compossibility and reciprocity as ethical principles for plural values, as a good relation of different values. Their harmony is often an advance and sometimes required of all values taken together. But meliorism can disrupt a previous harmonic order and creatively actualize a novel one, as in the ascent of contemporary art and music, which superceded that of the romantic and nationalist eras of the nineteenth century.

4. Ontological Theories

Perfectionism: A theory of value closely related to *pros hen* is perfectionism. Perfection can be the perfection of being or substance, and thus refer back to substance. However, it can also be perfection of human nature using reason as the model of human nature, the Stoic view echoed by the Enlightenment: the perfectibility of man. Both Stoic and Aristotelian forms of perfectionism involve a kind of essentialism, in which there is an essence that has to be perfected. Perfectionism as a kind of essentialism posits certain attributes that are taken as essential to being human and some judge that they ought to be perfected as essential to well being. They are universal in potency. Both Plato and Aristotle argued for the perfection of character and Aristotle believed virtues are potential modes of being in everyone. In particular, humans were essentially distinguished as *zoion ekhon logon*, the rational animal, and the perfection of reasoning in humans was judged an (intellectual) virtue. This essentialist model also involved the *pros hen* relation as a referring to a perfected state.

For the Stoics, human nature is essentially rational and perfection of this character is the ultimate goal, so their view is strongly intellectualist. In both the Stoic and Aristotelian cases, an essential nature is assumed which is then perfected through development, from a potential of a being to full perfection. Perfection involves developing into full form, the excellence of a nature. Thus it is often idealistic, since perfectionism involves development into a perfected form, idea or character. Perfection of character is perfection of the being of the person. Since perfections of character are virtues, the ethics of perfection of character includes the recent unfortunate rubric "virtue ethics." Perfectionism originates, then, in an intellectualist value hierarchy, since perfectionism consists in an instance developing into a full form, that is full participation in a standard of perfection or development.[12]

Some might argue that "perfectionism" is a straw man, since the perfectibility of humans, the Enlightenment project, is not identical with ancient notions of virtue. Although there is a "family resemblance" between both as involving development, achieving virtue in the sense of excellence is not quite the same as the perfectibility of humans. My response is acknowledging that they are not identical, but that the Enlightenment project built upon ancient notions of virtue, as anyone who has read Rousseau soon realizes. The Enlightenment took this project to extremes, perhaps, at least as a projection of an ideal type as a goal. However, there were ancient analogues of perfectibility and the difference between virtue as excellence and as perfection is slight.

Perfectionism posits a static model of virtue, the perfected character. One problem raised by this model is the contradiction between its ostensive naturalism and its perfectionism. How can "natural" tendencies or impulses need perfection?[13] If human being or nature is unchanging, it cannot stand in need of perfection. If humans have an essence, it cannot need perfection, since it would be part of our nature: it would be innate or inherent. Perfection of it would be superfluous. If it must be developed artificially, as any sense of perfection of character would be, it cannot be "natural," that is, it cannot be our human nature. But if perfection is not part of our nature, then using reason or feeling as a standard is either arbitrary or an evaluation of one potentiality within humans over others. That is, the standard is based on a value judgment, not the basis for one. Conventional culture selects some of these over others as (ostensibly) better, but they cannot be "natural" if they must be perfected. Perfectionism implies an ethos or culture that sets the standards of perfection: the virtues. That the virtues can differ among cultures is evident and an argument against their naturalness (this is not an argument for cultural relativism). Moreover, cultural variety argues against a human "nature." The counterargument would be that human nature "exists" as an abstraction from the cultural varieties, for example, as *zoon logikon*, or *Homo sapiens*. But *logikon* and *sapiens* are culturally mediated. The further question is the "existence" of such abstractions. They are surely not *per se* existents.

Aristotle or other perfectionists might argue that development is involved in natural processes, for example, in the growth of an oak from an acorn to a mighty tree. However, this growth is, by his own account, "self-caused," as the tree does not need outside agencies to grow. Thus the growth of the tree is not development in the same sense as that which a culture artificially molds persons. "Natural" impulses would be essential and would develop without any external agency. Any character that is selected may be rooted in potentialities, as when certain favored characters are chosen in a crop. But it is not "natural" to select some of these and perfect character in their light. Perfection of human nature also presumes a norm of human nature. But this cannot be as perfected, since such a model would not be human nature, but human artifice, especially if culturally variegated. But if we always did act according to our nature, we could not be perfected: nature would be destiny, as it is for instinctual animals. This model would be a denial of morality and the opposite of the perfection of character.

What of the argument from excellence, that is, that excellence should be aimed for and other goals would enshrine mediocrity? This view unites otherwise quite different sorts of philosophers, namely, Plato, Aristotle and Nietzsche. This argument is valid up to a point; the issue is whether development is equivalent to perfection. Excellence cannot mean perfection for finite creatures. The issue, however, is whether excellence can be natural, and obviously, artificial development of it or culturally variegated versions of it are not "natural." Again if perfection of character constitutes a single model, it cannot permit the diversity of talents and other excellences of which humans and even animals are capable. Plato himself gave arguments for specialization; many talents require such specialization for development. This implies that different sorts of lives may be excellent, that is, evidence against one model. Is excellence then a genus with many species? But then excellence is manifest in the species, which form the different sorts of excellence. Excellence is indeed an appropriate goal, but those who act morally need not do so perfectly. Morality can be good, not best. Even mediocrities can be moral. "Perfection belongs to the Gods."

Creative actualization involves bringing about a value as a goal, and thus achieving the end. The end may be but imperfectly achieved however, while still achieving the good. The good may never be fully achieved by action: value can be finite, ambiguous and imperfect in relation to or in comparison with the imagined good. Later improvements and developments are ameliorations; "perfections" may later be superceded either in part or in kind. To be valuable in itself, a value need not be perfected, but only revealed as good in itself, that is, distinctively valuable. Values are inherent as they differentiate instances that have been evaluated as worth differentiating. The distinct value of the instance is not a general predicate, that is, a universal in the form of a species or genus under which an instance is subsumed in the categorial proposition. Nor are inherent values perfectionistic, that is, perfections of being. By contrast with the ontology of perfection, Kong fu-tse emphasized striving after virtue was a continuing exercise; and Dewey that it involves process and growth. Virtue is ongoing and involves changing circumstances.

Perfectionism would limit variety and thus freedom. The freedom to choose alternative lives would be ruled out. Humans would be treated as means, not ends in themselves, since they would be molded according to an external standard, forsaking autonomy. Thus perfectionism is morally deficient. The question is not only whether one standard should be applied to all, but whether we need to subject children to an unnecessarily high standard of perfection. The point is to develop character and talents and this need not be absolutely perfect, only morally and categorially good. Moreover, perfection is the means, not the end and positing it as the end is in effect the denial of value to a person without such perfection. The person is devalued.

Perfectionists might also argue that all cultures mold their children in the process of acculturation and this development of what is considered a good character is a part of childhood maturation. But character is formed by culture in terms of some standard. This implies that character is derivative. Character types reflect cultural values, not ethics. No "types" of characters necessarily commit bad acts: the relation of character and action is contingent. Nor does any character type guarantee either perfection or virtue. A person of apparently good character may prove in the crunch to lack excellence or moral virtue. Character is set as duration, the being of virtue, its substance. But character has no place except as a history of action. The action is good or bad not the character, since the character is revealed in the action. Character alone is not the "reason" or ground of an act, as acts are context laden. Where character is relevant to perfected actions it limits them, but is not their essence. A trustworthy person would not step over the line and steal. But it is the action that is right: character only limits the action to good actions. Thus character is not the substance of the act but its limits, a reading of character contrary to Aristotle's.

The idealist notion that our duty includes being virtuous, which is a doctrine common to Plato, Confucius, and Kant, perhaps sets too high a standard. We can have no duty to be something we are not. Since courage may in part involve feelings, this would constitute the duty to have certain feelings. But we often cannot help what we feel. Nor is there any moral need to: we cannot always control what we feel, but we can control our actions. Our duty is not to "be" courageous, but only to protect our family, our country and perhaps our honor. Again, sometimes we cannot change what we "are." The life of character is a guideline not a duty, although it may well be good: a guide to either the good life or to good character. The good life is the positive part of ethics, but does not include norms for duty. I am not arguing against the development of character here; a good character is a worthy development. But it is a means not an end, and in terms of theory, character requires creative actualization, that is, neither ontic nor part of human "nature." Insofar as character is good it is a mode of creative actualization, actualized by a culture in creative ways. Similarly, virtue requires creative actualization: actions speak louder than words. "Feeling" certain sentiments to achieve virtue is not enough. Generosity is not a "sentiment" but consists in actual deeds. The affective theory of the virtues in such authors as Thomas Hobbes (1994) and David Hume (1970) is a fallacious reduction. Feeling is not a virtue, since we cannot help what we feel. Moreover, virtue may consist in overcoming our feelings or acting in spite of our feelings, for example, courage in spite of feelings of fear.

Perfectionism has entered public affairs in two forms. In the historical version of human perfection, humans are means of realizing some predestined end of history, which can be grasped only by reason. Present generations would be sacrificed to future generations to realize an ideal. This sacrifice would be unjust to the present generation, for their lives are treated as means and not as equal in worth to others.[14]

In the educational form, certain capacities are arbitrarily subordinated to others. The "perfected" do not participate but are molded and so play the passive role appropriate to a "tabula rasa" model: perfectionism becomes indoctrination. Further, an arbitrary identity is imposed via education on what is considered a "tabula rasa" child. The child is not considered a person by themselves, treated as having autonomous goals. The standard is also unjust to the personal differences among individuals. Doing justice to differences implies that excellence takes different forms in different individuals. The development of persons should take the participation of the person into consideration: it ought to be a reciprocal activity involving input from the child. Autonomy of the child is required; self-reliance a goal, not passivity and dependence. This precludes perfectionism, as the last is hierarchical: a single good to which all must conform. Development as "self-realization" necessarily includes a self, not just an a priori notion of realization. Room should be allowed for autonomy and thus should include choice.

I am not arguing against the development of character, the perfection of certain cultural virtues or especially striving for excellence. I am only noting that such a view is both axiologically and morally incomplete. Neither perfection of human character nor excellent actions exhaust value theory, since they do not account for inherent value, instrumental value and other modes. As for the morality of character, character cannot justify an evil action and lack of character does not detract from a good action. On the contrary, a good act from someone without a sterling character is that much more commendable. This does not make character morally irrelevant so much as instrumental. If perfecting good character aids in doing what is right so much the better. But it cannot justify right. An act is not right just because it proceeds from a good character; this would be as morally and axiologically arbitrary as voluntarism. Character may be an "element of an ethic," but does not constitute an ethic. "Virtue ethics" is a only a part of ethics, since it is not required, although a good character may be commendable.

5. Intellectualism

What of reason as the highest good of humans? In one sense this theory is subjective or psychological, since it takes a psychological state of humans as the *summum bonum*. However, reason has been associated so completely with perfectionism and the ethics of character, due to Socrates' equation of virtue, wisdom, and knowledge, that it is best treated under that general heading. Socrates argued that those who do what is apparently right may simply be ignorant of what is truly good, which Ferdinand C.S. Schiller dubbed intellectualism. As a logically or ontologically necessary distinction of humans, such a model forms a hidden hierarchy of value with rationality at the top. Intellectualist values as a universal model of human life, as the specific difference of humans, forms a model of the human essence. Reason as the ultimate court of appeal becomes the authority for all matters. Our destiny is to be rational, and other values are subordinated under intellectualist values. Subsidiary values receive their value from their compatibility with reason or as "reasonable." Everything must be "reasonable" to be fair, to be just, to be legitimate, to be a form of knowledge, and so on. All social values are at the service of intellectualist institutions: science, technology and the gathering of information. Education is for the benefit of intellectualist values. Since reason as a species difference is not seen as voluntary, this *summum bonum* would work against human freedom, for our destiny would be limited to values compatible with reason, and rule out both alternatives and plurality. The value would be imposed upon oneself as necessary, giving reality to the identity. It would be instantiated in different acts over time judged "reasonable."

In this model the inherent value of a distinct good is subordinated to reason and thereby treated as a means. The pleasure people derive from certain activities, notably human sexuality; the beauty in contemplation of art and landscapes; the satisfaction of creative actualization; social relationships and prosperity are subordinate in importance in the intellectualist view. Justice is not done to their differences, only to what they have in common: their conceptual identity, which reason alone can grasp. However, the other kinds of values in most cases are independent of reason and thus of distinct kinds.

We are not devaluing intellect or reason by arguing that non-intellectualist values often add as much to human life. It is not elitist to note that some people do not have and never would have, under any circumstances, an inclination to "do criticism in the evening." Helvetius' vision of so many Shakespeares walking the streets of Paris, given proper education, has not been actualized despite universal education. The elitism is on the part of those who assume that their own intellectual values must be held dear by all. In truth, most people would rather sing, dance, play in sports, gossip or loaf than hold Socratic dialogues on the nature of beauty, especially if the more serious conversations begin to exceed their limited attention span by getting into details. Baseball and astrology are more popular than Shakespeare. It would be foolish to assign this to "environment" or the "corruptions of capitalism" when capitalist society itself rewards the educated and bankrolls massive educational structures. The truth is that of the diverse values possible for humans, differences of personal values are possible; evidently intellectual values are the values of a minority. Since they are values, and the values of a minority they can in no sense form an essence. Pluralism of values recognizes that humans pursue diverse ends, no one of which is essential to being human. Justice must be done to other values if knowledge is to serve human life, and not vice versa. Each must be given its due. Such distinct ends require reciprocal adjustment; pluralism follows from reciprocity.

Intellectualism, like other category theories, is less a theory of values than a theory of the highest good. Such theories are monistic, and therefore not value theories. Value "monism" should not be confused with value theory. A value theory can argue that all values, instrumental and intrinsic, economic and moral, and so on, have one, univocal character or "nature." Value monism is any value theory which argues that all values derive their value from one source, for example, God, the highest good, and so on. The different kinds of good derive their value from the one source, and would not have value without this source. Value monisms base all good on one category of value, in this case reason or knowledge, and thereby reduce value. Thus values which are "akin" to reason: knowledge, wisdom and the like are rated highly in an intellectualist theory. However, a capability that must be perfected, like reason, cannot be part of human nature. If it were, it would not stand in need of perfecting, for it would be perfect at birth, as an instinct. If the ability to reason is instilled by culture, by education or by practice it cannot be innate and thus is not natural. The empiricists and others (Hobbes) who deny that reason is innate, but argue that it is perfected in humans have created doubt about "innate ideas," which means at the least that there is no consensus on this issue. Rationality as the essence is an open issue.

Like other category and *summum bonum* theories, intellectualism confuses value and evaluation. Rationality as a critical tool for comparing different hypotheses, options, desires, and so on belongs to evaluation and norms, not value. The value that is demonstrated as the outcome of following what is evaluated best cannot be demonstrated *a priori*. We cannot ground our actions only by what stands to reason, since reason has no content. Proof does not consist in reasoning itself but in testing our reasoning with trial and error, experiments, and the like. Reason can be of value as a prelude to such activities, but cannot be substituted for them. Reason is usually connected with model construction and the correlate simplification needed for analysis. Thus it constructs an ideal whose value has not yet been proven. Since this ideal is not yet actualized, it is utopian: an evaluative construct. This activity is prior to creative actualization, and thus in the realm of the normative and evaluative. The value is projected or potential, not actual. Nor can reason tell us the value of a value; it is not a substitute for evaluation as such either in its critical, comparative or deliberative functions. Reason may tell us how to fulfill a desire but cannot tell us the value of a desire, whether relative or abstract. Evaluating this value is the role of wisdom, which, contrary to Socrates, is not equivalent to knowledge, however defined.

Socrates equating of virtue and wisdom (Plato, *Gorgias, et al.*) is even more fallacious. Aristotle wisely separated moral and intellectual virtue, since he recognized that intelligence and virtue are hardly coextensive, although he was enough of his teacher's student to call the illicit use of intellect "mere cleverness." Intellect can be put to the service of evil (Mengele). Another ground for this was technical: someone could have an excess of moral virtue, for example, rashness instead of courage. However, an excess of knowledge would be a form of excellence. The structure of moral and intellectual virtue differs. Socrates was also naïve about the source of evil: evil is not confined to the ignorant. Intellectualists might argue that evil people are ignorant, but this will hardly do when highly educated physicians are known to experiment on humans (Tuskegee, Auschwitz, the Japanese biological warfare experiments in Manchuria). Education has not invariably produced wisdom and often has produced folly. Reason is used as a means to enhance individual wealth at the expense of the society: an instrument. However wrong or misguided this may be on the intellectualist scale of values, the instrumental use of reason is not the result of ignorance, but on the contrary, high intelligence. Knowledge has not invariably brought moral virtue in its wake, but often a scale of values with means promoted as ends: economism; and reason demoted to a means.

A more important point is whether reason can understand evil, since the motives for evil are often extra rational and frequently involve the subordination of reason as a means, the problematic of the opening book of the *Republic*. An act that cannot be understood would not fall into an intellectualist value schema. Slavery in the United States and the Holocaust may constitute such incomprehensible evils. It might be said that a bad act has a cause and hence can be understood. But by intellectualist standards this cause is either irrational or adopted as a motive on extra rational grounds, which defeats the theory. Intellectualism cannot explain evil. Acts, as opposed to understanding of acts, may not be rational.

A more technical point is whether a freely adopted choice of evil is determined. Understanding actions in terms of motives or causes requires complete determination by them; but the choice of evil may be idiosyncratic and thereby undetermined. Indeed, if it is determined it cannot be bad. Both of these consequences are incompatible with our notions about the Holocaust. An intellectualist, then, can only "describe," not "explain" bad acts. Even intellectualism must admit an irrational element in choice, for example, desire or appetite in Plato, which following the Pythagoreans is treated as infinite and thus not rational. A determinist might argue that choices are ultimately determined by chemical reactions in the brain, whether they attempt to reduce mind to brain or not (identity theory). This issue is slightly far afield for value theory, but it can be said that brain chemistry as such does not enter consciousness and the causal chain of the motive to the act; that speaking of chemical affinity as a "reason" is not intellectualist; and the causal chain is broken by interaction of mental and non-mental.

The theory is also axiologically deficient. According to intellectualism, a computer is more valuable than a person, for it is more "rational." To say that other human qualities are not equally good—love, affection, happiness, care, and so on—is to eliminate many humans and their activities from "humanity," in a reductive move consistent with the reduction of value to intellectualist values. A rational life not for everyone, as Plato and Aristotle were forced to admit.[15] Moreover, rationality ultimately has no content and consists ultimately of empty form, since the rational in the strict sense means valid inference using universal form without content. Rationality is ultimately dependent upon other values for its content. Thus a rational life would not be a self-sufficient life and like subjectivity in general requires a relation to some other of value to be complete.

The issue for value theory is whether reason should be anything but a means. Reason indeed has instrumental value but to value humans insofar as they are rational is a perversion of the relation of reason to humans. Reason is properly a tool, of service to humans, not their master, their destiny. Reason is but one species or mode of value; its main use is as an instrument in which it is an aid to other values. Again, we are not devaluing reason by pointing out its instrumental role. On a personal level our values are often non-rational. Even a fondness for vanilla over chocolate is not "rational." Nevertheless, reason can narrow down value decisions by showing that some values are mutually exclusive, for example, the absolute right to reproduce and a superior quality of life.

Reason does not distinguish the human as such and thus trap human destiny in an intellectualist web. For reason requires creative actualization through practice. Nor does it separate humans from their world, since reasoning is learned in interaction with the world. Reason is not autonomous, but requires a relation to the world. Descartes' mind would not be able to reason even the *cogito* unless in the world and interacting with the world. The devaluation of the world by intellectualism is axiologically deficient. Even if reason were the essential mark of humans, however, its use as a *summum bonum* would constitute the subordination of the world, other species, and habitats to a small part of human life, a devaluation. Intellectualism is thereby revealed as an obstacle to moral progress.

6. Life

A second set of objective theories are life-centered theories, in which value is value for life or to human life. Life as the center of value is the theory of Albert Schweitzer, Buddhism, and Jainism as well as certain environmentalists; human life as the apex is the theory of Arthur Schopenhauer, Friedrich Nietzsche, and the pragmatists.[16] Life as the center of value in the form of value as what "preserves or enhances life" is a *pros hen* to life and thus a metaphysical theory. I criticized this view in chapter two. Life as such cannot have *absolute* intrinsic value for life kills life. Life as such is the bearer of value, but certainly cannot be intrinsically valuable, since prey relations render life precarious. To argue against the predator-prey relation is Utopian and devalues the actual. Life is a condition of value and even has inherent value but life as such cannot constitute intrinsic value. Life is also the agency of value and specific instrumental and categorial values may be more measurable in terms of life. But ultimately a completely life-centered theory is impossible.

Life-centered value as what aids life in the form of "survival of the fittest" (Spencer), survival as such, is not the highest value. Mere survival is not improvement or amelioration, so it is inferior to the improved world of what enhances life. Survival is a form of duration, but actions come to an end when their goal is reached. They do not endure although the goods as values or creative actualizations of which they are means may endure. Survival value is an ongoing value, not a means: an end in itself, since what is creatively actualized is that which survives. In this form the theory is actually a species of life-centered value, which is also hierarchical, since a "natural" hierarchy is the end of evolution. I will address this point below.

As for human life as the *summum bonum*, this constitutes a *pros hen* to one species: in John Dewey's words, a mere "conceit."[17] To place human life at the apex of life, as the top of the food chain, ignores the conditions of human life in the environment. More, it gives a transcendent meaning to evolution and thus a goal to the process of evolution, since all of evolution has human life as its goal: human life is the final end, the *pros hen* of all evolution. However, it is unlikely that evolution has a goal: since the process is "open-ended" in the words of Prof. Reuben Abel. Further, the theory is no longer life-centered but speciesist, that is, centered on one species ("humanism"). Life as such is not the center or apex of value but humans. Thus whatever in the world fulfills our life is of value. However, this also is a hierarchical view and subject to all the arguments against hierarchy.

Value as life-centered is a species of naturalism, that is, the view that value can be based on nature, a view close to Dewey's.[18] Dewey is correct that context or situation is vital in defining parameters and the framework of a creative problematic. However, he errs in his attempt to reduce value to the "natural," for creative possibilities go far beyond what is "naturally occurring." He does recognize the interaction of possible to actual in what he calls "bringing about" and thus anticipates creative actualization as a theory of value. Value is brought about or creatively actualized by action. Value is not a mere wish or "interest": actions are grounded in the world as result or they are useless. The results of pragmatic activity are actual in the world, but are inherent in the instance.

Finally, life must creatively actualize itself to preserve the value of life. Such self-actualization is the mark of intrinsically valuable agents. However, it indicates the degree to which theories that equate intrinsic value with absolute value are off the mark. The intuitive idea is the absolute value of each individual life, or individual human life.[19] But no one could consistently hold all of life to be of absolute value, for the killing of prey by predator is a condition of all of life. A prey that has been captured has failed to actualize itself, to self-actualize its life or survival. Self-actualization as the character of intrinsic value is a limit on intrinsic value. Another is death, since this brings life-value to an end, even if death is not due to a predator.

The subordination of all value in the *pros hen* and hierarchical relations to humans reveals the hidden egoism and tribalism of Western philosophy. Everything has value for me or for my kind. Multiple values and inherent value are reduced to instruments of human mastery of the world, creating a humanistic hierarchy in which all the value of the world reflects humanity at the apex. The interrelation of parts of our environment to each other in a complex whole, and human dependence upon this environment are ignored. The environment is the condition of humans, not humans the condition of the environment. The environment changes, of course, but remains a condition. Hierarchical value theories that ignore this condition are blind.

7. Needs

Still another theory that could be interpreted as objective is the "needs" theory. This theory has not been explicitly advanced in philosophy as a theory of values, but is bound up with a well-known form of the labor theory, namely, some interpretations of Marxism. Nevertheless, the notion of needs could be taken as a theory of human values, and has been treated as such by social scientists of various schools, for example, by Abraham Maslow (1959, ch. 10) and Erich Fromm (Fromm in Maslow, 1959, ch. 12). Needs are objective values of human beings, required for them. It may be interpreted as subjective when it is psychologized, that is, when psychological needs are taken as values or fulfillment of them is taken as good. In either case the theory is still within the Cartesian problematic, since needs as objective presume a Cartesian universe of subjects and objects, and these subjects have objective needs. Regardless of the metaphysics of this theory, however, the theory ascribes value to fulfilling needs, and thus is closely related in form to the conative theory, which ascribes value to fulfilling desires. In another form, needs are based on an organic model of human nature, in which the different parts have distinct needs, which must be fulfilled. In this form the theory creates a *pros hen* back to the subject as the substance whose needs must be fulfilled. Finally, needs can be closely related to a vitalist theory in which the needs of living creatures or human life form the ground of value.

The main problem with the theory is that necessities are confused with values. Any needs past the animal stage, including stone-age activities, are values, not needs, since they are not requirements of survival, but enhancements of life. They are valued, but not required. "Needs" as the basis for ends and goals is a sort of pre-cultural naturalism and thus defective as a model for civilization, for "*homo faber*." The life consisting of fulfillment of "needs" is at best a description of animal and plant life, not of human association and culture. If the theory were treated as a comprehensive theory of value it would ignore the many creative actualizations that do not reflect "needs" at all. An automobile is hardly a necessity, even in a remote area, since humans survived for centuries without cars and still do in poorer countries. An automobile may well be *good* or of *value* as a mode of transportation, but is not required. Cars are needed in rural areas only if specific values are presumed, for example, speed, a way of life, and so on. To interpret needs as values is to bring in the mode of the necessary to a sphere of values in which it is not "needed": a category mistake.

Needs implies requirements, and most values are not required. The mode of (practical) necessity is limited to basics, for example, food and sufficient warmth to prevent freezing. Even clothes and shelter are enhancements, since some cultures forego clothing and all humans get exposed to the elements to some degree. Obviously I am not arguing that clothing and shelter are not goods, but that such goods are not needs, in the sense of necessities. To confuse needs and values is to confuse what is distinctive about each. If the use of the word is a proposal to change the meaning, then it would still not mean an equation of two coextensive concepts, since such prescriptive needs would be dependent upon values. Biological necessities, that is, those that are solely related to the well-being of the organism, would then have to be distinguished from cultural, economic, and other "needs" peculiar to cultures.

Karl Marx, following Georg W. F. Hegel, argued that at a specific stage of historical development what were once considered luxuries become by a magical process of dialectical transformation, "needs." Let us consider this argument in relation to the Rolls-Royce automobile. Will the Rolls-Royce ever become a need, instead of a luxury? I find it extremely implausible that any society would regard the Rolls-Royce as a necessity of life, as opposed to a specific valued way of life. But even if they did, the confusion caused by the use of the word "need" would remain: the auto is certainly not a biological necessity! Just how then is it a necessity? How is it needed? Some wealthy individuals might perceive the Rolls as a necessity in their social circle. The British or Saudis would not expect their monarch to drive (or be driven) in a compact "Mini." But this use of the word is an extremely contingent, narrow, and indeed arbitrary form of "need," and has little relation to food and adequate heat. It is contingent on a way of life that is itself not required, whether for biological or any other reason. The way of life of the wealthy is not a necessity of life, and indeed may involve "conspicuous waste" instead of need. Many other "needs" of consumer society fall into this category, and it is basically silly to treat them as needs in any serious sense: the "need" for perfume. Even the monarch does not "need" a Rolls-Royce as such but only a generic luxury automobile, suitable to his/her station.

The whole notion of "needs" for humans stands in "need" of a critical evaluation. No needs are actually "biological" necessities of life (or "natural") as they are post-cultural, that is, not defined in terms of animal life. "Needing to urinate" is normative and linguistic in form; association has transformed biological "needs" into conventional ones (we urinate for "health"). Urination is not a need for a wild animal, just a function. Seeing it as a need presumes a civilization of deferred gratification, whereby such acts as urination are controlled and subject to rules of propriety. We do not expect animals to "hold it in," or ask to be excused to go to the bathroom. When we speak of an animal's "needs" in metaphor with human needs, we transform the term. Needs are defined by the person in relation to and as embodied in the person. One person "needs" wealth in her life as much as another "needs" beauty. But it should be noticed that this constitutes an argument against an actual need for either.

To say all values are needs is like saying that doing without them would be life-threatening. Life enhancing values are hardly life-preserving values, however, and reveal the absurdity of the historical accretion of "needs." Doing without the latest technology or capitalist goodie would hardly be life-threatening, and might actually enhance a person's life, since they might discover genuine values. Indeed most of the highest values are not biological necessities at all. Abstract knowledge, refined beauty, spiritual bliss, and sexual pleasure taken as an end in itself are not necessities of life, not "needs." They are not biological per se, but instead values. Some might argue that once achieved they may be psychological necessities. However, this too stretches the meaning of the word "necessity." We may not want a life where we have to do without them, but might find ourselves in a situation where we must. The inmates of the concentration camps survived on hope alone. Further, the emergence of new values is creative, not based on the needs of survival in the jungle. Creative actualization is based on imagination and evaluation, not necessity, despite the implications of the phrase "necessity is the mother of invention." Even wealth is evaluative: the pursuit of a better standard of living than one actually needs.[20] I will address this point below.

A "need" is such only in relation to an end, that is, it may be the only means to an end in specific circumstances. An automobile is a luxury unless one lives fifty miles from work and there is no public transportation. Then a person may "need" a car. (Those who use a car in their line of work, for example, taxi drivers or salespeople are an exception). The "need" of a commuter for a car is based on his/her other values, namely, that they desire to live outside the city instead of within it. The need is not basic but contingent on other values. Since this is the case for almost all the so-called needs, the theory is incomplete as a theory of value, and indeed presumes some other theory of value. Need is not a basic category, but derivative from other values and generally contingent upon freely chosen ends which are not "needed." These may include life-style, the kind of life it involves, the place one lives, and so on There can be no system of needs only a system of values. Despite the critique of the "domination of instrumental reason" in critical theory, the school's own evaluation of values as "needs" confuses means with ends and thus, by treating necessary means to unnecessary ends as ends in themselves, treats what should be an instrument, a means, as an end in itself. Thus it ends with the domination of instrumental reason, that is the logic of needs, which are means only since they are "needs" only in relation to specific ends.

What would Maslow and other intellectual descendants of Hegel, knowing and unwitting, respond to these points? Would they say that there are other "needs" of creatures beyond the merely biological, which contribute to their overall well-being, for example, the need even for simple affection, benevolent nurturing, and other psychological needs? Even music may contribute to the well being of humans. Are there medical needs for any organism if it is to be healthy? I think these arguments support the point that such needs are in relation to values. "Well-being" as more than a biological minimum is a valued end, not a need. I am not disputing its value, but its necessity. After psychologists came to the conclusion that music did contribute to human well being, managers proceeded to pipe music into many public places, especially workplaces. Unfortunately, not everyone liked the same music, so managers selected the music with the lowest common denominator. Tasteless musak became standard. But how anyone could posit a need for tasteless musak is incomprehensible. "Health" is more of a borderline case, since a wounded organism requires "medical" aid, which is closer to a biological than a cultural need. But curing a life-threatening condition presumes the value of life, while achieving a state of optimum health is much more than absence of disease. The "need" for love, cherishing, others of one's kind, community, belonging is *hypothetical*. Even proven therapeutic value only demonstrates pragmatic worth, not need.

Values and not needs constitute the normative identity delineating any general schema which includes the diverse goods people make, purchase or exchange; actions in pursuit of goals; functioning; personal, social and cultural differences, and moral norms. History argues that different cultures and epochs, as well as different persons value not only different goods, but novel goods. Novelties cannot be "needs" in any biological sense, since such goods were not required prior to their novel creative actualization. Such goods have been created: actions were generated by the evaluation that the effort required for their actualization would be worthwhile. In truth the theory confuses values with needs by extending the "notion" of need to include what are obviously not "needs" in any sense, much less a strict sense of necessities of life. Values are the basis for human actions, not biological requirements; values regulate most activities, not need. Humans need much less than they actually pursue. At most, needs are a small subset of all such values, not equivalent to value. But then need cannot provide a general theory of value, even if the mediating role of civilization in defining needs were ignored.

One reading of Hegel is that society is like an organism with needs and that the function of society is to fulfill such "needs." Needs expand as society progresses. But the "needs" for which society is "organized" are values, which are or are not achieved by a particular society, as institutions fail or succeed historically. Cultures fulfill values more or less well. Society is not organized to meet needs, since association is not required to meet needs: animals meet their needs in many cases with minimal association. I would respond that, on the contrary, the advances of society are advances, that is, improvements. Novelties are creative actualizations of new goods, those that are added to the inventory of goods in the world. They are above and beyond the wild life of animal needs, and an improvement on life in the wild. The division of labor does not meet needs but allows new ways and valued ways of life, above and beyond the hunter-gatherer stage. Without it we would be in a world of needs where everyone fends for themselves. With it, the means have been creatively actualized to achieve variegated values, that is, new and more diverse goals.

Needs in the strict sense could not provide a theory of historical progress, since needs are fixed. What is needed to fulfill needs is fixed, not dynamic. Needs would provide an unchanging standard, not a melioristic one, since needs would be fixed by biology. The advances of civilization could not be accounted for, contrary to some readings of Hegel, since "needs" would end history with biology. All the novel goods of the world would never have been creatively actualized if humans stuck to their needs. "Needs" as a model would stop history and result in institutional stagnation, discouraging improvement. To take values as "needs" would potentially freeze human development, for it precludes development of new values. Needs are "ontic" in the sense of enduring requirements of all persons over time. Values are not needs precisely because they are historical novelties, an advance on previous standards. They are invented or discovered, not fixed. Institutions designed to meet valued ends have been historically established: they are themselves novel creative actualizations. Just as the inventory of goods in the world can increase so can the institutions. But if they do, they do not reflect needs, since the needs were not met prior to their creative actualization. Institutions designed to meet "needs" are soon superceded, since they become superfluous with the advance of history. The epoch of "needs" is pre-civil, pre-technological, belonging to the age where humans were wild. "Needs" would be fixed and thereby a-historical, for they would not be a factor in any historical developments. As soon as humans learned to control fire and use stone, the epoch regulated or determined solely by needs was over. To view institutions as creations of value is to realize that values can change and thus institutions can change. Value is the source of change, particularly historical change.

Even if needs are defined historically, just what is a "healthy, happy, life-supporting" need? Is preserving the wilderness a greater need or more necessary than powerful automobiles for ultimate human well-being? Or only preserving each species, as in zoos? Is a bare minimum subsistence, a hedonistic maximum or some standard in between the optimum "needed" for human well-being? If it is a necessity should it be a right guaranteed by government or allowed to evolve of its own accord? What is telling is that only some categories of value are treated in terms of "needs." Are there cases, then, where the value of the category is not considered a "need," for example, in an educational curriculum where funding for the arts is the first thing to be cut? But doesn't this reveal the arbitrariness of dividing values up in terms of necessities and luxuries? During the Cold-War, especially after the space race, the science and math curriculum was revalued from one aspect of education to its centerpiece. But this reveals the degree to which it reflects cultural and other values. Any hierarchy of needs reflects cultural, political, and personal values, not needs. Categories of value qua categories, qua kind, do not derive their value from another category, or they would not constitute a category. Beauty cannot be reduced to prosperity or prosperity to beauty. But then what of a reformulated theory of needs which gave equal weight to all the categories of value? As I argued above, these would not be needs, but values. An axiological theory of ends would be superior here, since it does not arbitrarily divide up goals into "needed" and "superfluous" or "luxuries." Similarly, personal values cannot be reduced to any schema of needs, since they are personal. If needs were treated as universal ends of society, social justice would demand that certain necessities of association precede personal desires. Personal values would be reduced, since if they did not fit the model of needs they would be considered unnecessary. But since any schema of needs reflect the values of some, such a schema would be unjust. Institutions would not serve individuals.

A counterargument might consist in the view that categories ultimately reflect individual or personal needs, that is, a reductive view, and thus that needs are ultimately based on individual judgment. But the categories could still be social, as a general end pursued by encultured individuals, which predates and survives any one of them. Social justice would demand that necessities of association precede personal values. Institutions of association serve individuals also, but call for a different ethic: sacrifice over greed, work over parasitism, even the beauty creatively actualized by playing a piece of music as written and not "as one wishes." Personal "needs" reflect individual evaluations of different goods and a resulting scale of personal values. This would not result in any scale of needs, then as possible, not necessary goods.

As for economic life, people value a higher standard of living and are willing to enact the means that best achieve such an end. A higher standard of living is not needed, but valued. The means to obtain it are not "necessary," but evaluated as superior for achieving it. A higher standard of living is a value not least because it is a goal, not a cause. The higher standard does not precede its creative actualization, it follows upon it, a true end, since it is not actualized except as a result. As with other purported "needs," true "economic needs" are actually biological, that is pre-economic. Most people are willing to work to raise their standard of living above the level of the wilderness, of wild creatures. This process is not necessary: what impels or attracts most people is not necessity or "needs," but ends evaluated as worthwhile and the prospect that they will be creatively actualized. Thus any higher standard cannot be determined and economic life in general is in the realm of freedom, not necessity. A standard of living, unlike a billiard ball, does not "push" people into laboring for a better life; instead, they imagine such a living standard, evaluate its worth for themselves, and decide if it is worth pursuing. A superior standard of living is aimed at because people evaluate it as better than other possible lives or present circumstances. Similarly, new methods are adopted not out of necessity, but because they are superior at achieving the end: the tractor to animal labor on the farm. (In hypothetical cases in which only one technique is available to achieve a particular end, this particular means is "necessary" but only if the end itself is necessary. To use the language of Kant it is hypothetical: if we pursue the end, we require such a means.)

"New needs arising" disguises (relative) improvement in living conditions behind the façade of "changing material conditions." "New needs" implies a rising standard of living in which more basic needs have been met, and new values arise that can be pursued. Several things formerly rare, or thought to be beyond reach, come to be expected as a concomitant of a way of life. In other words, economies change, and no model of needs can explain such changes. In the "needs" model, luxuries and necessities are distinguished, based on the organic model of the division of labor in which a specific minimum is "needed" to maintain survival. Yet economic practice goes beyond any such model, for example, in any system of exchange, where diamonds or gold can be exchanged for food. But surely diamonds and gold are not necessities: not "needs." Again, is "Coca-Cola" a need? If Coca-Cola had a monopoly would we be compelled to buy from them? Or is a preference for Coke over other brands and other liquids a value, a freely chosen alternative? The model of "needs" does not reflect practice, since in actual economies no such distinction between luxuries and necessities is made; both are economic goods. Economies are not institutionalized among organisms.

Marx's "necessities of life" in the wide sense are treated as a model for economic goods in general by some orthodox Marxists, a bourgeois way of viewing things indeed. Bourgeois culture has separated out economic "needs" as categorially distinct. In many other cultures, biological necessities are fulfilled in relation to local resources, for example, that of the Inuit, who hardly would have regarded bourgeois luxuries as requirements for life in the Arctic prior to their contact with European culture. Such comparisons cannot fail to reveal the fallacy of mistaking an "artificial" need for some commodity, created by clever advertising, with biological requirements for survival. This fallacy is one some forms of Marxism share with some economists who are not Marxist, namely, that the market "supplies needs." Few of the goods on the market meet genuine needs. Indeed the model is technocratic, a view which Marx shares with non-Marxist technological determinists. Humans have needs, technology alone can fulfill them, therefore humans require these technologies. The issue here is whether imposed needs are not oxymorons: if they are imposed from above, they can hardly be "needs." On the contrary, advertising creates perceived needs, sometimes playing on anxieties, sometimes on sex, for example, in the "need" for mouthwash.

Needs have also been used as a basis for rights. The "right" to decent housing, food, clothing, and so on as rights to "basic human needs" has been advanced as an argument for positive rights. But the model prescribes just what humans do need and mixes biological needs with improvements due to the escape from a merely biological life, for example, "decent" housing. Decent housing involves the evaluation that some housing is superior in value to other housing. But this cannot be justified by a scale of need, only a scale of value. Values are surreptitiously inserted as the ultimate basis of rights. My quarrel is not with the latter, but the model of needs as the basis for rights.

This model would ultimately reduce rights to the biological, and thus eliminate instead of expand rights. It would certainly not justify expanding rights beyond the minimal to the decent.

The theory that values are based in need exaggerates one motive, need, as if it could explain all value. What is valid about the theory is that values can be determined apart from subjective whim, presuming we correct the theory by substituting "value" for "need" except in the case of some basic biological norms. Values are not subjective in the Cartesian sense but can be determined accurately by social science. The entire technical civilization embodies such values and reflects them as manifested: as creatively actualized in continuing novelties that separate our civilization from those of the past. The values held by the society are embodied in the economic system designed to fulfill them.

8. Other Value Theories: Action

Another group of theories of value center around action as the focus of value. Value is revealed through action: value as the product of labor and thus of one form of action; or in the form of what works, or practicality. (Needs could also be treated under this heading since in this theory value reflects need, and thereby generates action to fulfill need.) The focus of these theories is on the action as such, not the being of the value, the subject or the object. Their ancestor is Aristotle's separation of acts done for their own sake from acts done for the sake of some further end, although not all action centered theories argue for the intrinsic good of the action itself. On the contrary, although action is essential, it is often a means, for example, when the product of labor has value, not the labor itself, although the value of the labor is recognized and even enhanced by the product: the "dignity" of labor. Action in such theories is a condition of the valued product of action, but the product is not considered an "object" so much as a distinctly novel value. Action alone creates value. In this sense the theories anticipate creative actualization but err by limiting value to one mode (labor), or to one criterion (practicality).

9. Labor

John Locke, Adam Smith, and Karl Marx argued that labor creates value for the sake of fulfilling needs. However, the labor theory is independent of the needs theory and was advanced by Locke and by Adam Smith as an economic theory before Marx adopted it. The theory is also nominalist, that is, in the Lockean view, my individual labor both creates value and gives value to some product. In this form, the model presumes an atomized, not a communal society, that is, a model that ignores the cooperation involved in the division of labor. The theory derives from a model used to justify an individualistic, property-oriented society, not one where many different individuals may labor on the same product or may exchange products not made by them but that still have value. The labor theory of value is nominalist: any individual labor gives value to something (cf. "socialized labor" in Karl Marx, which implies cooperation.) Thus the model is within the atomistic view of society formulated in the British version of the contract theory in Hobbes and John Locke. Labor creates value for Locke in the state of nature, which is prior to general association. The theory is originally bourgeois.

As is well known, the theory was rejected by economists on the ground that labor as such cannot create value without quality of the product and demand.[21] To use Prof. Lee McDonald's (1968) example, we may labor mightily on strawberries which are nevertheless of inferior quality and which no one will buy. The theory also concentrates more on the supply side than the demand side, that is, insofar as labor explains some of the costs of production it validly points to one factor in pricing. But it does not explain the interaction required in exchange, that is, the demand side, or why are some items purchased not others. What is the ground of purchases? Why do some products succeed and others fail in the market? As I noted in the last section, "needs" don't do the job required of them. Why, for example, do the fickle contemporary markets prefer good-looking but tasteless fruit to tasty if slightly tarnished fruit? Since labor is one cost of production for all economic theories, it is not as if labor has absolutely nothing to do with economic value. But then the theory is connected with price, not value, since economic value must reflect both sides in an exchange. Labor alone cannot create economic value: I may cultivate weeds for sixteen hours a day but the flower markets will never purchase them.

The theory is also deficient as a theory of value. Humans tend not to labor on something that has no potential value, either for its own sake, or as valuable for some further end. More labor will produce nothing of value without direction: without purpose. Value is given to the labor or accrues to the labor in relation to its end. In other words, labor is a means, not an end in itself. Labor is a consequent of value not its ground, even where we work on something of little value, like a sandcastle. The result is valued even where satisfaction is gained from the process. It may well be that work should be, can be or is enjoyed, but this qualification requires its proponents to bring in hedonism or some form of the affective theory. Since labor is a means not an end, proponents evaluate it too highly by ranking it as the real basis of historical change. Invention and successful creative actualization are as important to prosperity as labor alone. Increases in prosperity, the standard of living, do not reflect endless toil, but new means of production and their successful actualization. Labor does not create the object of value it only makes and shapes it. The imagination, (invention), and trial and error in view of some end have first created it. Inventions, contra Althusser, are not a copy of production. Imagination is a condition of progress, that is, non-material factors are not only essential but the ground.

Marx's theory of history ultimately recognizes this, since changes in historical epochs involve changes in the means of production. What he does not acknowledge is that this undermines the labor theory. As Hannah Arendt (Arendt, 1958) correctly noted, labor is tied to the cycles of biological life, that is, to becoming or coming to be. Progress breaks the cycle. Humans labor on what has proven value or the prospect of value. Labor is a condition of some types of creative actualization, then, but is not their ground. Labor is bound up with exchange value because labor in an economic context aims at some economic value, not only something of personal value only, without exchange or market value. Without a distinct economic end in view, the labor might be *useless*, its product might not sell. More strictly, the theory lacks a theory of worth, of what is worthy of creative actualization. Such a theory belongs in the realm of evaluation, that is, evaluation of competing projects and of their potential worth. Labor follows such a project as a means, but this does not assure value, since the labor may fail to achieve the end and the evaluation may be invalid to begin with. No market for widgets can be created despite millions of dollars of advertising spent in the attempt to creatively actualize demand.

To hold that labor is itself the end is a perversion, since it creates a *pros hen* to work: to travail as a good in itself: an incredible view! It would justify the workhouse and the slave labor camp. In this formulation, the right to property, in the form of what has been labored on, would be absolute, and all other rights would be subordinate to it: the right to life and liberty, for example, since all value would be created by labor. The lives of all would be subordinate to their laboring to produce their property; life and liberty would be subordinate to this property qualification. In other words a bourgeois society would prevail, in which the needs of property, which reflect the embodied labor would be paramount and create a hierarchy in which only what was in accord with property would be allowed.

And how are we to evaluate the value of an age in which the highest value is labor? We hardly work for its own sake, especially hard physical labor. On the contrary, such work is a burdensome necessity, which humans cease as soon as machines can replace it.[22] In some of its formulations, this defect of the theory is recognized since value is regarded as the end of labor: labor is the condition of value, just as desire or interest is the condition in the conative theory. But then the theory fails on the grounds discussed above. The theory as it stands lacks a theory of ends, or what ends are worth laboring on, of the values pursued by means of labor. Adam Smith and Marx supply, respectively, the required conative or affective theories, or the theory of needs. But then the theory cannot be a theory of value, only of instrumental value, or of price.

The theory also ignores the moral values that are the condition of the division of labor, to speak nothing of the condition of all association. The division of labor is not grounded only in economics, but requires moral norms.

The ascent from "nature" was not simply a process of increasing prosperity but also increased cooperation and the development of character requisite to such cooperation. "Productive man" as such would still be in a wild state (bees, ants, beavers, and so on). The "promises men live by" are forms of life which allow economic activities and exchanges of several kinds. If contracts were never honored, the goods they aim at could not be achieved. If we cannot depend on the products we purchase, faith in exchange would collapse. The theory also absolutizes property value over the right to life and liberty. My property rights to the product of my labor supercede anyone else's right to life, since they have no right to my property, even if they are starving. But this prioritizing would make a means of life, property, the end and life, the means. Unless property serves life, it has no value whatsoever. Property is a means to sustain life, not life a means to sustain property.

This labor model for value justifies economism, of economic values as basic, a view that has been uncritically accepted far outside Marxist circles. Indeed it may be taken as the unofficial view of capitalist society. But if economic goods do not reflect needs, the view is fallacious. Economic values are one species of value, not the cause of other values. Economic goods require creative actualization as much as other goods. Economic values are indeed a distinct type, then, separate from political and other kinds. But economics are specific, not general and require a general moral condition of the larger society. Economic values require a general theory of value that will separate and limit "realms" of value and do justice to each.

10. Practicality

Practicality as a theory of value has been connected with utilitarianism, since the utility of something is its usefulness; and pragmatism, since practical consequences are the test of meaning and, derivatively, value. However, practicality can be treated by itself as a common sense theory of value, that if something has a practical use, or functions well in a situation in which it is needed, it has value. Practicality is also the unofficial theory of value of the engineers and technocrats. This theory is flexible, since anything that can be used or which can function in action or a situation has at least temporary value: in the absence of a hammer we may sometimes use a rock for the same function. The connection to labor and needs should be evident here; value is measured in terms of functioning well at a task or with respect to the needs of a situation. Functioning well is also a test of creative actualization, so the theory is at least partially connected to our own. Practicality may be said to have multiple criteria, then, namely, functioning to achieve an end, useful for a purpose, or practical by contrast with theoretical, that is, a workable scheme. All of these criteria share an instrumental orientation, that is, what is practical is such as a means. I will discuss three aspects of the theory in this section, namely, whether what functions well has value, whether it is sufficient as a theory of value and the relation of practice to theory.

The theory, like the labor theory, lacks a theory of ends and is thereby deficient as a general theory of value. The issue for value theory is what does something work *for*? Is it life supporting or life enhancing or not? An atomic war would certainly "work" at destroying the enemy. It would also destroy the attacker, the neighboring countries, and perhaps all life on earth. The concept or norm "practical value" implies a gap between means and ends: practical for what? Useful for what? Functional for what end? The theory can provide no answer that does not go beyond the theory, and thereby negate it. Either something is practical for humans, or for economic ends, or for all of life, or some other more general theory of value. But more, without such a supplemental theory, practicality cannot be measured at all. Practicality must be for a purpose other than practicality itself or the theory becomes absurd.[22]

The unstated premise is that "practical" is what works for humans. But this creates a false dichotomy, between humans and the non-human. The implication is that what humans can extract from their environment through practice is thereby of value. However, different elements of the non-human can function perfectly well in relation to each other without human input or interference. Medicines can work just as well for other organisms as for humans. Functioning is a larger theory than the anthropocentric implications allow. By hiding the implied ends for which the theory is used, the non-human is tacitly excluded. Value theory, however, must cover all value, not a limited class of value. By ignoring ends, the theory can be used as a rationale for domination, instead of a general theory of value. However, its limitations as a theory of value are supportive of this use, since it does lack a theory of ends as it stands.

Practice should not determine value, for practice alone is uncritical. Does something "work" which is destructive? Does it have value? Nietzsche, *inter alia*, has argued that creation of new values requires the destruction of older ones. But he overstates his case: the old ones may simply fade away, like horses and buggies, which are now largely confined to the Amish. If something works economically and is destructive of human spirit does it have value?

What if it goes beyond moral limits? By what standard of value is the workable then to be measured? That something "works" is not necessarily or always an indication of its *value*. Cars may work but also pollute. An explanation may not be a good explanation (babies arrive via storks). What "works" must be evaluated and tested for value against moral, categorical, cultural, and even personal forms of value before anything can be considered a good device, procedure, explanation, institution, and so on. Functioning is one thing, functioning well another. Value can then be distinguished from functioning as such. The workable is only the beginning of a good and thus only satisfies the lowest level of categorial value. Common sense practicality recognizes this factor, since no one would argue that one should use a rock where a hammer is available, but only that in the absence of the tool, if a rock works, it should be used. But this argument already goes beyond mere practicality to improvement, that is, amelioration.

The quality of what works is always an issue: that is, a standard other than what works. In other words, "practicality" has often been used by "hard-headed realists"—that is, those who would exploit the non-human without limit—to avoid amelioration, specifically with respect to environmental damage. But by their own arguments, practicality presumes amelioration and is thereby subject to critical evaluation. Isn't value more fundamental than mere practice, in the light of these comments, that is, the value of practices as problematic?

The moral deficiencies of the theory are even more decisive against it. Concentration camps "work." The Nazis succeeded in slaughtering six million Jews, thousands of Gypsies, Slavs, gays, clerics, "mental defectives," and others. The theory is amoral as it stands: any functioning is good, including functioning for evil ends. Thus it is not only deficient with regard to ends but morally. Again, chain saws work better than axes in destroying old-growth forests. But the disastrous consequences of such destruction are beyond the theory. The limitation of the theory to means provides a rationale for evil, since ends are open for the theory. Again, "functioning" may include a society that functions by exploitation: the American South in the days of slavery. Slavery had a function, a use value in this society. Functioning and utility do not imply equality, freedom or even the value of life. Gladiatorial contests no doubt had a function in ancient Rome in keeping the masses content by providing them with bread and circuses (compare TV). Again, mistreating minorities may have a function for a particular society. The society may acquiesce in violence at its margins if it sees a use for such violence, for example, kkkeeping minorities in their "place" by lynchings.

As for utility, economists use the Phillips Curve as a justification for throwing people out of work to "prevent inflation." Throwing the least skilled, most vulnerable, and least able to afford unemployment out of work has the "use value" of serving the greatest good of the greatest number with the non-market solution of raising interest rates at the Federal Reserve Bank. The control of the money supply works to keep the wealthy, privileged, and powerful prosperous: useful for those who already have too much to sacrifice what little the poor have. The utility is measured by the preservation of wealth of those who are already wealthy, by keeping inflation down and thus preserving the "real value" of the currency. Overpaid corporate executives can thus increase their wealth by hiding behind governmental collusion while using "free enterprise" as a rationale and even decrying "governmental handouts." It is obvious that "utility" in this sense is incompatible with the ideals expressed in the Declaration of Independence.

Some philosophers also use practical as a dialectical counterfoil to theory. Plato, Rene Descartes, Johann Fichte, Charles S. Peirce, and William James each took a practical orientation as basic in philosophy and defined theory in terms of practice in some respect. The Aristotelian tradition (including Thomas Aquinas, Georg Hegel and possibly Gottfried Leibniz) argued for the autonomy of theory/contemplation and its superior value in relation to practice. Materialists have split between these two orientations, some favoring contemplation (Epicurus) others practical consequences (Hobbes and Marx).

However, they disagree over the paradigm of practicality. The practical is the political for the Sophists and Machiavelli, economic or technical for Marx, religious for Augustine, science for Descartes and others, virtuous activity for Aristotle, and moral for Kant and Fichte. The issue for a general theory of value is whether and how any of these paradigms either has practical value or reveals anything about practical value? Is there only one practical value? Don't these variegated paradigms make practice problematic as a basis for practical value? Does the "practical value" of some doctrine presume that the value question has been answered first, namely, what is of value? Is the distinction between value and practice more basic than practical value?

Practical value is a theory of instrumental value of limited value. Practicality is a mode of value instead of a mark of a general theory of value. While feasibility and functioning well are indeed one element of creative actualization, they are hardly exhaustive of it. Utilitarian value theory, insofar as it is interpreted in a functional or practical sense, as most economists do, and not from the perspective of its theory of value, that is, hedonism, shares this defect. A theory which analyzes good in terms of "good for…," that is, useful for, implies useful for something else, namely, that for which the utilitarian value is a means, as it is evaluated only for the sake of something else. The theory thereby undermines itself, since its expression implies its own limitations. Utilitarian value cannot ground: it cannot provide an ultimate justification of value, for it must end in intrinsic value. What practicality and utility do reveal is the "objectivity" of instrumental value, as I argued above.[24] That is, instrumental values are actual and in the world and do have actual consequences in the world, apart from any subject. Since these consequences are the ends, ends can be actually in the world. The theory also upholds the means-end relation, while failing to specify ends within the theory.

Finally, practicality does not apply to non-economic categories. No culture measures religion by practicality, nor art. *A fortiori*, practicality cannot provide any sort of moral guidance, especially with respect to the destruction of the environment by humans.

Six

CRITICAL EVALUATION OF SUBJECTIVE VALUE THEORIES

What is the significance of the *summum bonum* and *pros hen* treatment of value in ancient philosophy for more modern, subjective theories of value? Generally *summum bonum* theories have been grounded in the subject in the modern period; all of reality is instrumental to the intrinsic value of the subject. Only what is valuable to a subject has value; the human subject is foundational for value, in that the chain of means to ends has subjective gratification as its ultimate goal. Human good is defined in terms of what J. Baird Callicott has called "subjective states" or "psychological capacities," reason, will, desire, feeling, emotions, or attitudes. A relation to internal states of the subject is necessary for value. All tools are used as instruments for the ultimate end of "happiness,"[1] or other internalized states, and this creates a hierarchy of goods with happiness (or reason, feeling, will, and so on) at the apex. Thus there is a further reduction of value from world to human and human to subjective psychological states. Reduction of all value to such subjective states is also a reduction of all values of the person to such states, and thus a devaluation of other personal values, including those connected with the body, a relation noticed by both Sigmund Freud implicitly and Michel Foucault explicitly. The theories of this sort reduce all value to capricious psychological states, quite arbitrarily.

Many philosophers still cling to the psychological model of value, in spite of contemporary critiques of the subject in metaphysics, and of psychologizing in logic. The subjective turn in philosophy initiated by Rene Descartes was tied to a solipsistic ego. The value of what is valuable belongs exclusively to a self; as solipsistic, the ego can refuse to acknowledge an autonomous other. This movement has revalued slave relations (Epictetus) and made them metaphysical. The slave has no control over the outer world. His/her own autonomous value as an other, as an agent cannot be granted by the other. The slave's value is contingent on the master's (Hegel, 1979 "Master and Slave") since, as Nietzsche argued, the master creates value Nietzsche, 1887/1969, 1). The slave is not granted intrinsic value, but only instrumental value as the tool of the master.

The climax of this movement is Friedrich Nietzsche's denial of all intrinsic value in the "chaos" of the flux of forces as "will-to-power." All value is an instrument of the master and the will of the master. The master refuses to grant reciprocity or recognize the autonomous value of the slave despite the dependence of the master on the slave in his subjective solipsism. The slave is sub-ject to a subjective ego bent on subjecting the world to its will. The subjectivity of the slave disappears in subjection to another subject. The world of value is lost to the slave in his subjection, since the slave has no rights in the world, but is subjected to the whim of the master, who can use the slave as a tool. In turn the slave follows the commands and does the bidding of the master in subjecting the world for the master. The world becomes a tool of the egoistic master, who treats its value as an instrument. This model has been generalized, despite the inconsistencies of universalizing an exclusive model, by modern philosophies of subjective value.

1. Subjective Value

The good of objects refers back as an instrument to a first in value, either human good or the intrinsically valuable human subject on which its claim to have value is based: a hierarchical relation. Value is always in relation to a subject. Value is attached to an object for a subject: all good is subordinated to human good, the anthropocentric view. The reduction to subjectivity is the ultimate basis of human domination of the environment, for the good of the world apart from its value to humans is not considered in the subordination of good to the human subject in a relation. The problem of the (intrinsic) value of nature arises in metaphysical subjectivism, especially if combined with anthropocentric species ethics, such as an ethic based on rationality as the specific difference of humans.

Subjective value in general is foundational and episodic, and involves a fourfold relation. The first is hierarchy, a *summum bonum*, in that all value derives its value from the subjective *summum bonum*, for example in relation to the amount of happiness it brings. The second is the *pros hen* relation in which all goods become means to the end of subjective enhancement, "referring back" to the justifying end. The third is the compositional relation of psychological states and their value, that is, happy feelings or proper attitudes. The feeling has the internal relation to happiness. Finally, there is the relation of life as a whole to random feelings that compose it axiologically. A good life is one with many happy feelings: one composed out of "good" atomic elements, just as a body is composed out of material atoms. Such theories then require relations of value, that is, a world of relations. There is bound to be a tension in the conglomeration of relations that mark subjective theories, especially hierarchy and foundation; and the relations with the reductionism implicit in them. But more, the relations argue against the theory as a type, since it argues against any solipsistic subject.

For subjective value, value is grounded in the "given," that is, such psychological capacities of humans as pleasure, valorized feelings, desires, and the like. However, value in such theories would also be episodic, since such states as pleasure come and go randomly. Such theories originated in a form of materialistic determinism, in which psychological states were caused by preceding physical states, a passive anthropology, as Karl Marx noted. Increasing valued psychological states requires activity. The ethical or practical task is to increase such states, that is, to increase good as defined by such theories. This approach conceals a theoretical contradiction, since if value is given, it should not need increase. If subjective states have value, any attempt to increase them goes beyond the given, and requires a rationale beyond the given.

The fleeting actualization of such states requires creative actualization. Otherwise the implicit meliorist project is defeated: the states remain random and cannot be increased or improved. In other words they cannot be perfected. However, this conclusion would go against both the thrust of such theories, and actual practice. On the contrary, organisms try to increase valued states and decrease devalued states. They attempt to creatively actualize a happy life at the expense of a miserable one. Thus the different species of foundational subjective value are not sufficient to explain value: for valued subjective states require creative actualization. Insofar as they are goods they are species or kinds of creative actualization. They must be enhanced. They cannot constitute a ground, since they require grounding as a project. Even personal values like happiness require creative actualization. Since subjective ethics requires creative actualization, it also requires the norms connected with creative actualization, namely, reciprocal relations, compossibility, pluralism, and particularly meliorism.

Subjective theories identify the good with a faculty, capability or other psychological bearer which is enhanced by creative actualization of valued states: *good* feelings, *good* will, *proper* attitudes, *happy* moods, and so on.[2] However, value is distinct from such psychological capabilities as such, since there can be feelings and attitudes which are not good, namely, bad feelings, moods or emotions, improper attitudes, malevolent sentiments, and the like. Value is a "quality" of such capabilities, not their "essence." The ability must be enhanced to achieve good, and this involves creative actualization of it. However, the theory is correct insofar as it denies that feelings and other psychological states are value free: a denial that there can be a "substance" to feelings with value as a mere "attribute." Feelings have a value tone and attempts are made to enhance those deemed superior, worthy or desirable, especially happiness, that is, to improve the quality of such states.[3] This improvement requires creative actualization. Our argument is that happiness is creatively actualized and thus a species of good, not the genus, insofar as happiness can be enhanced at all. Moreover, creative actualization of such feelings is the active overcoming of what is given in favor of what is better.

Subjective value theories have generally failed to distinguish value from evaluation. All subjective value is in the realm of evaluation not value, except personal values. To desire some feeling or pleasure should be subject to evaluation: to evaluate whether creative actualization of a good would be an improvement. The feeling remains in the realm of evaluation, not value, since without creative actualization the project is either no good or else Utopian. The subjective is not actualized in the world and remains a mere wish, not a value. The possibility has not yet proven valuable in the world. This point applies to theories based on interest, liking, feeling, emotion, attitude, and other psychological factors. Feeling, emotion, sentiment, and attitudes have to do with critical evaluation of some ends over others in the realm of potential ends, potential projects. They rarely determine such ends.

Martin Heidegger made the same mistake in his judgment that to assign value is to devalue. Assigning value is an evaluation, not a value, since value is measured against a standard, or is quantified. Any creative actualization has value as inherent to an actuality. Assigning value is to concede inherent value while also bringing the value into a relation with the larger whole of values, an evaluation. Heidegger overstated his case. Assigning value cannot be devaluation, since value, after all, is assigned. However, a good may indeed be *under*valued in an evaluation and thereby devalued.

Subjective theories are self-refuting in this light. Not everything felt (pain), not every emotion, or other psychological state has value. Thus such theories are immediately qualified by extrinsic factors. As soon as subjective states are identified as the source of value they are qualified: feelings, but not bad or painful feelings; attitudes but only "proper" attitudes, and so on. Such subjective states, then, are not the source of value, since some of these states have value and others do not. The value of such states can and must be critically and morally evaluated: pleasure as a valued feeling and pain as bad. Thus value is independent of such states, as their value must be evaluated by an extrinsic standard: not feeling as such but whether the feeling is a good one. Value is not identical with feeling, pleasure or any other psychological state or capability, since some of these may be disvalued and an attempt made to increase other states. This applies to all psychological state theories. Psychological theories of value are, then, erroneous as theories of value, since they are connected with evaluation, and at best as the locus of value only, not its value element. Reductionists have confused the relation of value to the feeling or other subjective state (happiness, attitudes), with feeling as such, as solely a psychological state. But the relation to value is crucial to give the subjective state value. Thus the reduction fails.

Moreover, we do not value attitudes as such or even feelings as such. Instead, good feelings are valued and these require creative actualization. Value is actualized through action, and the feasibility of a project has nothing to do with subjective wishes. I may whine all I like but whining alone will not get the plane off the ground. More, objectives of subjective states that are critically evaluated as inferior do not even have potential value as a project, for example, flying unassisted. Creative actualizations of many feelings and objects of desire are passed over as projects, since they are judged as not worth the bother. Thus value cannot be an "object" of feeling, or other subjective states, since many such objects are evaluated as not worthy of pursuit. Even personal values are separate from evaluation since they have been creatively actualized and are exhibited in actions undertaken by an agent for personal reasons or grounds. Actions speak louder than words: action is the criterion of such personal values, although not their measure. Personal values require prior evaluation but are not values only as evaluations.

Subjectivity shares the problem of inherent value with other hierarchical theories of a *summum bonum*, absolutism, perfectionism, and so on. If an object has the inherent power to creatively actualize specific valued internal states, then it has inherent value (Clarence I. Lewis, 1946). Value belongs to the object, if only as an instrument of such valued states. Thus formulations of intrinsic value which require an object for subjective states are relational, not entirely subjective. A purely subjective theory would only involve a relation of value to a subjective state: good feelings, proper attitudes, and the like. However, almost no theories do not require an object: almost no theory is entirely subjective. Thus objects can have a "value of their own," apart from subjective agents. At the Grand Canyon we see the beauty. When we go away we don't, except in memory of the sight. If beauty were only subjective, we would always see it, as it would be part of or posited by the subject. The beauty of the Canyon would not require a Canyon. But could one value the Grand Canyon without an actual Canyon?

A purely subjective view of value could not take pleasure (pro attitudes, liking, and so on) in an object or it would be relational. That is, if it took pleasure in the object it would require such objects to arouse such states. But the relational view requires inherent value in the object: that some quality inhere in the object or that the object as a whole has the inherent power to cause valorized subjective states, whether pleasure or other valued feelings. Otherwise it reverts to the solipsistic subjective view. Since it requires the inherent value of the object, this inherent value is necessary. Since the inherent value is necessary for such subjective states, it meets the criterion of "objective" in Immanuel Kant's definition. Value is inherent in the object whether any subject wishes it, is feeling it at a particular moment, or has attitudes about it, pro or con. Inherent value cannot belong to an object if it depends upon a subject, for then it would not be inherent.

We use a hammer, not a paper clip to drive nails into the wall. The superiority of the hammer is not a mere subjective whim or attitude since its superiority can easily be demonstrated. Inherent value means that the value of an object is independent from—autonomous from—a subject. The subject must demonstrate value first to ground the choice. Subjects would not choose objects unless their inherent value was already exhibited, or potentially feasible.

Subjective intrinsic value belongs to esthetics and the refusal to acknowledge an autonomous other, a point stressed by Emanuel Levinas. The psychological origin of egoism lies in the attempt at personal self-sufficiency, which began with the Sophists but was emphasized in Hellenistic ethics. The Stoics championed the self-sufficiency of virtue; the Epicureans, of happiness. Later this was incorporated into Christianity as the "ascetic ideal," as Nietzsche noted. The Protestant version of the monastic life was asceticism within marriage: Calvinist guilt regarding sexuality, which requires the other as an object. This tradition was criticized by Hegel ("Stoic consciousness") as based on an extremely restricted sense of freedom, the inner freedom of free will (Immanuel Kant), which "avoids responsibility for consequences" as John Dewey argued. Inner value, the modern view, restricts the realm of freedom, and stands in contrast to freedom of action in Hellenic ethics. It also implicitly denies intrinsic value to other agents: the origin (rationalization?) of self-interest. The value of what is valuable belongs exclusively to a self: subjective value is egoistic as connected with a self that values and confers value, an ego. Anything beyond the ego becomes reduced to an instrument of the self and derives its value from such a subject. The value of the other derives from the self. Anthropocentric subjectivity effectively gave a metaphysical basis for egoism as the solitary subject was identified as an ego, and its world-view is necessarily egoistic. Egoism has been taken as self-justifying by modern ethics in the form that self-interest is defined as "rational." However, this view constitutes a denial of the inherent, to speak nothing of intrinsic value of the other. We cannot judge an object as the other valuable if it derives from the subject, for then it would not have inherent value, but only conditional value. Thus "objective" theories which start from a subject are fallacious. Samuel Alexander (1933/1968), for example formulates intrinsic value (what I call inherent value) as "contemplation" of the object "for its own sake." But inherent value cannot belong to an object if it depends upon contemplation. Alexander is in fact assigning intrinsic value, reductively, to contemplation.

How can any object or act be assigned value if value is not inherent in what is referred to? In the judgment, "this is good," value is assigned to a bearer as inherently good, not conditionally upon a subject. Otherwise the judgment is without reference and thus incoherent. "This is good," is a value judgment that assigns value to a referent, a bearer, not to either a subjective state or some other, third referent, a *summum bonum*. The judgment is *not* "this feels good," but "this is good." The argument that "analysis" "clarifies" the "meaning" of the judgment by reducing it to feelings is little more than an agenda disguised as an epistemology.

The attempt to replace judgments of value in a bearer with emotive or other subjective value is reductionistic. "I like wine" and "wine is good" are equivalent in this view (DeWitt Parker in Lepley, 1949). Liking the wine and the goodness of the wine are equated. This leaves out of account the goodness of the wine as such, as a distinct creative actualization, a distinct good in the world apart from other goods. It leaves out the success in making the wine, of avoiding vinegar, for example; as well as gradations in the quality of the wine and comparative quality. Such a reduction is the attempt to ignore value in all its differentiation and plurality: the kinds and modes of value. Only one mode of value is recognized in subjective reductionism: pleasures, specific feelings or attitudes. All the rest of the world is without value in this view, since value is reduced to random, episodic inner events. Value is confined to arbitrary feelings, pleasures, and attitudes. Such theories have only one value: feelings, emotions or other subjective states. Beauty is not valuable only pleasure; health is not valuable only pleasure. There is no criterion for difference in such monistic subjective theories. Beauty, health, and knowledge cannot be differentiated as values by such a reduction. But as the example of using a paper clip to drive nails indicates, such a reduction is pragmatically impossible. "I like" to use paper clips to drive nails belongs in theater of the absurd, farce, or to the inmates of an asylum for mental cases. Such a reduction would be *ungrounded in the world.*

Similarly, subjective theories of value ignore socially derived values and values peculiar to a kind or category of activity. Culturally shared values that direct action and are inculcated into subjects in the process of maturation are socially derived. We may act not out of a desire for pleasure or feeling but under cultural and social norms, for example, norms of courtesy for those we don't like. I will return to this point below. Values in distinct categories of activity are pragmatic: the actual value of means within a category, in the popular sense of pragmatism: does it work? Playing the violin well will not result in coal being mined, no matter how much we feel it will. A thimble is no good as a means of draining a lake, despite whatever attitude we have about it, proper or not. A steam shovel is no good for playing Beethoven, even if it may increase our pleasure. Subjective states are simply irrelevant to the value of such means. I hope that this constitutes a *reductio ad absurdam* of the attempt to reduce value to monistic subjective states.

Value theories based on the subject are monistic value theories, which recognize only one actual kind of value. To evaluate any such theory we need only imagine a world with just such a value and no other to realize its limitations. A world of perfect rationality would be a world of calculating machines, not of humans with mixed capacities and plural goals. A world of desire without fulfillment would be a world of continuous frustrations; one of fulfilments of desire would lack variety, since only one value would be consistent with monism of value. A world of feeling without cessation would be a world of bland satiation without the contrast of feelings that give different values to different states.

Such theories are modeled on atomic foundationalism in its subjective form, grounding all good in random episodes of an ego. In this view, we are not responsible for such states, which come and go randomly. Humans are the passive recipients of different psychological feelings, emotions, and attitudes. Value is arbitrary since its advocates would connect it with the occurrence of random psychological states. As others have noted, the problem with such a view is that it ignores the "active side," the creative actualization of good through activity and the attempt to ameliorate the human condition. John Dewey remarked that we attempt to increase the good and decrease the bad. Humans are not passive victims of random psychological episodes but active world builders, who act to "bring about" determinate and favorable results.[4] Humans increase good in the world with diverse activities aiming at plural goods. Further, the randomness of such states cannot ground any attachment to an object, since such states do not occur regularly. Their attachment to an object of value is contingent upon their occurrence and thus the value of the object is contingent.

Because pleasure, feeling, attitudes, and other subjective states are private, they cannot establish a publicly confirmable good. However, the value of health aids for animals and plants is publicly confirmable, to speak nothing of nutrition, specific medicines, and so on (veterinary science). Withhold water from any organism and ultimately it will die. Give it sufficient water and it will not, *ceteris paribus*. This shows the fallacy of private, subjective theories, of identifying all value with private value. Not all value is private, in a subjective ego.

In subjective theories, value is attached to interior states of the subject, who is a self or ego. Pleasure, feeling, emotions, and attitudes are those of a self or ego. However, the identification of self with such interior states is problematic. Why is "a" desire "my" desire? In what sense are random episodes of desire and feeling constitutive of a unity, a self? If "I" have parts, what makes me "one" self? My unity as a self can only be that I act as a unity, the unity of the actor. But then what is "mine" must be connected with action, not random, atomic psychological events. Unity of self is grounded in unity of action, not random episodes of psychological events over which "I" have no control.[5]

If value is subjective—feeling, pleasure, emotion, liking, or attitudes—then *the world is devalued*. The subordination of all value to human value, to the human subject, is the hidden egoism of Western philosophy: everything must have value for *me*. The beauty of the wild, its scenic landscapes, flowers, and the complex interrelation of species within it are without inherent value for subjective theories. No objects or actions have inherent value in this view. They are devalued from inherent to instruments for episodic occurrences of an ego in a value hierarchy. Only feelings, emotions, and attitudes have a value of their own. The entire world in its vastness is, from this perspective, only the instrument of the enhancement of such petty feelings and random emotions. Thus if there were no such psychological states, there would be no value, since, according to such theories, the value of the world as an instrument derives from its instrumental value in achieving or enhancing such states. Hypothetically, if everyone caught a rare condition or disease that ended all feelings, they and the world would have no value. Their life would have no value, since there would be no feelings, emotions, and so on. Subjectivity is the devaluation of life.

The objects of feelings and emotion are reduced to instruments of gratification of such subjective states. Thus having such a feeling or emotion would be as valuable, or more valuable, than having an object which is instrumental to such psychological states. According to this reasoning, "feeling" the value of beauty without any object of beauty would be a more valuable state since it alone has a value of its own. Moreover, the theory of the value of means in subjective state theories also requires objects that are instrumental to such states, that arouse pleasure, some esthetic feelings, and so on The means also require creative actualization.

Subjective theories locate value within individuals: they are nominalistic. Nominalistic subjectivism also inherits all the philosophic baggage of nominalism. The ground of value lies within individuals. But then value is instrumental to such individuals and relations between them become a problem. Metaphysically, the reduction to the subject is the attempt to deny the actuality of relations, including social relations—that is, society, culture, family, and the enculturation of individuals during childhood in the learning of cultural norms, including language. Nominalists want to deny cultural anthropology and sociology in the denial of the relation of individuals to each other. Such theories are incomplete as they stand since they ignore the larger world of society, culture, practice, and action and the different categories of value, to speak nothing of the world. All these kinds are subordinated or reduced to subjective states of the individual. Yet social, cultural, environmental, and other extrinsic and external forces may constitute individual values. The individual treated as solipsistic is fallacious.

Nominalism of values would mean, if consistent, that no one should ever impose their values on anyone else, since inherent value attaches to each individual subject with their own feelings, emotions or attitudes. There could be no bad or false values in this view, in contrast to good or true ones, since value derives from individual psychological states and these have inherent value. However, this would be the end of all value, since value is in part a contrast to what is without value or is bad or evil. If any feeling, desire or attitude has value, then none could be good or bad. Racist feelings would be as good as tolerant or loving feelings. Judging that some feelings, emotions, and attitudes are good and others bad is to have a standard beyond such feelings, emotions, and attitudes as such, namely, good feelings, emotions, and attitudes. It requires evaluation of such psychological states. The development of character in children also requires imposition of cultural values (Freud, 1962).

Even the value of the other is dubious in this view, unless we use another instrumentally for the enhancement of valued subjective states. The autonomous value of the other as an other, as a self, is not consistent with egoism. Thus such theories are relativist and anti-equalitarian at the same time. From the perspective of the ego, the relation to the other is as a means of self-gratification, a tool (Hegel, 1979). Consistent egoism is the refusal of reciprocity and thus the autonomy of the other by a self/ego. The other is not equal to the self, since the other is devalued from this perspective. Thus ethical egoism is ultimately anti-egalitarian, since self and other are not equal. However, the other regards one equally as a tool for egoistic gratification. Thus good is relative for all subjective theories.

Ethical egoism is an oxymoron. The value of the other cannot consistently be derived from a self. Similarly, confining morality to a subject is itself normative: an unjustified moral imperative, since it creates a norm from a singular. Ultimately, others are devalued in such a theory since they are instrumental to self-gratification. The intrinsic value of the other, having a value of their own, would be contingent upon the value of the self and enhancement of its valued interior states. More, the value of the world is subordinated to an ego. In this view, the world had no value before the solitary ego, nor will it after it.[6] The world only acquires value during the life of the solitary ego. The world as having value for others or a value of its own does not enter this perspective. Value is confined to human subjects for the subjective theories, as only humans are considered subjects. It is difficult, if not impossible, to see how the intrinsic value of species or environments could be grounded in subjective theories: they are reactionary.

Subjective theories, echoing the metaphysics of nominalism, would like to dispense with social relations altogether in value theory. However, the distinction of subject and object already involves a relation and the value of any object for a subject—water to quench thirst—requires such a relation. A *purely* subjective ethics is unreal: for relations to the world are required: instruments of pleasure, feelings of beauty, and so on. To be out of touch with the world is catatonia or other forms of schizophrenia. No purely or absolutely subjective theory of value is possible; all are relational and thus require relations in some form.

Subjective value (feeling, attitude, desire, will, interest, and so on) as relational presumes value in the world: the value of the inventory of the world with which it can interact and draw value. Otherwise feeling for or attitudes toward the world would be in vain. Most subjective theories distinguish between valued internal states and disvalued, for example, happiness and pleasure as opposed to pain. The attempt is made to enhance such valued internal states. If the subject is not in a circumstance where such states have been achieved, they attempt to change them. Frequently changes in circumstances involve changes in the world, since the circumstances are those that subjects find themselves *in*. These are adjudged bad in relation to other arrangements in the world. The world thus has a certain value or disvalue which is the origin of the action to alter it. The subject is only the agent of evaluation, not the source of value. Value as located in the subject as the "origin" of value is only conceivable in an unreal separation of the subject from the world.

The needs or biological requirements of the subject are an argument against the self-sufficiency of the subject and thus the subject and subjective states as the source of value.[7] A need is a need for something external, for example, for water. Value is what is needed to enhance specific feeling states and moods at the expense of others. The subject, far from being the source of value, is dependent upon the environment and the values of the world. The subject derives its values from the world, not the reverse. We require the world: it does not require us.

Similarly, the subject depends upon life, and thus life is required for subjective states. Subjective states are evaluated the highest good in psychological theories of value. For example, pleasure is the highest good for hedonistic theories. Thus life becomes an instrument of subjective states. However, all subjective states depend upon life, since they take place within a life: they are states of living creatures. A good life is one with a greater number of pleasures over its course than pains, as John Stuart Mill admits. Life is a condition of such states then and thus required for such states. What is required is obligatory: thus life is the value required for subjective states: a more supreme value, since agents require life before subjective states.

Moreover, all subjective theories are guilty of the fallacy of composition in their haste to reduce value to such random episodes. To argue that an episode of pleasure or sentiment is good is not a basis for all value. An episode of value in a life, even granting the value of such an episode, is not the same as a happy life as a whole. Nor are a disjunctive series of such episodes taken collectively, for they are random. Pleasure also requires the body, that is, it could not be completely subjective in the metaphysical sense. The independent value relations among living things in the larger world have nothing to do with the subjective states of an individual ego *a fortiori*. To say that all these many forms of value derive from one ego is as ludicrous as arguing that the whole universe depends on our ego.

Subjective theories of value ignore socially and culturally derived values. In this view persons are autonomous after a period of dependency and development in which they are acculturated and socialized (Mead 1964, VII-XII). Individuals do indeed pursue egoistic ends, but this reflects a culture that encourages such pursuits. The values of the subject do not come from the subject but originally from their culture and society. "Feelings" and "pleasures" are culturally approved. Thus the subject is not the origin of values. Indeed, moral education and perfectionism both presume that such values can be instilled; for example, John Stuart Mill thought it requisite that "moral sentiments" be taught to the young in Utilitarianism. If the subject is perfected in any respect, with development or training in moral virtue, social rules and duties, even intellectual, artistic or other socially derived values, they cannot be the original source of such values. Education and acculturation are the origin of norms of what is acceptable to feel, have pleasure or have attitudes about.

Subjective theories also ignore *particular* cultural values: "patterns of culture." Specific values distinguish one culture from another, for example, an emphasis on individuals over the community. In every culture some values are emphasized at the expense of other possibilities and such values are instilled in the individual. Culture is an intermediate realm between the individual and the world and cannot be reduced to either, as the source of many of the values of the individual person, which are internalized as subjective.

Again, subjective theories ignore the pragmatic-practical element in value. Individuals choose the tried and true, not their whimsy, when acting. They go with what works, functions and is good in the sense that it does function, not arbitrary attitudes or feelings. Those who go with feelings instead of what works or functions soon end up with feelings and attitudes they did not desire, namely, frustration, anger, and irritation. Value is not enhanced but diminished.

There is also a strong bias in all subjective theories toward individuals. Since ethics is primarily social and has to do with relations with others, subjective theories cannot provide a consistent ground for an ethics. As evidence, consider Utilitarian theories, which attempt to combine an egoistic theory of value based on specific feelings with an altruistic ethic. If my feelings are the ground how can I possibly care about others where their good conflicts with my happiness? How can any obligation or duty to the other be derived from an ultimate concern with my self, based on my feelings? The theory of obligation is inconsistent with the theory of value in this and all egoistic theories; personal feeling states can not be the ground of obligation to others, even if random episodes of feelings for others occur.

It might be argued that morality is in our self-interest since the benefits of society and cooperation are superior to strife and conflict (Hobbes). Since our self-interest lies in following morals, morality could be established on an egoistic basis. However, moral principles themselves could not be based on egoism or individual feelings. Rather, once moral principles have been established, individuals might conclude that it would be in their interest to conform to them. Further, I am inclined to think that self-interest is culturally mediated and thus constituted within individuals. A moral framework is required to cultivate self-interest as a norm for individuals.

If feelings, emotions, and attitudes are not in our power they would not be subject to moral suasion, perfection or regulation. Some schools of psychology believe just this and argue that we can or should control our actions and not seek to control our emotions. If we cannot control our emotions then we cannot be expected to treat them in accordance with moral principles. But then morals and emotions would be independent from one another. "Exploring feelings" might be valuable in therapy or as a topic for discussion between intimates. But whatever its value, personal feelings are not morals and cannot be the basis for morals. Morals involve choice: whether and which principle or rule to creatively actualize. If we cannot choose our feelings, we cannot base choice on them.

What then of value as "objective?" Value as objective still implies a Cartesian universe of subjects and objects. An object is an object for a subject: the concepts belong together. Value as objective implies that value belongs to things, objects, existences, "out there"[8] while subjective states are internal. However, both internal states of the subject and the objects it adjudges external are in a larger whole, of world. Subjective theories further presume that the subject can be abstracted from such an environment. The subject does not exist and cannot exist apart from an environment: the separation is artificial and Utopian. The environment, as Dewey pointed out, does not stop at the skin but permeates our entire body. The subject is entirely dependent upon such an environment and is entirely within it, as are their subjective states, even if these are private.

Values are not "objective," as they are not attributes of beings. Values are the cause of actual goods, and thus not an attribute of a prior "substance." Values must be successfully enacted: they are not present as "beings." Similarly, values are not subjective ideals, for creative actualization successfully "realizes" them as accomplished goods. Value is indifferent with respect to subjectivity and objectivity, the Cartesian problematic of consciousness, since value mediates subject and object in the larger world. This problematic is only part of a larger whole of value, one difference, one normative contrary among others (true/false, good/bad, reality/illusion, and so on). Both the subject and the object are derivative and informed by values: through acculturation, education and development into "subjects," and creative actualization of an objective, that is, goods. The goal is pursued through culturally sanctioned actions (instead of a "subject") which creatively actualize a good (instead of "an object") in an environment and thus brings the so-called "subject" and so-called "object" into relation in a larger whole in the world.

Even where value is the object of a *personal* feeling, attitude, interest, or emotion—a personal valuation—value is independent of these as their objective, and thus not contingent upon them. Values are neither subjective nor objective, since values ground the relation that makes the distinction possible. They are beyond such a distinction as its ground.[9] The subject is formed by acculturation and thus creatively actualized as autonomous through development and education. The object is made by creative actualization of worthwhile ends. Objects are pragmatic products of action, which have proven value. Differentiation itself reflects evaluations and the worth of distinguishing moral, categorial, cultural, and personal values.

Is there no element of subjectivity in values, then? It might be objected that creative actualization of value reflects subjective states in some respect, since we may personally desire to bring about a new good, and thus work on it. Whatever place such personal motives may have, they must be evaluated as worth pursuing before they are acted upon. Action, as even Mill noted, is end-governed, not motive governed. Action aims at results in the world, not in the psyche, except where some purely internal states are aimed at, for example, by taking anti-depressant drugs. My motive is beside the point in the achievement of value since the best intentions in the world will not make a plane fly. Desire, feeling, and other subjective psychological states have nothing to do with such value in the world. A good car runs whether or not I wish it, feel it or display "proper attitudes" about it. Mechanical qualities are standards without any reference to psychological states whatsoever. Car repair manuals are not therapeutic. The same arguments apply to art works, prices, health, and other kinds of values.

Creative actualization disproves all subjective theories. According to hedonistic theories, to take pleasure in the contemplation of a faster-than-light spaceship or a time machine gives these objects value. (Or to have strong feelings or correct attitudes about them, desire or take an interest in them, and so on). But such "objects" have no value, for they can never be creatively actualized, if general relativity is correct. Thus objects of this sort that are desired, felt, and so on are not values but whims, for they are "no good." They can never be creatively actualized.

Subjective theories are reactionary as they stand, since they are against moral progress: the expansion of rights to larger, not smaller considerability. Subjectivity devalues the world with its adaptation of the *pros hen* relation, the referring back of all value to such states, including the value of life, of other species, and the world. Thus it would be the opposite of expansion of rights to include a larger field. Subjectivity would also consist in an intellectual retreat away from the public world as the arena for improvement to the Stoic inner world: introversion of value. Subjective theories are pessimistic, since they have given up the attempt to transcend the inner. This strange internalization is belied by the diverse improvements in the quality of life over the centuries.

Subjective theories of value can be treated under three main headings, affective or feeling theories, including hedonism, conative or desire theories, and intellectualist or rational theories. These headings correspond to the main psychological capabilities, or "parts of the soul" of humans that Plato mentions. Many varieties of such theories have cropped up in the literature of value. I will discuss the main affective theories below, and include criticisms of some of the more prominent varieties, for example, hedonism. Rational-intellectualist theories were covered in the last chapter with perfectionism. Conative, interest, and other theories associated with desire will be covered in the next chapter.

2. Affective Theories

One subjective theory is the affective theory, for which value is connected with feeling. The affective theory is often closely connected with the conative theory, since Mill and other hedonists often define pleasure or happiness as the object of desire, but pleasure is itself a feeling, so we can safely classify hedonism as a species of the affective theory. One main form of the affective theory is the "moral sentiments" theory of David Hume and Adam Smith. A closely related approach is that of specific forms of feminism, in which the role of feelings and emotions in moral life is stressed. Others include the esthetic theories of Alexis Meinong and his followers (Pepper in Ferm1950, Werkmeister, 1970) and perhaps the agapism of Christianity. Still another form the affective theory has taken is the "emotivism" of Alfred J. Ayer and Charles Stevenson, which I will examine below. George E. Moore could also be interpreted as within this tradition. Affective and conative theories could also be combined as one theory.

Virginia Held's argument (Held 1990) that Western ethics has been patriarchal in its persistent rationalism, whereas women tend to emphasize feeling, overlooks the ancient debate in ethics over the primacy of reason vs. feeling from Plato on. Hedonism, which has a long history from the Cyrenaics and Epicureans on, is a species of affective value theory since the feeling of happiness or pleasure is the *summum bonum* by which everything is evaluated. There is also the notion of "moral sentiments" in David Hume, and other figures from the Scottish Enlightenment. Held's position, far from being radically new, places her within the anti-rationalist school for which "reason is a slave of the passions." This holds even if she were a non-cognitivist: the extreme irrationalism and relativism of A. J. Ayer and Charles Stevenson: "emotivism." Feminism of this type is less "transformational" than supportive of the anti-rationalist school, in value theory.[10] Its innovations are in its view of the social development of the individual and its gender relativism. Held's view is widespread among a particular school of feminists.

The question here is what is the justification for calling all such affective theories one in kind? "Feeling" is an ambiguous or equivocal term and can have many different meanings. These include sensation, as when I feel the texture of a garment; emotion or mood as in a feeling of anger; attitude as in what someone feels about a public issue; or refined feelings, such as may be instilled in a sensitive reader by a novel or other art work. Apart from the fact that they posit or presume the psychological or bodily ability to feel, what unites such "feelings?" I would argue that the theories themselves have more than a mere "family resemblance," since they center around emotions. (I am equating Hume's term "passions" with the more modern "emotion." I will argue this point below. However, this does not "affect" the present argument, since if passion is a variant, like attitude, feeling as a generic term covers it.) The theories do not dwell upon feeling as sensation, except hedonism, or if they do, they consider it as a part of emotions or play on the ambiguity in the word "feeling." Refined feelings are refined emotions. Attitudes can mean opinions, but they are based on feelings, since they are not intellectualist. Otherwise their source would be reason. "Proper" attitudes, like refined feelings, are cultivated emotions.

Despite some differences in their treatment of feelings, these theories can be treated together since they are similar enough in the relevant respects. Emotivism is a type of affective theory since it holds that judgments of value are expressions of emotion or feeling; no strong distinction is made between emotions and feelings by these authors. Thus any feeling has value, which is to confuse feeling and value. Since value judgments are the expression of feelings, value is located in feelings. Other psychological states are either subordinate to emotions and feelings or without value entirely, if this theory is consistent. Reason and desire are either instruments, as in Hume, or lack value: they are value neutral. The feelings are spontaneous and may be contradictory, but still have value or express value according to the theory. Value has its locus in random and episodic events in a subject who feels. Since feelings and emotions may come and go, value does not endure but is random and contingent upon having feelings. No feeling in a subject, no value.

Affective theories could take the form of a relational theory but also a completely subjective theory, in which valued feelings alone bear intrinsic value. I will examine affective theories first as a relational theory, and then as a radically subjective theory, which emphasizes the value of specific subjective or psychological states.

Like other subjective theories, affective theories are hierarchical. There is a double hierarchy involved in the relation of the outer to intrinsically valuable inner states of the feeler and also of other psychological states and subjective capacities to the feelings, for example, desire and reason as the "slave of the passions." Indeed, even feelings are ranked in a hierarchy, from bad to good feelings. With regard to good feelings, desire and reason are the means to the end of attaining satisfying feelings or increasing positive feelings: they have instrumental value. Indeed, all the good in the world is an instrument for satisfying or increasing (positive) feelings. The value of the world derives its value from whether it satisfies or enhances such feelings; the world derives its value from whatever feelings it evokes, satisfies or enhances. Nothing in the world has inherent value in this view since the world derives its value by virtue of participation in a feeling state.

As with conative theories, affective theories are egoistic, whether they are in a completely subjective form or relational. Value refers back to an egoistic subject who has or pursues valued feelings. Feelings are feelings of a self, my feelings. Feelings are an "attribute" or at least state of a subject, and thereby their locus is removed from any public or social context. Feelings are isolated within an individual and have no social or public element.[11] Thus they cannot be confirmed or known as such. Someone can report their feeling states but no one else can share the original feeling states of another. Thus value is atomized in affective theories in internal states of individuals. The individual derives their value by virtue of having such states or as a participant/ bearer in such states, instead of as the end or purpose of such states, since the individual as such does not have intrinsic value. If such theories are consistent, only the feelings of the individual have intrinsic value. Since value is atomized in random episodes of feeling, affective theories are foundational. Value is based on such episodes of feeling in a *pros hen* to their occurrence whose repetition gives value to the feeler.

Feelings are also episodic and random, since they come and go without deliberate planning or choice. We do not control how we feel, we only "have feelings." They come and go without conscious input or deliberation. Nor do they come in any order: they are random. However, the attempt is made by affectivists to increase positive feelings by external actions of different kinds. Thus there is some attempt made to control feelings and enhance some at the expense of others. Whether this is consistent with the theory is problematic, since if feelings have value any attempt to enhance some at the expense of others requires a standard other than feelings, by which some are adjudged positive and some negative, some good, some bad.

The first issue that is raised by such theories is whether "feeling" is a valid norm either as a human capability or as a foundation for value. I noted above that feeling is an ambiguous term, which is also fairly modern in philosophic usage. Plato argued that the three parts of the soul are the rational, the appetitive, and "*thymos*" (the latter has been translated as "spirited" part). Where are feelings in this tripartite model? Are feelings equivalent to "*thymos*?" Or are "feelings" a modern norm? Whether *thymos* can be interpreted so as to include feeling, and emotion is problematic. Again, Hume spoke of reason as the slave of the "passions." Are "passions" the same as "feelings?" This is moot; it could be that passion does not mean emotion or that emotion is a more precise term of which "passion" is a primitive forerunner. Again what of moral "sentiments" in Mill? Are these equivalent to the modern "feeling?"

One problem here is that since feelings are within the private world of the subject that the question of whether what I experience is what another experiences is problematic. "Feelings" are problematic, not a "given." They are problematic both epistemologically and even as psychological actualities. The epistemological problem is precisely what led to behaviorism, since "reports of inner states" could not be scientifically validated. Nor can they ever be compared: I will never know directly what another experiences. Thus the conceptual norm "feeling" is problematic, since it may cover a host of unrelated inner events.[12] What these "are" may be quite different, which is why philosophical psychology has changed the terms used to designate them over the centuries: from "spirited part," it later posited "passion" and "sentiment," and still later "feeling" and "emotion." Philosophical psychology is historical, and the normative understanding of human (or even animal) psychology is superceded in later times, just as Aristotle's notion of nature was superceded. However, "feeling" is even more problematic than nature. Since we cannot compare inner experiences, how can we either determine if they are feelings or evaluate them? How can we determine whether a feeling, emotion or pleasure of one person is similar or comparable in type or in value to another's? Or even of the same kind or intensity? Such a problematic norm as this cannot provide a foundation for anything. There may be no foundation there at all, since what I call feeling you may call passion and there is no way to compare the experiences in order to determine if they are similar. More, my feelings are mine, not yours. My feelings cannot be the basis for yours since they are mine. Nor can our feelings be the basis since we cannot compare them to determine if they are the same. Comparing them with a third person's may be even more problematic to speak nothing of larger groups. Thus they cannot form a unified standard of value, a *summum bonum*.

There is also the question of the validity or actuality of "feelings." Thomas Hobbes reduces "passions" to desires in *Leviathan*. The question is the actuality of "feelings." Similarly, the "moral sense" theory posits a new ability to the human psyche or subject. But is this a positing of an actuality to account for morals or a genuine part of human psychology?

Another question is whether what is private can be taken as a "principle" or norm. Internal emotions cannot be compared either to each other or to those of another person to determine if they are enough alike to constitute an identical element or principle. Indeed the atomism of this model makes it problematic if we can even compare our own feelings to one another. According to behaviorism and some forms of physicalism, so-called emotions are as much behavioral as internal: which of these are to be taken as paradigmatic? Psychology is historical and the categories used for psychological description are historically relative to some degree. Thus psychological categories such as emotion and feeling are problematic as a basis for values. More, they ignore the social dimension entirely, the way values and ethics are learned, taught, and inculcated. Individual emotions are often the result of social influences, forces, conditioning, conflicts, and so on. They may reflect class background, parental hectoring, peer pressure, and other social and cultural forces. To posit them as basic ignores the social origin of both the form and the content of emotions, and thereby social relations in their origin and cultivation. Nietzsche would have placed this under the "absurd overestimation of consciousness," as if emotions could be autonomous from the world.

The other problem is the axiological one of the relation of such states to value. How can one ever know what is felt is truly valuable? There is no way to examine such feelings, since they are private; no way to test their worth as subjective. Even if some feelings or emotions are deemed intrinsically worthy, how do I know if I or anyone else have such states or whether they measure up to the standard? They are not only incomparable to each other but to a standard of feeling.

Only specific feelings are valued. As I noted above, this creates problems for any theory of value based on feelings, since a standard outside of feeling itself must be appealed to or used to determine which feelings are to be valued. Not everything felt or for which we have an emotional response has value, including feelings themselves. We value specific feelings but not others: elation and bliss but not depression. The feelings themselves become problematic in value. Should I value whatever feeling comes along or only elevated feelings? Should I repress feelings that do not accord with the standard, for example, feelings of hatred in agapistic theories? Since some feelings are problematic, feelings as such cannot provide a basis for value. Enhancing specific feelings at the expense of others is an argument against feelings as such and appeals to a standard beyond feeling.

Affectivists recognize these problems and often immediately qualify their theories with standards beyond the theory. Good is first identified with the psychological state and then this identification is immediately denied through qualification. Not feelings but only good feelings. Not attitudes but only proper ones. Such theories then refute themselves, since both the occurrence and the creative actualization of hatred, depression, and other "negative" feelings is generally discouraged. The value of such states must be evaluated even according to their proponents, since only some feeling states have value. Value, however, must be independent of such states if we used it as a standard to evaluate them.

The resulting propitious evaluations, for example, "proper" attitudes, and feelings, "appropriate" emotions, and the like, are extrinsic to the theories, and bring in a normative element, propriety. Thus the theories require some non-psychological standard for completion, in order to constitute a value theory. Further, they implicitly presume perfectionism. Feelings must be perfected through cultivation of desirable feelings like "positive feeling quality" and avoidance of negative feelings: hatred, depression, scorn for art, and so on. Feelings must be molded to achieve value. Otherwise they remain random and good feelings have equal value to bad. But if they are equivalent in value, the theory cannot determine what good feelings are, since all feelings are equal. Thus it could not be a theory of value, only an apology for random feelings. However, if the theory appeals to perfection, that is, the cultivation of some feelings at the expense of others, as in the ethics of virtue, then a standard outside of feelings are the actual determinant and criterion of value. More, the atomism of the theory is anti-perfection, and this creates a tension. Should the feeling as it comes be valued or should it be perfected?

As in most subjective theories, affective theories confuse value and evaluation. Even if our feeling about some object of feeling informs us of its relative value to us, it cannot tell us of its inherent value. As with conative theories, worth must be established prior to any feeling of value in an object, or it would not generate a feeling of *value* but some other feeling. We do not value something whose independent value has not been established. We can rank objects in a hierarchy based on the strength of feeling it arouses in us or even on the quality of feelings, as in Mill. But this would constitute a personal evaluation, not a value judgment. The value of the object apart from our emotions, its inherent value, would not be covered by such an evaluation. The value of some object like water for quenching thirst does not change when our feelings about it change. Nor is this value of water in any way contingent upon our emotions. If we come upon a pool of "bad" water in the desert, our emotions will not change the poisoned or undrinkable water into good water, no matter how thirsty we feel.

An emotive or affective theory cannot in principle tell us the value of action, events, things, and so on since none of these have inherent value. Only emotions and feelings have value, including contrary emotions. Feelings are not just logically but spatially distinct from a good or valued "object." Such objects can have no value at all according to this theory since they do not have feelings. Our emotions toward them are not relational, but entirely subjective. The world is entirely devalued. Our emotions about them cannot extend value to the objects, since only feelings have value according to this theory. Thus if objects have value, it cannot be as an object of feeling or emotion according to this theory, for value is confined to feelings, in the sense of emotions, or perhaps to their expression. However, the theory is inadequate even as an affective theory, for we often do have feelings about something in the world and these feelings arise because of the object. They involve a relation to an object that precedes the feeling, for example, the love between parent and child. Thus the capability of the object to excite the feeling must precede the feeling and thus its value in exciting the feeling or emotion can not derive from the feeling, since it preceded it. Further, the object does not change when our emotions change. If the thing does not change, how can the value of it change? The object can be good in the same respect in which it aroused the emotions, for example, beauty, even when our feelings change. Only our relation to it has changed: additional evidence for a relation and thus all the location problems of a relational theory. Something in the world can be good without arousing any emotions in us at all, for example, the Grand Canyon or the ceiling of the Sistine Chapel for aesthetically obtuse individuals.

Similarly, emotions and feelings can tell us little about the value of actions. Feelings about an action, apart from the feeling itself, would be relational and require that value consist in more than subjective feelings. However, actions are subject to evaluation as successfully achieving the end or falling short. If the ethics of virtue has any validity, some actions, namely moral actions, may be valuable in themselves. This would be impossible for a consistent emotivism. Thus the entire literature of virtue would be fantasy. However, as some linguistic philosophers have noted, many virtue words are descriptive as well as "commendatory," for example a "brave" action. Bravery can be evaluated apart from our personal feelings; Native Americans were said to have applauded the courage even of enemies.

There can be no intrinsic value at all if value is reduced to emotions; the same emotion can be valuable at one moment but not at the next when a new feeling arises in which the old is devalued. Thus value does not belong to the feeling but the moment of feeling. Intrinsic value is generally used to cover ends valuable in themselves, instead of mere instruments. The question is how a reductionistic theory can extend intrinsic value to persons. If value only attaches to the moment of feeling, how can it extend to the person? The feeling—indeed the moment—has value, not the person; the person is an instrument of the feelings, and presumably would have no intrinsic value unless it had feelings. But how could we know whether others have the requisite feelings? If they did not have these emotions, why would they have intrinsic value? And if the emotions of a person were negative about herself how could it be wrong for the person to commit suicide? For in this theory neither life nor the person has value except as the instrument of the moment of emotion.

If feeling is made a standard for evaluation it would still not be equivalent to value. What is evaluated superior or worthwhile must still be grounded: it must be creatively actualized. We may feel we ought to do something and still not do it. The action has not been brought about: the feeling has not been creatively actualized or resulted in a dutiful action or the like. Thus it remains in the realm of potentiality, that is, of evaluation. Insofar as feelings are in the realm of value at all they have to do with evaluation of some ends as superior to others, that is, as contributing to enhancement, preservation or achievement of specific positive feelings. For the most part, these feelings must still be creatively actualized, thus can only be potential feelings if they are not produced, preserved or enhanced (happiness). Insofar as feeling lies in the realm of evaluation, however, it would not have intrinsic value, since the goal is to enhance it. The goal of creatively actualizing the feeling would be value, not the evaluation that one would be better off if one had specific feelings which one lacks and whose enhancement is evaluated worthwhile. (This argument does not apply to feelings one has and is trying to preserve.) Thus feelings are not necessarily either loci of value nor inherently valuable. However, whether feelings can be a standard of evaluation is problematic in any case, since they are private and internal. Feelings themselves require critical evaluation.

Even if our feeling were to lead to creative actualization, the value of what is created must be established separately from what we feel about it. Otherwise rotten vegetables which we feel ought to have value have as much value as fresh vegetables. A plane that we feel ought to fly but does not is different than a plane that does fly. A plane that fails is no good: it has no value as a plane, and cannot unless it meets the non-affective standard of a good one. To argue that feeling could be the origin of creation, that a feeling that the value of something that has not been actualized might be valuable if actualized is to confuse value and evaluation.

Evaluations are not confined to personal feelings. Scientists themselves make such judgments of worth and can back them up with considerable evidence. The medical and nutritional sciences can give publicly confirmable means to ends, that is, foods and medicines which have been proven good for health or recovery, for example, those with more vitamins. Penicillin is better than or superior to candy as an antibiotic. The health of an organism, the end, is not emotive for such sciences, but can be measured by publicly confirmable standards, which can be established for animals as well as humans. Similarly, engineering can establish whether particular materials are superior for construction to others—better. Steel is better for the framework of high-rise buildings and bridges than wood, to speak nothing of paper. Its inherent qualities are superior for heavy construction. Aluminum is better than lead for particular parts in planes, since it weighs less. Glass is better than opaque materials for windows. These judgments are not based on feelings or emotions, but trial and error, calculation, function, and other standards. Such judgments and evaluations are reasonable precisely because they are not based on emotions. We are more likely to trust a building that has been built with materials that are evaluated superior based on impersonal standards than the emotions of the engineers.

Affective theories confuse value with feeling. In the first place, goods are not feelings. Feelings are not just logically distinct from goods but actually distinct as well. A good can be such without exciting a feeling in us. A bolt or a nail may be in excellent condition—good bolts—without creating any feeling in us whatsoever, much less positive ones. If we are not building or repairing something we may not care in the least whether the bolts we see at the hardware store are defective or not. We usually have no feeling at all about them. We can have a feeling about something that is not good, an emotional investment in a project that we feel ought to work but does not. We can have a feeling about goods that are not our own project and thus the feeling is not directly related to the good. The feeling about the good is a variable independent of the good.

Goods retain value apart from anyone's feelings about them. A good does not cease to have value when a feeling or desire passes. Gold still sells at a specific price, tends not to tarnish, and is ductile whether or not I make any emotional investment in it. Flowers are either fresh or withered; art works fine, primitive or amateurish. Actual goods do not change due to feelings; thus how could their value change? Its personal value may change to one individual, but its inherent value as a good, as non-defective and functional, has not changed. Only our personal relation to it changes with feeling changes. A plane still functions well even if we hate flying. Our feelings are simply irrelevant to whether a hammer functions well; a tool is valuable whether we have feelings or proper attitudes toward it, or take pleasure in it or not. Our feelings cannot restore a wilted flower. Since they are irrelevant to the inherent value of such instrumental goods, what reason is there to think that they are relevant to intrinsic value? To ends in themselves?

The identification of feelings and value is an arbitrary relation. The relation of the feeling to the object of feeling may be totally arbitrary, and often whimsical, since different people may have different feelings about the same work of art or other object of feeling. We may feel one thing today and something completely different tomorrow about the same event, person, good, idea or any other object of feeling. Which of these feelings is correct or proper? Which is right or good? Feeling alone cannot be a guide, since which of the feelings we should use as a standard is problematic. Feelings may be illogical: we may hate something which is perfectly good for us, medicine, and like something bad for our health: tobacco, excessive alcohol, junk food, drugs. How can arbitrary, shifting, personal and relative feelings tell us anything about the value of the plane? A good car operates well whether not I have good or bad feelings about it, regardless of my attitudes toward it, proper or not, and emotional reactions. Mechanical qualities are standards without any reference to feeling states whatsoever. A good mechanic does not consult his feelings to fix a car but the manufacturer's manual. The mechanic does not worry if his emotions are good but whether the car runs well. The case is similar with art works, price, health, and other kinds of goods. "Touchy-feely" has little place in the practical world of goods. Feeling and other psychological states have almost nothing to do with value, with the exception of valued personal states. The latter, however, are not just a distinct kind of good but entirely different than goods, since they are internal, private, and subjective. They are also selective.

Affectivists might counter that feelings are only the ultimate source of value with respect to instrumental goods like a plane. Ultimately, such tools are instruments for our happiness, or other intrinsically valuable feeling states. The value of the tool may be inherent but derives this value in relation to our feelings. Without human feelings, the tool could not have value. Our point in the above examples is that feeling cannot be the source of value for such instruments, since their value must be inherent or they are no good. Feeling will not make a defective or broken tool work nor will it detract from the value of a good tool.

The language of value is not equivalent to the language of feelings, since if I say, "It is a good plane," I am decidedly not saying, "I feel it is a good plane." The two judgments are not equivalent and the attempt to reduce one to the other is fallacious. First, the judgment that "it is a good plane," may be a professional one and require extrinsic standards. We do not want a mechanic's or engineer's report on their personal feeling states when they are evaluating the air-worthiness of the plane. Instead, we want their professional judgment, which follows norms proper to aerodynamics, not psychology. Similarly, we do not seek personal feelings from a jurist, but legal precedents and the like in a judgment. Judgments are not based on feelings, but considered reasoning and normative evaluations.

In judgments evaluating actual goods, the language almost never involves feelings. Feelings get in the way of good judgment. If a carpenter claims that a job is "no good," s/he cannot allow his or her feelings to cloud his/her judgment. The judgment that a tool is "no good" cannot express feelings if the judgment is about a hammer. A judgment about the *feelings* a carpenter has about a tool would call for *different* language, a different judgment. An apologia for keeping an old tool around may involve sentiment, if a new tool is also available. "She's old but still works." If it were defective the judgment in which the virtues of the tool were expressed would be erroneous. But the old tool would never be used if it has no value as a tool, if it were broken, defective or unsuitable.

On a psychological level, the language of value is again distinct from psychological states, including feeling states. Reports of feeling states, whether anger, hatred, love, and the like, are distinct from the value and ethical terms "good," "right," "virtue," and so on. No such distinction would be made in language were the feelings equivalent to value terms. There would not be two languages, one of value terms and another of feeling states, since these would be equivalent. Further, we could not distinguish feeling states from value states, since they would be equivalent. Each feeling would carry its own value or evaluation. But this is obviously oversimplified. We often describe emotions without evaluating them at all. And we can evaluate actual goods in the world without reference to feelings. Further, we can also evaluate feelings apart from having the feeling. We can say "you shouldn't have gotten so angry at her," or "I feel guilty about not loving him more." This would be impossible if feelings and values were equivalent. Again, if they were, enhancing such feeling states would be either impossible or untenable.

"The language of morals" has little relation to the language of feeling, since moral rules "prescribe," "command," or state obligations. Whose feeling is supposed to be expressed by the Decalogue in "Thou shalt not kill?" God's? The author of the Decalogue? But the command says nothing about anyone's feelings, that of the author or how anyone following the commandment is supposed to feel. It does not state that we are supposed to feel that we should not kill. The commandment is not concerned with what we feel about killing. On the contrary, the commandment implicitly recognizes that feelings may be the context in which the rule comes into play: feelings of outrage and revenge. But these are not valid for putting aside the commandment. Actions are proscribed, not feelings. One may still have murderous thoughts but cannot act on them. But the rule covers actions regardless of feelings, that is, it covers cold-blooded as well as "hot-blooded" murders.

Some might argue that there is cultural evidence for the use of feeling as a substitute for evaluation, that, "I feel that" may mean a normative or evaluative judgment. In such a case, a feeling is not an emotion but is given a new meaning, equivalent to making a professional judgment. It could also be the case that the person making the remark has been influenced by affective theories of value and thus is attempting to creatively actualize a theory in a self-fulfilling manner. In this case their language is "theory laden." However, we generally learn morals and values with different language than the language of feeling. We are taught rules, not parental feelings. We are told, "don't hit your brother," not "I feel badly when you hit your brother." A parent warning a young daughter does not say "I feel you should not get pregnant," but "you'd better not get pregnant." A warning which expresses concern for the well-being of the child is expressed, not the personal feelings of the parent. Indeed, trying to match the feelings of the parent is impossible, and would be an onerous burden for the child.

We can read the affective theory in two ways, passive and active. Feelings as something that happen to an individual involve a passive model. Moods and emotions come upon us and there is little or nothing we can do about these feelings. We may be able to avoid acting upon them, but cannot suppress the feelings as such (or, in another reading, it would not advance our psychological health to suppress them.) A more active reading of feelings is that they can be manipulated for the better. Good feelings can be creatively actualized; bad feelings suppressed, overcome, or replaced by more positive feelings.

In the passive view, we cannot creatively actualize feelings, since the feelings we have are largely out of our control. Feelings are "there." If this view is correct, positive, valued feelings cannot be creatively actualized. There is no point in trying to increase happiness or other positive feelings, since our feelings are largely out of our control. If this reading of our feelings is correct, progress in value is impossible. Our feelings come and go like the weather and we cannot increase the positive value of our feelings. Such a model of feeling would disprove all happiness and hedonistic theories, since their premise is that feelings of happiness and pleasure can be attained through actions, changes in circumstances, and other efforts.

Random, isolated, internal events cannot give any identity to a principle that is the mark of any theory. Since the events are internal and isolated inside different persons, they cannot be publicly verified. How can one ever know what is felt, emoted or liked is truly valuable? If they are internal they are private and cannot be compared with a standard. Since they are random, we cannot even compare them with one another. Since they may be contrary, they cannot provide an identical principle. Since they cannot provide a principle, there can be no principle that identifies a theory or an instance of a theory.

In the more active interpretation of the emotions, we can increase positive feelings. But we may not always have such feelings. Thus they must be increased through our efforts. But then such feelings are a species of creative actualization, since the positive feelings must be creatively actualized. We are not merely passive creatures but can creatively actualize increases in happiness and other positive feelings. Since we can also creatively actualize other goods and both these and feelings require creative actualization if actual good is achieved, feeling cannot be basic. Just as perfection of character in the form of virtue requires the creative actualization of such perfections, so feelings require creative actualization or they remain Utopian, mere possibilities, not actual goods. The difference, as the passive reading of the emotions implies, is that we cannot require anyone to feel something. Feelings are not available on demand; we do not necessarily have a suitable feeling about something, that is, feelings cannot always be creatively actualized. But if we cannot creatively actualize these feelings, they remain Utopian, that is, possibilities or norms, not actual goods. If there is any truth to the passive reading of feelings, then, they cannot be goods, for they cannot invariably be creatively actualized. We may not be able to achieve positive feelings in our life, even if we have tried.

Nor are they as manipulable as the active theory requires: they are not always, or even frequently, perfectable. It may be that we should love our neighbor but sometimes we just don't. Thus we cannot require feelings of approbation or disapprobation, since our feelings may be independent of our best intentions. We may despise our neighbor despite our attempts to overcome such unlovely feelings. "Our" feelings are not always under "our" control. Emotivism and other affective theories would also mean that we are not free in our choices, since our choices are subject to determination by arbitrary feelings. Thus we could not freely choose the better of two arguments, proofs, recipes, medicines, and so on. Since evidently we can evaluate such affairs free from our feelings, it is equally evident that such decisions are not determined by feelings and thus that some goods are independent of our feelings.

Presuming we could separate worthwhile emotions from negative ones like depression and sadness, there is another requirement for the value of emotions. We must be able to creatively actualize such emotional states. Indeed, insofar as emotive theories are an ethic, they are optimistic, that is, they argue that positive emotions can be increased and negative ones decreased. But this is to argue that we can increase the worthwhile emotions. For unless we could the theory could not be an ethic, could not be normative. It would only be descriptive, a science: when people have positive emotions they have positive emotions, when they have negative ones they are negative. Emotional quality could not be changed for the better. If control of emotions were not in our power at all, they could not be in the province of ethics. Some emotions are not subject to our control, although our actions are. However, this is not the same as to say that such emotional episodes pass and new emotions arise. Now evidently at least some emotional states are subject to change and more positive ones can in some cases be substituted. All psychological therapies presume this, to speak nothing of moral and value theories. The thesis of therapy is that this change can also be an improvement. Thus the process of obtaining worthwhile emotional states is a process of creative actualization of such states, that is, bringing potentially better states into actuality in the person. Creative actualization of positive emotive states is inherently worthwhile.

As in other psychological states, feelings that must be perfected or cultivated require creative actualization. We must creatively actualize "positive feeling quality," perhaps through spending more time with art than at other activities. By doing so we shall have more of the good sort of feelings and less of the bad sort. But then the theory presumes creative actualization. If such "positive feeling qualities" are pursued but not attained do they still have value? Would a life of striving after good feelings but not attaining them be worthwhile? But then it could not constitute a good, since would never be experienced as good. Feelings that are valued but never actualized would be Utopian. The standard would not be actualized: it would remain a Utopian ideal, but never actually felt. No one would ever experience such feelings. If feelings are actual goods they must be actually felt. At best, then, feeling is one species of value, not value, since if we could not actualize "good" or "positive" feelings through enhancement, we could not attain the good. If feelings are values, they require creative actualization. That is, insofar as emotions are of value, they are one species of creative actualization. Far from all values being reducible to emotions, emotions are but one sort of creative actualization.

Specific feelings may be instilled or developed in us, just as advertising or other external causes may arouse emotions. Such feelings are not original, not a foundation, but derivative. They might be the "normal" reactions of a culture, which we are expected to feel, such as grief at death, and which have therefore been cultivated within us. They arise in an ongoing context of activity as consequent not ground. Thus they are not original to a subject, but are socially determined. Their creative actualization is considered socially necessary, and thus a means to social ends. In such cases the feeling is not an end in itself, but a means. More, they have been instilled as a conditioned response by long training: they are not individual feelings. We may doubt whether the feelings of patriotism, which make young men march off to war and probable death is "natural," since they conflict with the more "natural" instinct of survival. If such feelings are instilled by the society, they are not the origin or foundation of value, since we can trace the origin of the values involved to society.

Why does feeling alone confer value? What is so valuable about feelings that all of value can be reduced to feeling? The best way to examine this question is to ask whether all feelings are valuable? Would a strong feeling like unrequited love, which could not be reciprocated, be of value? or would depression? Unhappiness? It might be argued that even unhappiness may have value by contrast with happiness as revealing what is positive about happiness or if it leads to happiness. But then is it the feeling that is valuable or the *positive quality* of the feeling or a later feeling? Evidently, not only do some feelings and emotions occur that are not valued by anyone but they could not be judged valuable by any standard, for example, depression leading to suicide. But then feelings as a whole are not inherently or intrinsically valuable and affective and emotive theories cannot provide a standard of value, since not all feelings are intrinsically valuable. Not all emotions, feelings, and attitudes (racism) are valuable. If not, what distinguishes the valuable ones? Are all feelings and emotions equally valuable? Or are some, like happiness, valuable and others, like unhappiness, not? If some have value and some do not (depression), we can distinguish the feeling from the value of the feeling. Feeling, emotion, and attitude alone, without consideration or evaluation of their quality cannot confer value.

I argued above that if we can make evaluations separate from influence by feelings, we can evaluate feelings. Indeed, we must if we are to improve feelings and only cultivate, enhance or pursue "positive" feelings. We must be able to evaluate whether some feelings are *worth* feeling. For example if we are depressed, should we live through it or try to immediately overcome it?

Since not all feelings are valued or worth feeling, the quality of feeling is a consideration in evaluation of feelings. Are the feelings worth having? If we value only particular positive feelings, it follows that not all feelings have value. In striving for "positive feeling states," happiness, proper attitudes, and other select feelings we attempt to avoid their contraries and enhance the feeling of quality. But then feeling cannot be the source of value. Not all feelings are themselves valued or valuable and non-valued feelings cannot ground all other values in the world. On the contrary, such feelings may be the source of bad values. Some feelings may be inherently bad and also the source of bad means to such feelings as the end.

Indeed, it may be asked whether or how we can evaluate two different feelings which we have ourselves about the same event. Are these of equal value? Since not all feelings are worth feeling, nor are all in our control, a standard of quality of feelings must be derived either from feelings themselves or from some other source with which to evaluate feelings. But how could this be derived from feeling alone? Which of two contrary feelings is more of a feeling? Suppose that we feel we ought to complete some task that we feel we ought to do but have been putting off for a long time. However, we also have a feeling that we have other projects to do, or better, feel like loafing for a day. Watching a movie would be more entertaining than cleaning. Which feeling as a feeling should be used as a standard? Which feeling, pro or con cleaning, is intrinsically valuable? Which has greater value? But if one is more feeling-like, that is, involves more passion, emotion, or other feeling state is this a better feeling? Is intense depression of greater value than mild happiness? But if not, then the intensity of feeling cannot supply the standard. Neither can strength of feeling for the same reason. Feelings of hatred for other "races" may be stronger than feelings of compassion, but are the former better in value?

More, if two different persons have different feelings, which is more valuable? If we decide based on our own feelings, isn't this using an arbitrary standard, since we do not know which is better? What if our own standard turns out to be negative? And if feelings are private, how can the feelings of others be evaluated? Thus in any conflict we may doubt if we could use feeling alone as a standard to evaluate feelings. For which feeling is to serve as a standard? Any judgment that one feeling is to set the standard would involve some factor independent of feeling, which can be used to evaluate the feelings. This cannot be still another feeling without raising the same problems, including infinite regress. But then feelings do not and cannot provide a standard for evaluation of the worth of feelings. Feelings must be evaluated by a standard that is not itself based on feelings. But then the emotive theory is fallacious, since it cannot be used to attempt to judge or evaluate feelings for their worth.

Another consideration is that we must evaluate and judge the particular circumstances in which a feeling is relevant. Amorous feelings may be inappropriate at a funeral; gloomy feelings at a party and envious feelings at a tribute. If we can control our feelings, in the light of circumstances, then having feelings, or even positive feelings, may not be any sort of standard. (If "we can't help having such feelings," then the entire basis for enhancing positive feelings may be in doubt.) The feelings must be appropriate, not because of attitudes, but due to circumstances.

Feelings may change over the course of time. What is judged good based on one feeling may not prove to be good and the feelings may change. More, the feeling about something inherently good may change. We may feel fine about flying until an airline disaster and then our feelings may change. But the vast majority of planes in the fleet may still be perfectly good. Our feelings have changed without the value of the planes changing. Further, those who are afraid of flying may never have positive feelings about it in spite of excellent safety records.

I would argue that feeling and emotion have almost nothing to do with value in general, as opposed to the inherent value of some feeling states and emotions themselves. A feeling that strawberries have value is no indication of the actual value of strawberries, since they may be underripe, rotten, worm eaten, and so on. I may emote that you ought to eat them, but this will not increase their value as food. As I noted with respect to affectivism, a tool is valuable entirely apart from our emotions toward it. Our feelings are completely irrelevant to whether such a tool functions well. Again, a plane that will not fly has no inherent value as a flying machine, no matter how much we feel it ought to, or yell and scream at it in vexation for not flying. A good plane is such apart from my feelings toward it. Feelings cannot justify creative actualization, since "feeling" that a plane will fly is not enough. An aircraft company does not evaluate its planes based on passenger emotions about the engine but mechanical quality. Mechanical quality is based on standards without any reference to supreme states of feeling or emotion whatsoever. True, the cabin of the aircraft may be designed with consumer preferences in mind. Specific feelings like comfort may be one factor in overall evaluation. And some feelings may be valuable in themselves. However, to extend this model to all of value would be an egregious fallacy of composition.

If value is creative actualization, feelings are a variable independent of creative actualization: they have no direct causal link even to other feelings, much less all goods. We may be able to creatively actualize specific feelings even if other feelings stand in the way (happiness, depression). But regardless, feelings are peripheral to creative actualization of most goods. They may or may not accompany achievement of valued ends. In many cases they are irrelevant: the truth may be "hard," as Nietzsche says, that is, seeing the truth of the Euro-American treatment of Native Americans may not be a pretty picture and may arouse in us very unpleasant feelings. But this does not detract one iota from the truth value of that treatment. We may despise the achievements of opponents and rivals, but that is irrelevant to whatever creative actualizations have been accomplished by them. Indeed, feelings may get in the way of seeing the truth.

Feelings may also accompany creative actualization without affecting the good of the outcome. True, positive feelings may accompany the achievement of the project. But the project can have value whether such feelings occur and can have value regardless of the quality of the feelings. If the Wright Brothers felt little but let-down at the end of their project would that have made a bit of difference regarding their creative actualization of flight? We can feel positive feelings, negative feelings, no feelings or inappropriate feelings at the conclusion of a project. None of these are relevant to the value of the project itself. Even if feelings can tell us the value of our feelings, they cannot tell us the worth of actions, projects, and other modes of creative actualization. They may enhance the project for us internally, but only if positive feelings are creatively actualized, not negative ones. The emotive theory again fails.

Feeling "about" some object, objective, action, event, person or completed project shows the limitations of feelings as a basis for a theory of value. Such feelings require an "intentional object" of feeling, that is, a relation to something in addition to the feeling, which is the cause of the feeling. Feeling is both incomplete as a theory of value and a consequent of such a relation. It cannot justify value as a ground.

Affective theories also confuse feeling and obligation. Even if one has no feelings toward some human, one may still have obligations toward them. One may even have negative feelings toward someone: hatred, jealous rage, contempt, and so on. None of this affects our obligations, namely, not to murder innocent humans. Feelings are not equivalent to obligations since some acts are wrong whether we have feelings and regardless of whether they are positive. Nor does the strength of the feeling give a guide to obligation. The difference between mild anger and outrage is not a moral difference. Indeed, strong feelings of a negative sort may be what are restrained by moral principles. Feelings may be a motive for an action but cannot constitute an ethic as such. Strong feelings may impel me to act but cannot make the act right or good. The randomness of feelings and the requirement that they be morally evaluated separate them from both a mark of value in general and obligation. (More sophisticated versions of the theory argue that feelings are the ground of obligation: the good as the basis of the right. I will argue that they are not a ground for obligation below.)

Like other subjective theories, feelings cannot generate obligation, since they are egoistic. Feelings are *my* feelings: they are personal not altruistic. Despite arguments that we can have benevolent feelings, in Hume and Mill, feelings as such can never rule out bad feelings or sentiments, for example, sadistic feelings causing us to torture someone. Good feelings are outweighed by bad. Duty is other-regarding: duty to someone to whom it is due. That altruistic feelings of duty or benevolence are at odds with such theories is evident, particularly if the affective theory is hedonistic. "Ethical egoism" is an oxymoron; the ego cannot be universalized as singular to take other-regarding duty into its calculations. Duty to others is an external relation, not an internal relation of the ego's own feelings. Only if other countervailing considerations are introduced can the theory succeed as an ethical theory, which would undermine the theory.

The Nazis had strong feelings that what they were doing at Auschwitz was right. No doubt General Sherman had strong feelings when he declared that "the only good Indian is a dead Indian." Yet few would wish to propose that Auschwitz or the policy of concentrating Native Americans on "reservations," murdering those who refused or resisted, stealing their land, and giving their "private property" to whites are moral. Affectivists might argue that the feelings of the Nazis and American whites who wished genocide on the tribes argue for relativism. But this attempt to save the theory ignores the possibility that attitudes can be wrong as much as beliefs can. The Nazi's attitude about race was not only racist: it was scientifically preposterous. Above all, the attitudes and feelings of the Nazis and genocidal Americans are *morally* repugnant.

No moral guidance or basis for evaluation can be gleaned from such a view as emotivism: the theory is morally nihilistic. There are no good or bad feelings, just feelings. No other person can evaluate my feelings both because as feelings they are good and because they cannot experience my emotions. Since feelings are the feelings of individuals, the theory is egoistic in at least the metaphysical sense; like the affective theory or subjectivism in general it cannot generate obligation. The ego cannot be universalized as distinct: what separates the ego is precisely the difference between me and others. Thus other-regarding duty cannot be derived from the theory as it stands. The relation of feelings to the ego is internal not external: my feelings or emotions. Emotivism cannot go beyond relativism, then, since what emotions I value have no relation to anyone else's. It cannot go beyond "I feel" to "I ought." In this view, racist feelings are as good as tolerant or benevolent ones qua feeling. The moral error of Auschwitz cannot be established on the basis of personal feelings, since racism can be based on bona fide feelings. Thus the theory is suspect from a moral point of view, since if it were adopted it would justify immorality.

Further, feelings are sometimes morally problematic, for example, strong emotions which may cause bad acts, like anger. Moral principles are obligatory in spite of feelings and may even require us to control our feelings, for example, feelings of rage or revenge. Since morality is often meant to control such feelings, it cannot be established on the basis of feelings. Morals regulate feelings, not the reverse. What we feel may be morally irrelevant, to speak nothing of morally reprehensible.

However, the theory is wrong apart from such consequences. Morality is social, not egoistic or concerned with personal emotions. Morality involves duty and duty implies duty to someone, that is, duty to others. Duty is social and thus is not concerned with egoistic feelings; emotions are personal and experienced by a self. As I argued above, egoistic feelings cannot be the basis for social norms. Indeed morals are often a check upon egoistic impulses, feelings, and emotions, for example in learning to control our anger and its consequences: the violence that results from it. We are taught not to act upon such impulses because of their bad consequences for others. We must creatively actualize moral values by not acting on feelings, but instead, obligations to others. Of course morals also consist in a way of life and virtues that involve development of the self, whether talents, capacities or more strictly moral virtues like honesty. Morals are potentially universal in scope since every event has a moral dimension. But consideration of moral consequences as well as deliberations upon duty are primarily social in their moral purport. Duty is to others and consequences for others at least as important from a moral point of view as consequences for self. Otherwise moral principles would be reduced to self-interest and morals *per se* lost in a *bellum omnium contra omnes*.

Morality in the social sense consists in rules that are essential to social survival and order.[13] We learn morality in the form of rules, principles, taboos, strictures, and the like, as several rule-Utilitarians and others have pointed out (Frankena 1963, 3).[14] Even cultural relativism argues that we should follow cultural rules and values or in other words that the morals of a society are expressed in rules and principles. Such rules are not feelings, but impersonal social principles, commands or "proscriptions."[15] "Thou shalt not kill," says nothing about emotions, whether mine or others; or any person as the source of the command. As commands or rules they are regulative over actions, whereas feelings are subjective, internal, private, and personal. Thus to confuse emotions and moral rules is a category mistake. Whether a parent who gives a child a rule has a feeling is irrelevant to the rule and conformity to it; the child cannot generally know the feeling of the parent but still can conform to the rule. The basis of the rule may be the child's welfare, not gratification of the parent's feelings. Again, such rules may be cultural (the Decalogue) or impersonal. Emotivism ignores how ethics are learned. The concern of the parent for the child's welfare, which accounts for the emotive emphasis with which parental commands are expressed, is confused with rules in general.

If a parent warns a daughter before a date, "Don't get pregnant," and emphasizes this point with a loud tone or by shaking their finger, they may be expressing an emotion. But the value is not that of the parent but the daughter's welfare. The rule expressed here has been ignored by emotivism. To argue that emotions accompany rules in some circumstances, however valid, is different than the curious notion that rules express emotions. Do logical rules express emotions? But they are as normative as moral rules. "Don't contradict yourself" or "don't fudge evidence" can also be expressed with emotions but not be equivalent to emotions. By confusing moral rules and emotions, the theory makes a serious mistake about the force of such rules. We may have obligations even if we don't have any emotions in a situation. We may have obligations in spite of our feelings (forbidding rape, young soldiers marching off to war). Moral rules, principles, and virtues are not a handmaiden to the emotions: they are a check upon them in many cases. We may even suppress emotions in particular cases based on cultural norms, for example, the Roman Catholic injunction not to hate any person. Whether we accept this principle as a valid norm or not, the opposition between moral rules, principles, and virtues, by contrast with feelings should be evident.

From these and other considerations, we may doubt that feelings in general and emotions and attitudes in particular can justify all value. Feelings, emotions, and attitudes may not have value, may not be intrinsically valuable, and cannot provide any standard for value. Feelings do not justify values or moral principles in the normative sense, since feelings that are opposed to such values are also feelings. A theory of value, then, cannot be formed from feelings or emotions. Feeling a plane will fly does not confer value on the plane. We cannot judge that a plane is a good plane only because we internally feel that it is a good plane. The ground was littered with such "planes" prior to 1903. Having positive, happy or other good feelings is not coextensive with what is good. Indeed, the question is how isolated, internal random events like feelings could ever confer value on external, autonomous objects, persons, accomplishments, and other forms of creative actualization. Unless feelings somehow emanate from inside of us, value is independent of feelings, or feelings would not be internal. Our feelings do not literally project outward to confer value; the value of goods in the world is autonomous from such feelings.

3. Emotivism

"Emotivism" is the name given to a form of affectivism from the mid-twentieth century, in which value is tied to feelings and value judgments consist in a non-scientific, non-verifiable statement of such feelings or of emotions. The theory could be taken as the idea that value judgments are a report of our feelings about objects or that they are only a report of our feelings as such. The latter is the more radical move and is associated with Ayer. I will treat the first form of the theory in this section, as less of a deviation from affectivism, that is, than from the views of Hume, Meinong, and others.[16]

According to emotivism, including Stevenson's version, all value judgments are erroneous, if they are taken as in any way cognitive, for they claim to find value in objects but in fact only express emotions about objects. "I feel that…," was interpreted by emotivists to mean valuation, but actually means something more like "I think that…," as some feminists have implicitly noticed. Virginia Held in her critique of rationality in ethics argues that women often act on their feelings, not on reasons. The "I feel" becomes an alternative for "I think." Emotivism confused two distinct sorts of feelings.

This view has created an incredible number of misplaced efforts in ethics, including those of Ludwig Wittgenstein and many other analysts who have been led down primrose paths into blind alleyways by the confusion of the emotional and the valuetive. Equating "I feel" with "it is good" is fallacious enough; going from "I feel" to "one ought to" is impossible if facts and norms are distinct. How someone feels is a psychological and historical fact about that individual. If feelings are facts, they cannot be equivalent to values according to this theory, since facts are subject to "scientific verification," but values are not on the same theory. Feelings cannot be both facts and values if values are in another realm from facts, although their total bifurcation would itself involve a dubious assertion. Moreover, if we cannot derive a norm or "ought" from a fact, ethical and other norms cannot be derived from emotions and feelings.[17] From the fact of having an emotion we cannot derive any norms or values.

Among the questions that this theory raises is whether the value is in the feeling itself or the "expression" of feelings in the emotive value judgment? Since value judgments are the "expression" of feelings in this theory, its value cannot lie only in the judgment, unless this is more of a linguistic than a subjective or psychological theory. But clearly such judgments "express" feelings in the theory. If value lies in the feelings, then their expression is not the locus of value, and the theory of value judgments as expressions of value fails. Feelings that are not expressed have value, and expression is irrelevant to their value. Thus value judgments add nothing to value and must be analyzed apart from value. The judgment is not identical with the emotion.

The emotive theory makes no normative qualification, if I understand it correctly. That is, there is no hint of "correct attitudes" as in the theory of Brentano or of good and bad feelings as in Meinong. Thus there is no notion that some feelings are appropriate and some are not. Instead, the emotion as such has value. The relativism of this position should be evident; the feelings of Hitler and his victims have equal validity as equally feelings and thus equally valuable. Bigoted feelings are as good as unbigoted; rage as good as love; anti-intellectual as good as intellectual attitudes. Thus whether science has *value* is dependent upon arbitrary whims, not evidence. It should be clear from the positivists' own program that this conclusion is unacceptable; and this shows the category mistake involved in confusing feelings and values. Feelings are an independent variable from values: if doctors prescribe a medicine as good for us, they should not base their medical regimen on what they feel but medical evidence, professional judgment, and a case history.

Emotivism never questions the value of emotions: all emotions have value, if the theory is consistent. But this view is both morally and axiologically defective. Are all emotions worth feeling? Do feelings of depression have value or do not have value? Do we not attempt to change such negative emotions or get rid of them, and thus not value all emotions, but only good emotions? Do they all have value? Or do some emotions have value and others not? Does this not imply a standard of good beyond emotions with which we evaluate emotions? Of the quality of the emotions? A distinctive axiological component? But then the actuality of emotions alone cannot confer value; they must be evaluated. Some factor independent of emotion must be used or the value of the emotion used as a standard arises, leading to an infinite regress. Why does emotion confer value in the first place? Can emotion even account for the value of simple goods like water or food? The question also arises whether emotions have value as transitory. Does any present emotion have value, or only particular emotions? What gives them value? Would a strong emotion of unrequited love have any value? Or irrational jealousy because a person for whom I have feelings loves another? If two people have different emotions which has value? The critical question of the value of emotions is never raised by emotivism; the value of emotions is problematic.

The theory ignores the use of values in the sciences themselves; the theory is wrong even with regards to science. Emotivism ignores values in the social sciences, for example, cultural and social values distinctive to a particular society that differentiate it and are essential to describing it. Also ignored is the use of value in economics, both what is valued as worth something and its market value. Often the price can be estimated and judged by experts, completely apart from subjective feelings, for example, of art works, commodities, and so on, for purposes of auction or other sales. Such forms of value cannot be explained in terms of emotions and are ignored by emotivism, as is the actual scientific treatment of them by anthropologists, sociologists, economists, and other social scientists. This itself is a value judgment. It also ignores the attenuation of the fact-value distinction in such scientific treatments, namely, that the "fact" that a person holds such and such a value or that the market value is within a specific price range.

Another problem for this theory is that if value expressions are expressions of feeling, as Stevenson argues, then they must be relational, that is express feelings about the object for which the feelings exist and which are expressed in the value judgment. This involves a relation of feeling to the situation: a relation. But if the value component consists in feelings alone it cannot be relational. (The same problem arises for feminist forms of the theory, since they argue that feeling can be a guide in moral situations.) The theory is untenable as it stands, then, since a relation is the necessary condition of most feelings. Feelings cannot be the ground of this relation, since they require it; nor can feelings as such be of value without a context, situation, referent or object of feeling. Feelings represent our reactions to something other: another person, organism, good or situation. They are conditional. This point holds even for the less extreme form of the theory held by Stevenson, since feelings cannot be the ground or locus of value.

If the feeling of grief follows the death of a loved one, whether this feeling is instinctual or culturally instilled, the feeling has a determinate cause, that is, the cause arouses the feeling. Feelings, then can be related as effects to some external cause, contra the theory. Nor are they irrational or amoral; love of family is widespread; grief for loss is even exhibited by animals. Such emotions do have an approximate and determinate cause. As effects, they bear a relation to a cause in the world; they are not ungrounded. The theory is psychologically mistaken as well, since some feelings are closely and even rationally related to their causes: we do not feel grief if no one dies, but if someone, especially someone close, dies.

Emotivists might respond that people do not always respond with grief at the death of a loved one. They may secretly be relieved or may even be happy at the prospect of an inheritance. The grief may be for show. Emotivists could also argue that the theory does give a causal account of emotive states, that is, a causal relation between the object of emotion or the event that arouses emotion and the emotion itself. In Stevenson's view, value statements tell us what someone feels about some event or object and thus that the event or object causes the emotion. However, value is assigned to the feeling alone, not the relation. Moreover, this relation is irrational, since different people will have different emotional reactions; no one reaction is appropriate, suitable, rational, or moral.

However, this is not a causal relation in the normal sense. An effect does not invariably follow a cause here, but only contingently and variably. As the effect may not follow at all in many cases, where no emotion is aroused, the relation is not causal. Normally an effect follows a cause if specific conditions are met. But emotivism argues that such emotions are not determined by causes under such conditions. Thus that the theory as described can be causal is improbable. Variable effects, that is, different emotive responses to the same cause are not a causal relation in the normative sense, but one of unforeseen and unpredictable consequences. They are reactions, not effects. The relation is irrational and relative.

If a relation of our feelings to an event is irrational, then it cannot tell us what our feelings should be *about* the event/object. These feelings may have no logical, or moral relation to the object or event, since the object or event may not arouse them at all. They may tell us about ourselves, but not the relation. Indeed, an event may not arouse a feeling in us, but none at all. Then Stevenson has not made his case for value statements, since they do not always express our feelings about the event/object, but at most, our feelings.

However, this exposes the main flaw in this highly flawed theory. If feelings are psychological facts and reports of them are accurate statements of our reactions—or non-reactions—to externals, then they can be treated like other facts. But then statements "expressing" value judgments about some event or object, given the theory that emotions do somehow express values, are either true or false, correct or mistaken. If they are untrue, then they can be corrected. But then they are not irrational for if they do express feelings, they report the fact that someone is having such a feeling. And values are not non-cognitive, since value judgments report on the fact of emotional states. In neither case are they only expressions of irrational emotion. Moreover, the distinction between fact and value, a distinction the theory was meant to underline, is attenuated, since value judgments report on a class of facts, feeling states.

As some feelings are instilled or cultivated by a culture as appropriate, they are not up to the individual. They are consequent upon cultural installation (shame). When such feelings are aroused they are the effect of cultural causes: consequences with determinate grounds. However, even if such feelings are culturally instilled, they are still the determinate effect of causes. Moreover, feelings that are instilled by a culture that reflect cultural values, such as grief as a response to death, or in respect for the dead, are the consequence of value as the ground. Respect for the dead is instilled as a value by the culture into the youth; grief is a sign that such the value of such respect has been heeded. The emotion is the consequence of the cultural value, not identical with it.

Emotivism cannot tell us the value or the moral value of different emotions. One emotion is as good as another, that is, emotivism is relativistic. Hitler's hatred of the Jews is as good as any other emotion on this view and cannot be morally condemned. The feelings of satisfaction of greed of the drug pusher are equal in value to the feelings of the parent for their addicted child. The feelings of resentment of the mass murderer are equal in value to those of the families of the victims. Emotivism as such is morally nihilistic, since any emotion can serve as a ground of action. Since emotivism cannot go beyond relativism, it is morally suspect: grossly immoral acts have value for this theory. Feeling alone cannot establish the immorality of Auschwitz and the destruction of Native America. Nor do "strong feelings" justify bad or immoral acts. The spouse who flies into a jealous rage at evidence of infidelity is not justified in murder, no matter how strong his/her feelings. Emotivism is morally depraved in its cavalier, relativistic, and immoral model of values and morals.

Another question is whether emotions are in our power in the same way as actions, especially moral actions are in our power. We may not be able to love someone, that is, we may not be able to have an emotion. But we learn in the process of maturation just how to control emotions and not to act upon them. Hating someone may not be in our power but learning not to murder someone we hate is, and is something we learn as children. If we have it in our power, it cannot then be emotive. Moral commands cannot be equivalent to expressions of emotion, since emotions are not generally in our power. Indeed, emotivism is tied to the metaphysics of determination, the individual determined by feelings over which they have little control. Moral rules, however, are voluntary, and require a deliberate choice or decision on the part of the agent.

The theory also treats rules as emotions or the reflection of emotions. Again, this ignores the use of impersonal rules in science itself, for example, the scientific method. This methodology is a means to the end of knowledge, required for a successful outcome in many cases. In other words it has instrumental value, a point that echoes Mill. The value of creatively actualizing such rules in experiments in new areas is required, or science could not be separated from pseudo-science, for example, astrology. Moral rules are normative, like those of mathematics and science. Their actualization is required in social situations for the good of society. Positivism, then, is an ideology. Value is defined in this way to preserve a false model of science, and does not reflect any careful examination of morals as such.

Moral values are often public, observable, and based on factors other than emotions. The revival of "virtue ethics" or the ethics of perfection of character has focused attention on how virtue words are both descriptive and evaluative (Murdoch, 1970/1985). A courageous person exhibits acts of courage and is commended by being judged courageous. The basis of the judgment involves at least some evidence of acts of courage. Virtues are "public" or observable and this is the basis of the language of virtues. Virtues are ascribed to people who generally or invariably exhibit them, or exhibit them to a notable degree. Public behavior is as observable as any thing, and the basis for behavioral psychology. A virtue can only be observed situationally and after repeat instances; this makes them little different from other scientific evidence. Virtues are similar to predictions, as ascriptions of character: the "trustworthy" are those who we can rely upon in the future to act honestly, who will not steal from us, and so on. Further, they are often regulative over emotions, for example, Aristotle describes how keeping one's anger in check in specific situations is a virtue. We use examples of good or right actions in teaching and explaining ethics.

Morality should control actions, not emotions. Theft is an act, not an emotion. Ethics is in part abstract or general descriptions of actions, not only commands. Actions are public, not subjective and private. The separation of fact and value must be attenuated in both the case of virtues, and thus of actions, of which virtues are a species. The value of the act in such cases is inseparable from the act itself and the act must be described in terms its value. This norm applies to both virtuous acts and acts which achieve their end successfully. Again, carrying out a moral duty consists in acting upon it, not upon feeling it. The emotive theory cannot in principle tell us the value of such acts since acts are not feelings. Thus the theory is seriously defective.

4. Emotivism as Purely Subjective

One reading of extreme emotivism is that feelings alone have value, that is, that value lies in the subject alone. This theory is the most extreme form of value subjectivism, since it denies any direct, logical relation between the object and the subjective emotion or feeling. I will take this to be Ayer's view on emotivism (1936, 5). One oddity of this theory is that since it does not include a normative or axiological standard of emotions, the increase of good emotions at the expense of bad is not unequivocally better. It would only be an improvement if an agent feels it would be better. However, if the agent *did* feel it was better, they would need to figure out how to increase such positive emotions. They would have to creatively actualize the good emotions and appeal to a standard beyond emotion as such in order to increase the good emotions. The theory is thereby axiologically inadequate and requires creative actualization for completion.

If the theory in this form were valid, nothing in the external world would have value, since value would be confined to internal emotions. The amount of good in the world would be diminished, since nothing in the world would have value except such feeling states. It could not be maintained that objects which aroused feeling states would have value, since this would require their inherent value to arouse the feeling, as C.I. Lewis argued (1947, III), or at least their instrumental value, which would require a relational theory. Such a hierarchy would also constitute a value judgment that goes beyond the theory, since the value of non-emotions is adjudged either inferior or nil. The theory would thereby be contradictory, since the value of what does not have value could be evaluated.

A *strictly* subjective theory would devalue the world in a reduction. Value would be confined to episodic incidents within the mind of individuals, in which only emotions would have value, not objects of emotion. A "positive feeling tone" about the Grand Canyon would have more value than viewing the actual Canyon. Indeed, seeing the Grand Canyon would add nothing to the value of the emotion, if the theory is consistent, as having the emotion without the object which the feeling is "for" would be more valuable in the subjective view. Actually viewing the Grand Canyon would be irrelevant.

According to extreme emotivism, a life of emotion without connection to world is as good as a life that includes a connection to world. Having emotions is enough: whether there is a basis for them is irrelevant. This excludes the actual value of acts, of things of the world, of the person as a whole, and of life, all conditions of having feelings. The world is devalued in this theory, for it has no value: only emotions do. All the species of animals, plants, flowers, landscapes, unique habitats, even astounding human inventions are devalued in an instant. Destruction of the entire world would have as much value as its conservation, if someone felt it would. Even the body may be devalued, if feelings are described mostly in terms of emotions. As such, the theory is ungrounded, unworldly, for emotions would have value apart from any basis in the world, and indeed, without the world. However, to exclude the value of all the other kinds is at best an incomplete theory; at worst, a distortion. Emotions are possible in the life of a person in the world who is living and can appropriate the things of the world through action for survival and enhancement of their lives. Feelings take place in a person who is very much in the world. Indeed, emotions may be caused by events in the world and can be the cause of events in the world, as when a person acts out of depression.

Nothing exterior to the emotional state would have any value in such a reduction. If emotions alone were valuable, then the life of feeling alone, without relation to the events and actual goods of the world would have to be judged superior to that of accomplishing something in the world. The life of pleasant daydreams would be superior to the life of DaVinci. Civilization must be judged a waste on this ground, as people can have emotions without all the advances made through it. This conclusion is almost a paradigm case for the fallacy of composition. However, no one, least of all Ayer, act as if they truly believe this; most prefer actual technical, economic, social, artistic, intellectual, and other advances. Ayer prefers science to nonsense. To argue that this is because they produce new and better feelings is to argue that feeling alone is not enough: that quality is also a consideration, to speak nothing of meliorism and the relation to something external. Nor would a life that was entirely subjective be worth living, even if it were possible. A comatose person with good emotions would have as good an existence as someone who is part of the larger world. But would anyone want to live such a life?

Emotivism is more of a fantasy state than a serious theory of value: does anyone actually believe that we could have emotions without a world that supported the life of persons who were capable of such emotions? Would a subject completely removed from the world, have any emotions whatsoever, much less values? (Compare Ibn Sena's (Avicenna's) "flying man" and his contention that he would have consciousness.) Could they know of beauty, pleasure, love? Or other worldly goods? Would any feelings or emotions they did experience have any value? For that matter, would they have any feelings or emotions at all? We may fairly ask whether a blind person could even have a notion of visual beauty. How could a "feeling" of beauty or pleasure arise in a "subject" without an object of beauty or a source of pleasure to give rise to such emotions? Emotions cannot be entirely subjective even if they have a subjective component.

A further question is whether such unworldly emotions could justify value at all? The emotions would be entirely self-referential, and nothing outside such feelings would be referred to. As I argued above, emotions involve a relation to the things of the world, the objects of feeling. At the least these add to the number and quality of our feelings and thus would be better even if emotivism were valid. More to the point, they are the source of value. If such evaluations had no reference to objects of feeling, they could not ever tell us about value, since the actual value of the object of feeling or emotion would be excluded. To think that we could survive based on whether we *felt* food had value as opposed to whether it truly has value is comic, as if advertising for junk food has equal validity with nutrition. It would also be extremely strange to exclude all else for the value of emotions, which are often obstacles to value, for example, in some types of neurosis.

The *pros hen* of all value to such subjective states ignores the conditions under which such emotions are possible. Emotions, however private, are an event in the world. A feeling of depression or elation is in the world, even if such emotions are not part of the public world. We can even act on our emotions and make such feelings a cause in the world. Emotive theories also ignore the extent to which emotions are representations about objects. While emotions can arise spontaneously and without relation to an external context, often such feelings are context laden. They are emotional reactions to situations, other people, events, and so on. They would not have arisen except for such events. Love of another individual does not arise if such an individual never had come into one's life. The emotion may even be a representation in feelings of the other person. Again, is it plausible that most humans would feel cheerful when first hearing the news of the discovery of the Nazi Concentration Camps? Or seeing photos of the dead bodies? Or of the genocidal wars against Native Americans? A few racists might have a positive reaction, but I am sure that most people would be sickened and depressed at this evidence of human evil. But even the reaction of racists is evidence against the theory, for they have reacted to an event in the world, a relation; and this sort of reaction is morally depraved, constituting a good argument against emotivism as a basis for moral value.

Emotions are often generated as part of a larger context, for example, the beauty of the Grand Canyon is not a mere emotion, since the Grand Canyon must have inherent features which generate the response. They are differentiated from other kinds of feelings, that is, they are specific to the Grand Canyon. Since the emotions that are generated by the Grand Canyon require an actual Grand Canyon, they cannot be entirely subjective. Could one value the Grand Canyon without an actual Grand Canyon? Could one value its unique features and beauty without such features and beauty? The judgment of beauty is specific to the unique features of the Grand Canyon, not our feeling about the Grand Canyon. The occasion and basis of the judgment is the seeing of the Grand Canyon. If beauty were a subjective feeling alone, it could be expressed anytime, about the feeling of beauty itself. It would not require an occasion of visiting the Grand Canyon, or judging its unique features.

A judgment about the Grand Canyon is a judgment about the Grand Canyon, not our feeling about the Grand Canyon. Even the evidence of language is an argument against the emotive theory, since such judgments do not refer to feelings but the object, to wit: "the Grand Canyon is beautiful," *not* "my feelings about the Grand Canyon are beautiful." Emotivism distorts the actual language of value. In actuality, the language of feelings and emotions is distinct from that of values, and value language does not generally report feelings. To say "I love the view at the Grand Canyon" is not equivalent to, "The view at the Grand Canyon is beautiful." One reports or describes an internal feeling state. The other says nothing about emotions but reports or describes the beauty of the scene. The aesthetic statement says nothing about emotive states whatsoever. To "analyze" it as saying what it clearly does not say is a distortion based upon an erroneous theory of language and value. The theory in its haste to reduce value to feeling ignores the conceptual distinction between anger, hatred, and love—emotions; and good, right, and virtue.

As Hannah Arendt noted (1958, 113), affective theories in the subjective mode constitute a totally private ethic. As such these theories would devalue the world, since value is attached solely to valued interior states, to personal emotions without regard for other persons, life or the world. Consequences of our actions for the world would be morally neutral. Thus destruction of the world would be morally neutral. This would result in destruction of ourselves as well as others, but this consequence would be morally neutral, since only our emotions have value. But if we are destroyed, our emotions would cease—would this effect also be morally neutral? Emotivism cannot tell us, since it does not argue for the continuation of our emotional states as persons as a good, but only the emotions themselves. Emotivism is so restrictive it cannot even give us the moral tools to preserve ourselves, much less the world in which we live.

In sum, if extremely subjective forms of the affective group of theories, which includes the emotive, the attitudinal theory, and hedonism, were true, then it would follow that goods of the world would be irrelevant. If we could have good feelings, pleasures or attitudes without ever experiencing art, beauty, health, prosperity, and other external goods, then these kinds of value would be without value. Only my feelings and pleasures would have value. But if there is any value in external goods, then extreme subjectivism is fallacious. Creative actualization of new goods has expanded the goods of the world and these have been the cause of new feelings. More importantly, they are elements in a good life as a whole.

Actual emotions involve objects of emotion or feeling, an emotion about or toward an event, person or thing. Feelings that are aroused by a event such as visiting the Grand Canyon or meeting a lover cannot be separated in a subjective world all their own. They are contingent upon such objects, and thus not intrinsically valuable. For their value is not autonomous from that upon which they are contingent, but extrinsic, in a relation to this other. If value lies in the relation, however, it cannot be purely subjective, but involves at least the relation to the other. But the relational view requires the inherent value of the object or it reverts to the subjective view. The relational view requires some property of the object, person or event that will normally excite the desired feelings. It also requires a world with relations among its "furniture," that is, a world with more than just individuals.

If an object of value is required to arouse a positive or negative feeling, then the object is necessary to arouse the feeling. The object is then the origin of value, not the feeling (Grand Canyon). Feeling and emotion are the result of the object of value, not the ground. As with socially instilled feelings, such feelings are aroused by such experiences as viewing the Grand Canyon and would not otherwise have arisen. Feelings are again a consequence, not a ground.

5. Hedonism

One common form of the affective theory is hedonism, which unites ancient Epicureanism with modern Utilitarianism. The Utilitarians address the question of the implicit egoism of Epicureanism with the principle of the greatest happiness of the greatest number, that is, it brings in social considerations in the form of the happiness of others. However, they share a hedonistic theory of value in which pleasure is the highest, if not the sole good, and pleasure is equated with happiness. Hedonism is a prominent theory, especially in the twentieth century.

The separation by some writers of affective from hedonistic theories implies a critique of hedonism as a defective theory.[18] It implies that there is a category mistake, both in the confusion of all feelings, for example, aesthetic feelings, with pleasures; and also in the confusion of goods in general with pleasures. Pleasure is one part of life, not all of life; one value, not the highest value or the only intrinsic value. Instead of repeating all of the criticisms of subjective theories of value in general and feeling as a type, I will only note here that all the criticisms that I made of the subjective and affective theories in general apply to hedonism in particular, as a species of the affective theory, since happiness is accounted a (positive) feeling by its advocates and opponents alike, that is, a feeling of pleasure. Hedonism involves a feeling of pleasure and thus reduces all value to random episodes in a *pros hen* relation to random episodes in the subject who feels the pleasure, insofar as they feel it. The theory confuses value and evaluation, denies the validity of non-hedonistic value judgments, and thus reduces all value to hierarchies based on solipsistic pleasures. However, some problems peculiar to the hedonistic theory demand separate treatment. These peculiarities will be the topic of this section.

Like other theories that attempt to reduce all of value to one category of value, hedonistic theories are not so much value theories as *summum bonum* theories. To say that pleasure or happiness is the highest good is different than that all value is pleasure. Even hedonists admit that not all value is pleasure: going to the dentist for a tooth extraction is far from pleasurable, even with modern anesthetics. Hedonists thus argue for the "pleasure" of a dental visit either by arguing that going is less painful than the alternative or that we are "ultimately" happier by going, since once we have recovered from the visit, we reattain a happy state. Pleasure as happiness is the test of value: we act for happiness as an end, not for immediate pleasure. Happiness is the overall or ultimate end, the highest good for the sake of which other goods are pursued.

The question arises of why pleasure is to be adjudged the highest good, the value that justifies the rest?[19] Is pleasure equivalent to happiness or a part of it as in Aristotle's account? Is happiness a mere feeling or more? Consider the child who on a holiday like Easter is deluged with all the candy they could want. They eat and eat and finally no longer wish to eat more candy. Yet the candy still *tastes* good, still brings pleasure. But if they continue eating it, they will not be happy. At the point of satiation, then, more pleasure does not bring happiness and cannot be considered equivalent to happiness. Yet the consistent hedonist would have to argue that eating beyond the point of satiation will still be equivalent to happiness.

To reduce all goods to pleasure would seem drastic, but this is not Mill's ultimate position. He also includes the "pleasures" of intellectual, social, esthetic, moral, and spiritual life. The question can be raised as to whether these are engaged in primarily for pleasure or other ends. That pleasure accompanies such activities does not prove the hedonist thesis. Firstly, we might engage in such activities regardless of how much or little pleasure they bring. The activities might be evaluated worthwhile for their own sake, especially the most sublime human activities, intellectual, aesthetic, moral, and spiritual. For the hedonist cannot distinguish such activities by their pleasure: pleasure as a generic justification is identical for all pleasures. Why then should anyone engage in difficult and prolonged scholarship when they might receive just as much pleasure from gluttony? Pleasure alone cannot differentiate the value of distinctive kinds. For Bentham and other quantitative hedonists, this is not a difficulty: if gluttony brings as much overall pleasure it has as much value. But then quality of pleasures is sacrificed: a life fit for a pig is as good as the life led by DaVinci.

However, the expansion of the idea of pleasure, which primarily connotes sensuous pleasures even for Bentham, to include these others is arguable: what do they have in common that defines them as pleasures? Is the "happiness" which attends solving a problem in mathematics equivalent to or generically alike with the pleasure of eating? Or that of eating with sex? Mill himself argues for a rational component in his inclusion of "higher" pleasures in happiness. The hierarchy of pleasures is based on their superiority as pleasures, but obviously some people might rate different pleasures than those of high culture as superior, a position consistently held by Bentham. If, then, the criterion is not superiority of pleasure, but superiority of some other feeling than pleasure, whether (species) nature, reason, and so on, a principle other than pleasure alone might determine ultimate value, which would invalidate the theory.

Even if the intrinsic value of pleasurable feelings is granted, the reduction of all value to pleasure is dubious. First, a hedonistic theory of value would be hard-pressed to avoid relativism. What gives one person pleasure may leave another cold. One may prefer beer and another wine; one a life of ease and another of accomplishment. Which is to serve as the standard of the highest good? Also, what of the problem of knowing which subjectively felt pleasures are highest? If I like both eating pie and looking at the sunset, which is higher?

A division of pleasures into higher and lower does not escape this relativism. Another problem with Mill's argument for the pleasures of the more educated as higher and serving as a criterion of the highest good is that differences arise even among the educated. Some think that Richard Wagner's music is sublime; others that it should be scene and not heard. (Mark Twain remarked that, "Wagner's music is better than it sounds.") Which educated judgment is to be taken as definitive? What of differences between scientists and artists: Snow's "two cultures?" Are we to take the pleasures of art as the standard or the sublime happiness of doing mathematics? What pleasure is to serve as a standard of the highest good?

Because hedonism is a subjective or personal feeling, hedonism may also be a confusion of personal values with categories of value. The argument that pleasure is tied to taste, and that no one can determine another's personal tastes argues against a trans-personal notion of pleasure. I like chocolate, you like vanilla and that is the end of it. The value of pleasures of taste is completely subjective, and thus not a specific kind. Even if pleasure is a distinct kind of value, a distinct category of good, this cannot justify the arguments which would make of it a *summum bonum*. If pleasure is a distinct kind, then it can be differentiated from other kinds and thereby is not generic for all the categories of good. Pleasure may be highly valued by hedonists, but this is not a criterion valid for all categories of value. Hedonists do not examine value in general, but only one category of value. A hierarchy derives from evaluation, not categories of value which, as distinct categories, cannot be reduced to one another. The reduction also confuses value and evaluation.

Further, the value of a plane is not measured by how much pleasure it gives us or indeed whether it gives pleasure at all. Pleasure is a consequent to the inherent value of the plane for flying. Indeed, this pleasure may come from having completed the flight safely. No matter how much pleasure we get from the feeling aroused by a plane which will not fly, we would not attempt to travel in it: it's no good, it won't fly. The theory is defective as a theory of value, then, since it cannot tell us what is valuable about things other than pleasure. The *pros hen* to pleasure conflates pleasure with value, but many kinds of value cannot be explained in terms of pleasure. A tool is valuable whether or not we take pleasure in it: the pleasure we get from a tool is simply irrelevant with regard to how well the tool functions, its "utility."

The theory is reductive, but fallaciously. The value of a tool must be inherent or it has no value as a tool, but, according to hedonism, only as an instrument of pleasure. Like other theories with a *pros hen* to the subject, hedonism is the denial of inherent value, that is, the value which is inherent in "objects." Yet Mill argues that virtue, a means, may become part of the end, that is, have intrinsic value, once it becomes bound up with pleasure; and that feelings of justice are essential to Utility.

The theory also presumes creative actualization, that we can increase or enhance pleasures. Pleasure also requires creative actualization: it is not self-actualizing. Otherwise, pleasure is not a good at all, moral or otherwise, but a biological function. If pleasure is a biological function we cannot control it. Just as we cannot stop our kidneys or bladder from functioning, so can we not prevent feelings of pleasure which accompany specific biological activities, notably eating and sex (Aristotle, *N.E.* VIII and IX). However, if we can deliberately and purposively increase or enhance pleasures, we require creative actualization of them to do so. We cannot feel pleasure if the fine meal is not prepared. But then pleasure is one kind of creative actualization. This may undermine the theory from another angle, however, since pleasure is passively experienced. We experience pleasure from a sunset, a fine meal or from sex if we engage in these activities whether or not we wish it. But if we require creative actualization to achieve pleasure, value cannot be passive, but must be creatively actualized. Pleasure must accompany creative actualization, then, or it cannot serve as a theory of value: pleasure would be a biological function, passively experienced. But as a mere accompaniment to creative actualization, it is invalid as a general theory of value.

The fallaciousness of the reductionism of the theory is again evident if pleasure is considered in relation to life. If "conscious beings" have "intrinsic value" because they can "experience pleasure and pain," a modern sentience view (Norton, 1987) does value lie in the pleasure and pain or in the person, that is, the life? The feeling of either pleasure or of happiness cannot be separated from the life in which such feelings arise. Life is the context in which pleasure takes place and happiness would be inconceivable without a life. Happiness occurs in the life of a person as a whole, during stages of their life and inseparable from a life. Mill argues for a collective hedonism: a happy life overall in which the sum of pleasures outweighs the sum of pains. But then pleasure cannot be detached from life. The theory is a theory that the happy life is the highest good, not pleasure as such. Happiness is in a life overall and life is the condition of happiness. Otherwise the theory commits the fallacy of composition. The issue here is whether we should pursue something which may make us temporarily "happy," like drugs or cheap thrills, but is bad for us as a whole precisely because it will ruin our health or shorten our life. This includes activities like taking dangerous drugs or playing "chicken" with cars. If playing "chicken" gives us a temporary thrill but ends our life, is it truly good? Is not value for life a decisive test of happiness, not vice versa? Is a short life with a few dangerous pleasures better than a longer one with more accomplishments and fewer dangerous and dubious pleasures? Life has intrinsic value—pleasure only has inherent value since it requires life and is the pleasure in and of a live individual.

Hedonism, like other subjective theories, commits the fallacy of composition. Does pleasure alone have intrinsic value or the life which includes such pleasures? But is a person of less value when they are not feeling pleasure, for example, the victims of illness or of the Nazis? It might be argued that such person is living for the day when the pain will have ceased: living on hope. But were their lives of less value just because they were not happy during that period of their lives? The accounts of the lives of the working classes and the slaves from the nineteenth century suggest otherwise. Working twelve hours a day, six to six and a half days a week for subsistence wages is hardly a happy life. But these persons persisted and found reasons to live other than immediate pleasures.

Another approach to the relation of pleasure to life is to ask, would a person who was bedridden but had their brain hooked up to electrodes which caused pleasurable stimuli have a happy life? Or the same person in a comatose state? Awareness and activity are also elements of a good life. Health is important even if we do not always find it pleasant to exercise or take other measures to retain or improve health. People pursue many other elements in a worthwhile life other than mere pleasure. Again, pleasure is in the life of a person in the world, but the value of these is reduced to pleasure according to hedonism. But as conditions of pleasure, the life of a person in the world are necessarily bound up with pleasure, and thus cannot be means to pleasure.

As in affective theories in general, the theory, if consistent, argues that pleasures without objects are as good as pleasures with objects. Even feelings of pleasure that arise and are pleasures taken in nothing are more valuable than indifference to something in the world. If I have pleasure without ever experiencing art, the beauty of landscapes, intellectual achievements, the development of talents or other goods then these others would be without value. Only my pleasures have value. Pleasure taken in being a "couch potato" would be equally valuable with pleasure taken in the Grand Canyon and all other external goods of the world for such a purely subjective theory. But if art, the life of the mind, natural landscapes, and other goods do have a value of their own, then hedonism must be erroneous.

Hedonism is not completely wrong, since pleasure is indeed a part of life. What is implausible is the attempt to account for all human activity in terms of pleasure, to make a *summum bonum* out of what is only a part of life. Hedonism is only partially true as a theory of value. Pleasure is fine and sometimes in life we pursue pleasure: fine dining as opposed to mere nutrition. Here we enhance a necessity of life with more pleasurable fare. However, other times we don't pursue pleasure, whether from the duties imposed by employment, the necessities of health, hard physical labor, and so on. Pleasure is generally not the motive of most of our activities; we pursue other ends in life and pleasure is only one of these. Indeed, as the example of dining shows, we sometimes deliberately pursue pleasure as a distinct activity apart from other activities of life, just as we pursue other kinds of goods apart from pleasure. But the argument in Mill is that pleasure is the "only" end.

Hedonists might argue that pleasure "ultimately" justifies pain, and thus justify pain as by a *pros hen*, an ultimate reference back to pleasure as the source of justification. But the activity itself is not engaged in for pleasure. Pleasure is not the immediate motive. More, the two senses of pleasure, combined with happiness are vague or equivocal. The cessation of pain at the end of a day of hard labor, or at the termination of a visit to the dentist is not equivalent to the pleasure of eating a fine meal. The pleasure of eating a chocolate bar is not equivalent to the happiness that may accompany some grander achievement such as developing a musical talent or becoming a good scholar. Which of these is to serve as a standard of the highest good? This "ultimate" standard taken consistently would justify masochism: pleasure taken in self-mutilation would, as the highest good, overrule health, physical well-being, and other goods. Any appeal to a higher good, an overall well-being in evaluation of pleasures in a life would also be illicit. Pleasure and pain are the locus of highest value, if the theory is consistent, not the rest of human consciousness, life, or society. If drugs give pleasure, the life of an alcoholic crackhead is as good as Wolfgang Amadeus Mozart's. Pleasure states are more valuable than health, overall well-being, and the effect of pleasurable actions on society. A delusional pleasure or an artificial one from drugs would, if the theory is consistent, be more valuable than a non-delusional indifference or a state free of drugs which included mixed pleasures and pains. But in the case of delusion, we have not creatively actualized pleasure.

Hedonism also devalues the world. As a subjective theory, hedonism strictly speaking entails that all value lies in the random feelings of the human subject. The world only has value as an instrument for enhancing such states. Thus the world has no inherent value. Indeed, if destroying or defacing natural beauty were to give more pleasure than retaining it, then it would be justified. This is the actual reasoning that was used to fill in scenic canyons along the Colorado River in order to dam the river for water. If consistent, hedonism is seriously defective as a theory of what is good or valuable.

Mill's quantitative criterion of quality—that if the majority who experience it mark the pleasure as more desirable we may judge it superior—contradicts his qualitative argument that quality of pleasure is distinct from quantity. Also, as both Moore (1903, 65) and Dewey (1939) have pointed out, some further reasoning is needed to derive the desirable from the desired. Just because the achievement of some value is possible does not indicate whether it should be achieved or actualized, whether we evaluate the desire as worth achieving. That some end is desired tells us nothing about the worth of achieving it. The issue is not if pleasure is ever worthwhile, but which pleasure is worthwhile?—which is of the greatest quality? The gist of both Epicurus' critique of the Cyrenaics and J.S. Mill's critique of Bentham is a qualitative critique of quantity. Quality of pleasures may be at odds with quantity, just as quality is itself problematic. Diversity of pleasures is problematic and if the desired is made the standard, then quality suffers, as Mill was aware. If quality is made the standard, the rationale for pleasure is undermined, since most people do not pursue refined pleasures and quality may require extra-hedonistic criteria and standards. Human behavior cannot provide standards for the worth of pleasures, since humans are anarchic in their pursuit of pleasures.

Mill has been criticized on the grounds of both his theory of value and the connection between this theory and obligation (Copleston 1946, and Frankena, 1963). The theory of value itself is morally deficient. Unqualified, hedonism as such could be used to justify sadism and cruelty. If it makes someone happy to torture animals or even other humans, then the torture is good, if the theory is consistent. If it gives us pleasure to kill people we don't like, then "ultimately" we are justified. Hedonistic theories reduce morals to values: an action is right if it has pleasurable consequences. This theory of value as it stands cannot rule out immoral acts and could even be used to justify them. Happiness of specific types may indeed be immoral: the happiness of a sadistic torturer. If the theory of value were to be qualified to rule out such immoral actions, it would no longer be a hedonistic theory. Pleasure would be an element of value but not the *summum bonum* of value. I will return to this point below. However, a hedonist would argue, as Mill does, that the relation of values and morals is problematic for all theories of value; hedonism is not peculiar in this regard. What he overlooked in this judgment is that the Stoic-Kantian moral theory of values does not suffer from this problem, since according to the moral theory, what is immoral cannot be good. Moreover, the theory of value as creative actualization is reciprocal with moral values, since we must creatively actualize them.

Can pleasure or happiness encompass all goods? Can they justify all acts as their ultimate warrant? Pleasure and happiness are legitimate ends for humans; we pursue this end as a distinctive value. Since we actively pursue pleasure it is in one respect not merely subjective, although the feeling of pleasure is private, even if we can determine public criteria of the activities surrounding some pleasures. Yet pleasure is still a cultural and social category, a distinct good, and therefore not "the" good. Some goods do not make us happy or give us pleasure, and others are good regardless of our individual relationship to them. Water remains valuable when our thirst is quenched; water in other parts of the world has value even if we never drink it. Some pleasures are heavily regulated by society (sex) and some means to pleasures are socially shared, for example, meals, the enjoyment of beauty, and others. While the enjoyment is individual, this may occur in a social context, for example, in art museums. The pursuit of such pleasures and individual happiness is one goal of society, one category of good. Thus it cannot be the whole, the most general value. The exclusion of rape is the acknowledgment that duties to others may have a higher priority than pleasure.

Epicurus has argued that pleasure is a "natural" good (Diog. Laert., *Lives*, X, "Letter to Menoeceus"). But many of the pleasures pursued by humans are not based on biology, if that is what Epicurus meant in this context. An interest in music, or taking pleasure in music, in religion, even in wealth cannot be based on biology as none of these kinds of goods arose until after human biology was well established. Nor can we predict using biology what is outside the biological realm, as if genes could be teleological (socio-biology). It is unlikely that there is a gene for mystic bliss. As for the environment, these goods may indeed be pursued as a result of the social environment, but hardly as a result of the natural environment. Churches, orchestras and stock tickers do not occur in the wilderness. The happiness that such goods give to individuals cannot be predicted of any one individual given a specific biology. Which child of all those who are schooled will end up with intellectual interests cannot be known *a priori*. Such goods are social categories. The ability to feel particular bodily pleasures may well be biological, but this cannot account for others.

The criticism that desire cannot ground what is desirable extends to the relation of value and obligation as well: derivation of the greatest happiness principle from the desire for or feeling of happiness. The ultimate question of such a consequentialist ethic, based on a nominalist locus of value, is how social obligation can be derived from the theory of value. True, Mill defines obligation in terms of happiness, so happiness is their entire rationale. But the inference that we are obliged to any duties based on happiness is more difficult to draw: how does a feeling create or warrant an obligation? Are we obliged if we do not feel happy? Or if such obligations end our happiness? No obligation can be drawn from pleasure as such, not even the obligation to pursue pleasure or happiness. For this would constitute an odd theory of obligation. Hedonism as such, Epicureanism, would involve an *obligation* to pleasure! To increasing my pleasure! In the Utilitarian form it would mean to increase the pleasure of the greatest number! From the observation that people do pursue pleasure it does not follow that they should; from individual pursuit of happiness no universal norm can be logically inferred, especially if happiness is individually different, along the lines of matters of taste. Whether obligation to others can be derived from, in the sense of justified by, individual feelings is at least problematic. As happiness is a feeling within an individual, it does not of itself entail such an obligation, and may work against it.

Why is the obligation required, or necessary: why does it follow logically? Isn't an additional minor needed to derive a social obligation from an individual, internal feeling? Mill argues at length both for instilling social feelings and for some basis for social feeling in human nature, but a further premise is required to show that these feelings will make one happy or are based on happiness. If each person desires "his own happiness," and this is individual and thereby variegated, how can social feelings be instilled? Social habits are instilled by education and rearing, thus they are not traceable to the natural desire of the individual for happiness. As the good of others is not subjectively based, almost by definition, the sanctions of society must be based on the interests of society, not those of individuals. Mill might argue that the greatest happiness principle argues for a more social criterion, but as based in individuals, it cannot be social as such. Happiness is individually located. Further, even if a social sentiment could be introduced, for example, the happiness of parent in relation to child, Mill has not distinguished intimate and casual association, and the feelings generated by intimates (family and tribal loyalties) might not extend to casual acquaintances.

While obligation cannot be derived from the pleasure or happiness of individuals, the converse does not follow. Pleasures are compatible with the ethics of duty and the ethics of perfection of character, so long as the pleasures are within moral limits or in accord with a good character. Indeed, such pleasures may accompany the pursuit of duty or the practice of virtue, as Kant and Aristotle, respectively, argued. This argues both that pleasure is only a part of good but insufficient for an ethic. Right and wrong, moral values cannot be based on pleasure and happiness, but can encompass happiness.

Pleasure may deter us from excellence, but excellence may include happiness. Thus the ethics of duty and of perfection of character are morally superior, for they include strictly moral values, rule out immoral pleasures, and encompass more values.

Another problem with Mill's form of the theory is that the "greatest happiness of the greatest number" is majoritarian; the "greatest number" is a majority of a society. Thus it cannot be used to uphold minority rights and may even supercede them. If the Nazis had made most people in Germany or Europe happy by the Holocaust, would this make it right? The majority in the Confederate States of the United States were so unhappy at the prospect of the end of slavery that they were willing to die to prevent it. The examples indicate that making the greatest number happy, the happiness of the majority, may involve gross acts of immorality and even justify them. The use of animals in research for humans is another example. Hedonism, even qualified by the more altruistic treatment of Utilitarianism, can be used to justify the violation of minority and individual rights. Indeed, it could be argued that hedonism and Utility are immoral, since they can justify immoral acts.[20]

Utilitarianism is ultimately anti-equalitarian, despite the qualification Mill makes that each person's happiness is to count as much as any other person's. Mill also notes that the Utility principle is supreme in cases of conflict. What this implies is that the happiness of the majority overrides that of minorities and individuals. The happiness of some is worth more than that of others: the happiness of the "greatest number." If the majority is made happy by actions or policies that make the minority unhappy, their happiness is of greater weight.

Another consequence consistent with hedonism would be the use of humans for entertainment: the spectacles of ancient Rome. Feeding humans to the lions entertained the masses: the "greatest good of the greatest number." More, the suffering of innocents could also be justified. The good of society has been used as a rationale by death penalty advocates, who argue that even if innocent persons are occasionally executed, that society still benefits from the practice (Ernest van den Haag, in Cahn and Markie, 1998). If this principle were universalized, randomly killing innocent persons would be justified if the majority were happy about it. Obviously, such injustices would undermine the entire criminal justice system, which in theory attempts to separate the guilty from the innocent.

Other bad consequences follow from the hedonistic theory of values. The theory is allied to a deterministic view of human nature in which humans are "blank slates" that must be filled. Advertisers have embraced this model and view humans as consumers of economic goods, which are justified in terms of pleasure and happiness. If some "good" makes consumers happy, their happiness justifies it. The destiny of humans is as passive consumers of whatever trivial product corporations with the capital to advertise can sell them. This destiny is quite different from that implied by the pursuit of moral excellence of character or as practicing duty for its own sake. A society of passive hedonistic consumers is inferior to a society of character.

We may wonder whether hedonism can be the basis of a society at all, since a consistent hedonism would result in a "society" of individuals pursuing their self-interests. This "society" would be more accurately described as a collection of individuals than a society with common goals. Mill argues that only a "moral blank" could live their whole life on self-interest alone, but this ignores economic theory, including economic forms of utility, which envisions just such a life a "rational." Mill concedes these points and argues both that the problem of ignoring conscience is not confined to utilitarian ethics and that "control by reason" may be required (Mill, 1863/1973, 446). In this case, however, happiness is not a sufficient basis for ethics and rationality is again required, as in his qualitative value theory.

Apologists for Utilitarianism argue that the spirit of the theory is benevolent: how can wishing that people be happy be bad? Hedonism is wishing that people can be happy. The issue in a moral context is not this benevolent wish but whether this can ever justify immoral consequences. Since obtaining my own happiness at the expense of another, in the form of doing harm—torture, rape, and so on—can never be morally right, pleasure can never be a general principle or norm of value. Happiness as a value can be seen as a part of value, not the whole, and not the *summum bonum* or justification of all values. It does not fail as one value but fails as the only intrinsic value.

6. Attitudes

What may be still another affective variant is the attitudinal theory of Franz Brentano and Noah Lemos: the theory of values as reflecting proper attitudes or "correct" attitudes. These varieties are inadequate on the same grounds as the other affective theories. An attitude is not precisely equivalent to a feeling, but the theory is subjective and relational. Thus it has all the problems of subjective or psychological theories. These include: the value of attitudes themselves (the critical question), the devaluation of world in terms of a *pros hen* to a small part of the world (reduction) and the resulting worldlessness of such theories, the location of the relation of attitudes and that which the attitudes intend. Further problems are the confusion of such attitudes with other values, that is, the category mistake, the confusion of evaluation and value, the language of attitudes and that of value, of attitudes and action, the egoism of such theories, and so on. It is difficult to see what value the notion of "attitudes" adds to value theory even if it could be determined what the value of attitudes themselves are or the legitimacy of this norm. If steel is superior to paper for construction this has to do with the inherent character of steel, not attitudes, "correct" or not. Steel does not lose value when our whimsical attitudes change. Again, beauty has little to do with attitudes, and may change them, for example, the spectacle of the Grand Canyon may bring a change in the previous attitude that the long trip to see the sight is not worthwhile. Thus beauty may be the ground of attitudes, instead of an effect or reflection of them. But the Canyon is beautiful in spite of negative attitudes and independently of them.

Just having attitudes may be bad or evil; attitudes alone cannot make a bad action a good one or an immoral one a moral one. Moral principles may be the contrary of attitudes, for example, racist attitudes. We may be obligated to change our attitudes in view of moral principles, rules, and values. Changing such attitudes is often the purpose of moral exhortation: "correct attitudes" are instilled as a result, not a basis, and require prior moral evaluation. The social consequences of actions are more of an issue for morals regardless of personal attitudes. We may have duties to others irrespective of whether we have attitudes at all, even "correct" ones. Morals are in our power whereas attitudes may not be, especially having a "correct" one. Who is to decide what attitudes, as opposed to actions, are correct? Actions are public, but attitudes are not. Above all, even if attitudes have value, then their creative actualization as correct has value. The creative actualization of new art works is also required in order to have correct attitudes to them. At best attitudes, correct or otherwise, are a species of value, and at worst a devaluation of values.

The same arguments apply a fortiori to taste taken as equivalent or determining of value. The fact of different tastes in clothing styles, food, and so on argues either against the relevance of taste to value or of value to taste. A taste for high quality clothes cannot of itself establish the quality of the clothes. What we find to our taste may not even be of personal value, although it can influence personal values. Nor can it tell us what is the value of a value or whether we should value a value. The value of the taste is never at the level of universality and whether a taste is in good taste requires the distinction of oughts/norms from personal taste. Basic distinctions in value: duty, categorial value, personal value and taste.

Proponents of this theory would no doubt argue that the epithet or qualification "correct" is needed to complete the theory. Attitudes alone are insufficient; only correct attitudes have value. This qualification defeats the theory however. Attitudes must be corrected. Thus a standard of value or of right other than and beyond attitudes themselves is required to give attitudes value. "Correctness" cannot come from attitudes themselves since they need correcting. Attitudes are then superfluous for determining value, since any psychological state, actuality or action can be corrected according to the non-attitudinal standard.

7. Agapism

Something must be said about agapism, the ethics of Christian love, presuming that this type can be classified with affective theories. This point is perhaps arguable, since love of our fellow humans may not be equivalent to an emotion. Indeed, the word is based on the Greek term *agape*, which is distinct from other kinds of love *eros* and *philo-*.[21] Augustine defines love as "I want you to be (exist)." In this form agapism could be a form of desire. Agape may be more than a mere emotion, an all encompassing One, as mystics describe it. Love comes in different kinds, despite the single term in English. Romance, love of family, of nation, and of one's fellow human are distinct kinds, to speak nothing of love of nature or of wisdom. Love in English is generally considered as an emotion or feeling for others, however, and I will treat it in this form.

If this interpretation is valid, the arguments against the other forms of affective theories of value apply to agapism. Christianity ultimately rests upon an emotive or an affective theory of values with a derivative theory of ethics. Universal love is the basis of what is right, the "greatest commandment." Love is the greatest measure of value in this view, the ultimate criteria of good, and the *summum bonum*. Each soul has absolute value, because the soul is the prerequisite for all other goods, whether experience of them, knowledge of them, interaction with them or excellent development (virtue): perfection of the soul. The existence of the soul is a necessary condition of experience. Love is the recognition of the existence of the other: the criterion of good. We have a duty to love that which is of absolute value: to love the good.

Henry David Thoreau has argued that the love that lies at the basis of specific altruistic arguments can have negative consequences. In the name of love, people commit acts "for your own good" which may not be good at all. He warned us to keep away from such well-intentioned people, who impose personal preferences in the name of "our own good." In this view, love is not an unconditional good.

Christians might argue that we should not hate others and should love others. I do not know if this is possible, since I am not sure that we can manipulate our actual feelings that easily. The Freudians believe that it can be harmful to suppress strong emotions, so this may not be a worthwhile option. This theory seems to me to be contingent upon whether we have such feelings so the question that arises is whether we are bound if we do not. As a commandment it may constitute an obligation to love, however, and therefore be removed from the contingent and relative. The question then arises if we can be bound to change our feelings, that is, do we have a duty not to feel what we do feel (hatred, indifference) and to feel what we do not (love). Even if we try hard to change our feelings about someone, we may not be able to change them. Suppressing them is not changing them and may have adverse psychological consequences. Trying to make such changes is virtuous, but cannot be a duty. While duties can be consistent with such feelings, they cannot be commanded by them if the feelings are not there and we cannot force them to be there. I think the only thing that we can demand morally is that people not act immorally based on feelings, for example, murder out of anger or revenge. But this constitutes an argument against affective theories as the source of value.

Nietzsche attempted to classify all ethics as slave and master and to include Christianity in the slavish type in the *Genealogy of Morals*. However, his own ethical views have far more in common with Christianity than he realized, since he adapted the same hierarchy to categorize values. Hierarchical value theories, including his own, do not ultimately recognize inherent value, or the intrinsic value of individuals. Christianity represents hierarchy (praying to the "Lord," the patriarchal "Father").[22] Its value theory does not actually recognize intrinsic value, but hierarchical value, value as a reflection of supreme value. Value is ultimately contingent upon divine grace or the will of God.[23] Value emanates from the perfection of God to his creation. This model has priority despite subtle dissembling: the "infinite" worth of the individual. In any case, the worth of individuals is contingent upon divine grace. Individuals are of worth because they are creations of God, emanations of God in the divine image. Love is good because God commands it. Neither the individual nor love can have intrinsic worth, then, since these derive their value from divine command. The "infinite" worth of the individual is contradicted by the hypothesis of hell (cf. predestination). If God's love were unconditional, consistent with the absolute value of his creatures, he would not punish them, for he would love them in spite of all. If the individual soul had absolute value it could not be placed in hell. Nor would some be predestined to hell. Christianity in this sense is contradictory: if hell, no absolute value; if absolute value, no hell. Moreover, individuals do not have infinite value, as mortal. Their life is for a finite period.

Søren Kierkegaard's defense of the religious life, which is counterpoised to the ethical life, argues that Abraham is the "father of faith," since he was willing to sacrifice Isaac at the command of God. This makes the action right according to theological ethics: the top of the hierarchy has commanded it. Theological ethics argues that crimes are not immoral if allowed or commanded by God (Osama bin Laden). In effect the argument is that there is no morality, but only divine arbitrariness, what Kant rightly called the ethics of "divine whim." No one is blamable for anything at all since God wills it: if it happens, God allowed it. Moreover, we would not, unlike Abraham, kill our loved ones based on "divine command," so divine will cannot be the basis for right.

I am not going to comment any further upon agapism, since it has had many able defenders. By noting the contradiction between divine grace and absolute value, and also the impossibility commanding a feeling or emotions one just does not feel, we have critically evaluated the theory sufficiently.

Actions may be required despite feelings. We can expect someone not to murder, but we cannot expect them not to hate. It may well be better if they did not hate, but this is an evaluation, not a moral principle. Morals cannot be based on feelings; no more can feelings be based on morals. Morality in the narrow sense of duty has to do with action, not whatever feelings accompany or motivate actors. Nor is religion required for law or morals. Either something is right in itself, or no religious sanction can morally justify it. Morality must stand on its own feet.

I have covered most of the affective theories in this chapter. Affective theories are closest to a purely subjective theory of values. However, they frequently involve a relation to a bearer of instrumental value. The relational theories will be treated in the next chapter. It should be noted that some affective theories are relational and the treatment of them as subjective is not meant to imply that affective theories cannot be relational. Their inclusion here is as affective, and subjective, but many versions are not entirely subjective.

Seven

CRITICAL EVALUATION OF RELATIONAL VALUE THEORIES

The instrumental value of all objects in subjective theories is an argument against pure subjectivity of values, as the theories require the instrumental value of the object. Instrumental value ascribes some inherent value to objects that cause the feelings, desires, interests, emotions, attitudes, and so on, of a subject, so they presume the notion of inherent value. If values were not intrinsic to such subjective states or faculties, then value would not be entirely subjective. But then subjective theories cannot make reference to an extrinsic relation to an object of desire, will, interest, and so on, without bringing in relations. Yet almost all subjective theories shade into—if they do not collapse into—relational theories. To take hedonism as an example, the pleasures of sex and the table depend on external causes. They involve a relation to objects like food that cause valued subjective states. Some formulations of affective theories recognize the requirement for this relation; "approval of" or "feeling for" imply of or for an object of approval or feeling. However, hedonism is not directly relational, since pleasure is judged good in itself, although it requires objects with the function of bringing pleasure, such as tasty foods.

In this chapter I will examine the problems of any relational theory. As I argued in the previous chapter, the relational view requires objects to arouse such states. That is, the relational view requires inherent value in the "object": that some quality inhere in the "object," that is, the good that has been creatively actualized, or that the "object" as a whole has the inherent power to cause valorized subjective states (Clarence I. Lewis, 1946, P. III). Otherwise it reverts to the solipsistic, subjective view. A relational view requires, then, an inherently valuable property of the good or "object" that will normally excite subjectively valorized states. Since it requires the inherent value of the "object," this inherent value is necessary; as necessary for such subjective states, it is "objective" by Immanuel Kant's definition. Value is inherent in the "object" whether any subject takes pleasure in it, is feeling it at a particular moment, or has specific attitudes about it, pro or con. Inherent value cannot belong to an "object" if it depends upon a subject, for then it would not inhere in the object. If these consequences of such theories are correct, then real "objects," that is, goods, are required to complete the theory, that is, they are necessary to the relation, the condition of values in relational theories. If goods are required for subjective states, the latter cannot be autonomous from the relation; the subject is not intrinsically valuable apart from a relation.

But then subjective theories have the location problems of any relational theory. In this context the question can be raised about where such a relation is located: the "topography" of the relation. Relational theories have metaphysical problems in that the locus or bearer of the relation is problematic: it seems to be nowhere. Value is not in the subject or the "object," for it involves a relation of these. A relation is not an "entity" or actuality by itself; what then is this relation instantiated or "embodied" in? Is it in the subject? But then the relation is subjective and we are right back to narcissism. But if in the "object," the relation is independent of our feelings or desires about it. But the "object" only has value for a subject and for the value to be in the "object" would be a denial of a necessary relation to the subject. The locus of value, that bearer in which the relation is inherent, is problematic for any relational theory. Value is not in the subject or the "object," but is an abstract relation without concrete location, instantiation, or embodiment.

Moreover, if the relation were attached to a subject or "object" solely, then it would not be relational, but a quality. A subjectively based value relation cannot posit intrinsic value in both subjects and "objects": it must be in one or the other. However, if subjective states are *intrinsically* valuable, of their *own* nature, they cannot require a relation, for then they would not be valuable on their own. Further, the value of the "object" is a necessary condition of valuing in relational theories.

Relationists might demur that the relation is more holistic, in that it involves valorized subjective states, an object that excites these, and their immediate relation. But then the theories shade into holism, and the whole is required, undermining the subjective notion of intrinsic value.

Inherent and relational value may be logically incompatible. If a value is inherent or intrinsic, its value cannot derive from an extrinsic relation to something else, subjective or not. Relational theories of value are a species of a hierarchy theory of value, in which value derives from a subject. Value cannot be intrinsic or inherent in this view, as it requires an extrinsic relation. Inherent value means valuable in *itself*, not in a relation.

1. Conative Theories

There is a group of theories which combine desire, will, liking, and interest into one theory of value, the interest theory. Prall's relational theory, which will be examined below, provides an example. For the advocates of this theory, an object of desire or willing is also an object in which an interest is taken or which is liked. Since desire is a component of will, several forms of voluntarism may also be treated as similar to the interest theory in important respects. This theory may take a subjective form as in David Prall or a more behavioral form as in Ralph Barton Perry, although both of these authors have elements of each: the subject and behavior. In the behavioral form, value is an object of striving, so the theory has been also been called the conative theory.

However, each of these species of the theory could be a distinct theory of value, despite their connection by Prall and Perry. An object of desire is not necessarily an object that is liked and vice versa. With this qualification, I will treat the theories as alike enough to be treated as one, namely, that value is an object of desire, the modern formulation of the theory by Christian Von Ehrenfels and the Austrian "psychological" school of value.[1] I will refer to such theories as conative theories and desire will be taken as generic for the content of conative theories, including voluntarism. I will then briefly discuss each of the different components of the theory separately: will, liking, and so on.

Another qualification is that conative and demand theories can be objective or subjective as well as relational. Desire could be interpreted as a subjective state, a psychological condition of human or living agents. However, value as the *object* of desire, demand, will, interest could also be taken as either a relational theory or even an objective theory. In this form value consists in a relation between a subject who desires, wills or demands an object, involving a relation between them. Alternatively, value could reside in any object that fulfills desire or demand. If value resides in the object, the theory is objective, since value is inherent in some object of desire. In this chapter I will mainly discuss relational forms of conativism, since objective and subjective theories were treated in previous chapters. However, the ambiguity of this theory should be noted.

Like affective value theories, conative theories attempt to derive value from random, episodic, and involuntary psychological states. Unlike them it allows room for objectivity, since value may attach to the object of desire, instead of the mere satisfaction of desire. However, this makes the theory logically contradictory. We may desire something at one time and not desire it at another. The object of desire both has value and does not have value: a contradiction. Moreover, value is not dependent upon episodic subjective events, whether voluntary or involuntary. Objects of value generally retain their value after all desire for them ceases. A good tool may still function quite well after we cease to want it; even where we replace a hand tool with a power tool, the hand tool may still work quite well. Similarly, water retains the "property" of quenching thirst long after our thirst is slaked. It has no less value for that purpose because we no longer desire it. Desire per se, that is, isolated random internal events cannot reveal anything about the value of objects.

Any relational theory suffers from the problem of the location of value, but especially value as the object of desire. Desire is intentional: a desire for. Generally, desires are specific or determinate, that is, a desire for water, not for anything at all. Intentional states are not entirely subjective, since they necessarily involve reference to some object, source of satisfaction, or whatever will fulfill the desire—or better, anything that will cause the desire to cease. Desire is a lack. Thus it makes reference beyond the desire to the world. In one sense the desire or lack is within a person, but in another, desire refers to the world as the source of fulfillment. In this locational aspect, conative theories are self-refuting, since if value is located in desire, striving, interest, willing or similar sources, it can never be satisfied, for as soon as we achieve satisfaction, the source of value ceases. If it is located in the object of desire (willing, and so on.) then it cannot be located in desire itself.

Desire itself is equally problematic. "Desires" are revealed in actions and are intentional: a desire for something. Thus their status is problematic since they are experienced as intentional states in which not desire but a lack, need or objective is experienced, not desire as such. The experience is of a state denoting a lack, for example, thirst or an objective that will satisfy such a lack, for example, water. To say "we all have desires" is not enough if this is only revealed in action: the attempt to satisfy a specific lack. "I desire x" is shown because I pursue or obtain x or report a state such as thirst. The evidence is in action, and can never in principle be anything else, since we do not have access to other's psychological states. To say that desire, appetite, will and so on are abstract identities of all such kinds of states may be conceptually valid, but does not establish the "existence" of desire as such, nor the identity of mine and another's desire.

Does value consist in the desire itself (including will, liking, interest, and so on) or the object of desire? If value is in the object desired, then desire cannot provide value, but at best only the negative condition of value. If value lies in the desire, then desire alone would have intrinsic value if the theory were consistent. A desire for water would have more value than quenching thirst. The thirstier we got, the more value. The supreme level of value would be reached when we were parched and dying of thirst, for then we would have the greatest amount of desire. I trust that this suffices as a reductio ad absurdam of this theory. Wanting water would be as valuable as having water, if the intensity of desire were sufficient or greater. But to desire something is to lack the actual value. Thus such states are pragmatically dependent upon their objective, not the source of their value. Value is independent of such states as the object of desire, interest, and so on. If we desire to enhance our feelings of pleasure, we must creatively actualize such states.

I am referring here to conation as a subjective theory. In some formulations, for example, Hobbes, value is the object of desire. Such an objective or relational theory of value has desire as its condition, but not as its goal. The goal in an objective theory, however, is cessation of desire, which is also a subjective state. Similarly, "satisfaction" of desire is a subjective state. Thus conative theories are ultimately subjective in orientation, despite a concession to objectivity: value as the "object" of desire.

Desire as lacking something is as such not a source of value. Neither an object of desire nor that which satisfies desire are themselves desires. On the contrary, they are the cause of the cessation of desire. And this point shows how odd this theory is. What does not have value, a state that we try to overcome the more it persists, is supposed to be equivalent to value! Desires themselves do not have value: they lack value and may even constitute a bad state. How can bad be definitive of good? This is another contradiction. On the axiological level, desire as lack could not be valued as such: a lack or want that is unfulfilled. Could thirst that is never quenched be a valued condition? Instead, what alleviates, satisfies, or fulfills desire has value. In this respect, the conative theory can be reduced to the affective theory. Only what satisfies desire has value, that is some feeling of satisfaction. But satisfaction of desire may mean what fulfills desire not satisfaction in the sense of happiness or a positive feeling. Conativism also has the odd result that what we already have has no value. Desire is a lack; what we have or own we do not lack. Thus our property, our spouse or partner, our children, our lives lack value, for we do not lack for them if we already have or own them. But this would be absurd on any theory of values. Proponents might argue that we still want what we have and would want it even more if it were missing. But this in fact defeats the theory, for what we do not lack for can have value.

Would an unfulfilled desire have value? A craving as such? A strong desire which could never be fulfilled, for example, unrequited love? Or must not the objective of desire enter in—that which satisfies desire as having value? Would an unfulfilled thirst as such have value, a thirst that someone would have after three days in the desert without water—would this be valuable? Or instead the water? Consider "badwater" in the desert, that is, poisonous or excessively alkaline springs, with "good" water. This is hardly a subjective distinction, to speak nothing of its being based on whimsy.

Is desire valuable or only fulfillment or satisfaction of desire? Does any desire as such have value? Or only some? Or the value of the object in fulfilling my desire? And by bringing it to an end, actualizing a good that ends desire? Would a life of desire be a good life? What of a life of just desires, without fulfillment, where we only experienced desires and nothing else: would this be a good life? Obviously, a life consisting of unsatisfied desires would hardly be worth living. Fulfillment of goals, satisfactions or objects of desire, will, and so on are valued not desire as such. Desires themselves have little if any value: only its objects or satisfaction have value. Nor are desires *per se* values; we do not value desire as such or will as such, but on the contrary, often attempt to end desire as a disvalue. The conative group of psychological theories of value is not only erroneous as theories of value but as of loci of value as well. In this, conativism is even less plausible as a theory of value than affective theories, which are at least partly true. While we do value creative actualization of more happiness, among other goods, we do not value the creative actualization of more thirst, hunger or other desires. Further, such theories are immediately qualified by extrinsic factors, since not everything that anyone or everyone desires has value. Thus the theories refute themselves: not everything desired, willed or the like have value. To equate value with desire not only reduces the scope of value to whimsy, but begs the question of the value of the desires: of just what is the value of what is desired.

Desire for something cannot give it value; at best, the value is located in the "object," that is, the good that satisfies the desire, not the desire itself. However, this makes value as problematic as ever, since what gives such objects value or makes them of value cannot be answered in terms of desire as such, without a vicious circle. Thus so-called conative theories of value, which include value as an object of desire, subjective will, liking or interest are at best theories of action or behavior, not theories of value. The "object of striving" posits a ground, objective or goal for striving, but not the what or why of this objective. Value as such is neither defined not characterized for such theories: the "object of value" is an empty set or content: an "x." If value is the object of desire, there could be no value theory—no normative identity that explains values, but only isolated, private, internal events. Even if it were established that desire is the cause of action, the value of what is desired and even the objective of the action have not been established. Moreover, since desire is intermediate in satisfaction, it is a means and cannot tell us what has intrinsic value.

Nor can it be argued that desire is the condition of anything receiving value, the subjective basis, since the cessation of desire has value more than the desire. The value of the object of desire is in putting an end to the lack, to the desire. Indeed, what is desired is the cessation of desire, not the desire as such. A valued state is reached in the person when desire is satisfied in the sense of negated, not when desire is present. The quenching of thirst is better than the desire to quench the thirst. The desire has negative value that only its cessation can make positive: excessive thirst is a bad state not a good one. Value often lies precisely in the *negation* of desire.

What of the point that desire is a condition of value, even if desire is not of value as such? I would respond that desire is private and thus cannot establish value at all. On the contrary, desire, will, interest, demand, liking, and other conative content have almost nothing to do with value. A good plane is such whether I wish it, desire it, like it, "demand" it, or not. Mechanical standards make no reference to psychological states whatsoever. This applies as well to art works, price, health, and all other value categories. The value of life-enhancing and health-enhancing aids for animals and plants is publicly confirmable, as is the value of water for quenching thirst. This value is the basis of horticulture, veterinary science, human medicine, and other studies that evaluate the efficacy of means to ends for living things.

The attempt to reduce all value to private or subjective value, to "me value," involves a fallacy. Not all value is private in a subjective ego. Indeed, is easy to give examples of desires that are of no value or even bad, for example, the desire of a smoker with lung cancer for one more puff. While desire may be one factor in the evaluation of action, it is hardly a condition of value. It would be giving desire a blank check to argue for its being a universal or necessary condition of all value. But arguing from desire as a possible condition in evaluation of action to desire as a necessary condition of value is a *non sequitur*. Many desires would not be valuable if fulfilled; others have a further necessary condition, namely, a moral one and thus are contingent; others are vain, still others are absurd or Utopian. Above all, goals that end desire require creative actualization.

A better statement of this flawed theory would be that cessation of desire is the condition of value, not desire! However, it is wrong to argue *a priori* that desire alone constitutes a universal and necessary condition of all value. Something can have value without our desiring it or if we cease to desire it. The rivals of the Wright Brothers may not desire the success of the latter, their rivals, but this desire will not remove the accomplishment of the Wrights. Since value is inherent, as I have argued at length, desire is not a condition of value at all. When we are not thirsty, water still has inherent value for quenching thirst. This value remains with the water independently of our desiring water. It does not receive this value from the subject's desire, but from its inherent value for quenching thirst. The desire for water to quench thirst presumes inherent value of water for quenching thirst – not mere subjective desire. Otherwise, thirst increases: mere desire. The water does not change when we have quenched our thirst, only the desire or interest in the water has changed. The inherent value of a bearer of value does not cease to have value when a desire passes. Thus desire is not a condition of the value of the water, since this value remains whatever our subjective state. If we desired poisonous liquid or water from a poisoned well this desire would not change the negative value of the poisoned liquid or water. Desire or interest does not in any way affect the suitability of liquids to actually quench thirst in a healthy way. Taking an interest in carbon tetrafluoride or in salty water as a thirst quenching liquid will not give it value in quenching thirst, no matter how much we may wish it. The conative theory confuses the personal with the valuable: personal values with value, as well as evaluation with actual value.

Like other subjective theories, conativism is hierarchical, both internally and in relation to other values. Desires are the source of value; the entire balance of goods in the world becomes an instrument of the whims of a subject, contingent upon the random occurrence of a desire. Desires themselves are also placed in a hierarchy; some have more value than others. This hierarchy makes the theory either contradictory, or false. No one desire as a desire can be the source of value for other desires, a desire to desire notwithstanding. If it could the theory would be based purely on whim—or would have to use some standard beyond desire. But then desire is not the standard. However, even if such a hierarchy of desires were consistent, it would reflect evaluation not value. The theory, like the others has confused personal and subjective rankings of desire with value as such. Desires may say something about personal motives, but are never the source or ground of value. The place of desire in value theory, if indeed there is any place, is in evaluation not value as such. Desire, will, and other conative psychological elements have to do with evaluation of some ends over others as more highly desired: desiring one good over another. Such desires are in the dimension of the possible, not of actual value. But this presumes the value of what is desired: value is the ground not the consequent.

All subjective value is in the realm of evaluation not value, except personal values. To desire some value is subject to evaluation: to evaluate that creative actualization of a good would be an improvement. The desire remains in the realm of evaluation, not value, since without creative actualization the project is no good. The subjective is not actualized in the world and remains a mere wish, not a value. The possibility has not yet proven valuable in the world. This argument applies to theories based on desire, interest, liking, will, and other psychological factors. Value is actualized by action, and the feasibility of a project has nothing to do with subjective wishes. I may whine all I like but this will not get the plane off of the ground. More, objects of desire which are critically evaluated as inferior for a project do not even have potential value, for example, flying unassisted. Many desires are passed over as projects, since they are judged as not worth the bother. Thus value cannot be an "object" of desire, or interest, since many such objects are evaluated as not worthy of pursuit. Even personal values are separate from evaluation since they have been creatively actualized and are exhibited in actions undertaken by an agent for personal reasons or grounds. Actions speak louder than words: action is the criterion of such personal values, although not their measure. Personal values require prior evaluation but are not values only as evaluations.

That desires are concerns for evaluation and not value is evident from considerations that were argued above. Desires as such do not cause; instead, desires can be placed in a scale or hierarchy of potential values and then evaluated. Even a scale consisting solely of desires would be given an order of preference. Then the most highly valued is pursued. Evaluation is thus required for action, since desires require evaluation prior to action, that is, creative actualization of the goal of desire which will end it. Desires require evaluation, and are entirely in the realm of evaluation, since the desire has no inherent value and *a fortiori* no intrinsic value. Desires alone are never enough to initiate action. Even where a desire is acted upon, after it has been evaluated, it is only a factor in evaluation, not value as such, which gets rid of desire. We act primarily because a goal has been evaluated worthwhile. Creative actualization of the goal is evaluated worthwhile because agents evaluate it superior, not only if it will end desire. This evaluation is prior to action. Even then, the non-desire that fulfills or ends our desire only *may* be good: not good by necessity, but only possibly good. For it may turn out to be bad (tobacco, junk food, and so on). The goal may prove unsatisfactory even if successfully actualized: not worth it, for example, whimsical desires. Desires may prove unfeasible, that is, the goal that might satisfy desire cannot be creatively actualized: attempts to do so are no good. The desire may be impossible, for example, to fly unaided or to square the circle. Desire is not value: rather we pursue feasible ends which are evaluated superior or worthy, but which still have no value until they are creatively actualized.

In general, worth is prior to desire or other subjective states that may involve a relation to values. We do not value something whose independent worth has not been evaluated. The value of water for quenching thirst is the condition of our desiring it. As I argued above, desire is specific or determinate, not general, that is, a desire for water when we are thirsty, not for just anything. Gold or flowers, whatever their value, will not quench our thirst. Worth must be established prior to desire, will or other relational or conative subjective-psychological states, by pragmatics. Value is the ground of desire, not vice versa. Means are also evaluated independently of desire. Thus desire is not the condition of the value of such objects; instead, the value of such "objects" is necessary for anyone to desire them.

Does some good cease to have value when our desire passes? Does water cease to have the quality of quenching thirst? If it did we could not drink it the next time we are thirsty. Water does not change, so how could its value change? Only our relation to it changes.[2] Water is good for all living things whether we desire it or not. This value judgment is not anthropocentric: all living things require it. The value of what is desired is independent of whether it is desired. Otherwise, wanting one million dollars would be a more valuable condition than having one million, if the desire was sufficient of itself, that is, inherently valuable. Desire without the inherent value of what is desired, that is the capability or property of what is desired to fulfill or end the desire is vain: an argument for inherent value. Indeed, unless value is inherent in the object of desire, the object could not satisfy desire: it would be no good in the context of desire. What then of a desire which leads to a novelty? Desire for a novelty is based on potential value: if I had a million dollars (a beautiful woman, a prestigious job, and so on) I could do x, y, and z. Therefore I desire it. Desire is the consequence of some imagined future state, not the ground. Even if our desire leads to creation we must first establish pragmatic value by experiment, testing, and so on.

Philosophy from the time of the Sophists has often taken the model of wants as "natural"—including the model of primitive desires. Animal behavior is often interpreted with the model of desire as natural, for example, if an animal quenches its thirst it desires water. Thus desire is natural and basic as natural. Desire is attributed either directly or by analogy to animals. The question is whether we can ever know what an animal experiences. But this model also ignores changes in desires that indicate that desire could never be natural. Almost all desire for consumer goods is "unnatural." Such desires could not be basic if the good that is desired is not "natural" but "artificial." A desire for a chocolate bar is not "natural." Conative theories are not "naturalistic" if desires for "natural" things must be instilled through social pressures and parenting. History and culture are evidence for this: different cultures, epochs, and persons prize different goods. Novel goods, which are not needs in any biological sense, go beyond the "given" of value previously available. Such "desires" for new goods have themselves been created: the value of the novelties has been sufficient to generate actions to create such goods. This is not a *denial* of desire, but a denial that desires are basic or given. Values form the basis for any schema covering the diverse field of personal purchases, actions in pursuit of usable goods, and especially intrinsic value, not desires or wants.

Further, desire and liking are constituted. What is liked in our age is different from what was liked 100 years ago, if only because some choices available now could not have been made then. A desire for a horse and buggy was constituted by an age where automobiles had not yet been invented; the marketing of the autos in the twentieth century and beyond has constituted such desires for our age. The liking or desiring is relative to what has been actualized in an age, including cultural mediations, and prescriptions of and rules about what is proper. Civilization, society, and culture instill some "desires," regulate others, and prohibit still others. Such cultural values regulate what is available to a degree, one factor in technological change, although technology is also an independent variable to some degree. Both of these normative factors mediate desire and liking, and frequently constitute them; desires in the Victorian era were more circumscribed than at present, particularly sexual desires. The value of what is liked is in relation to the differentiation of what is valued, for otherwise there could be no ground for differentiation of liking. Desires are not foundational but created with respect to what has already been actualized and thus has proven its value. We could not distinguish a liking for an automobile if the machines had not been successfully actualized, since there would be nothing to like. The distinction of likings from one another is thus grounded in actualization, including mediation by cultural values. The subject is originally made or constituted, not foundational. The atomic, completely autonomous individual is a myth that ignores culture, parental and other kin relations, and socialization. Raising children is a socialization process in which some desires are filtered out from others: desires are regulated by social and cultural standards.[3] Desires are not basic as they are both constituted and informed by social values.

Conative theories attempt to reduce all the good in the world to one, the whims of a subject. Only the subject who wants something has intrinsic value. But this theory is absurd: the want as such has no value. Arguing that the subject "wants" argues for subjectivism, not conativism. These points constitute a strong argument against subjectivity of *intrinsic* value, given this theory. If value lies in something that satisfies a desire, liking, willing, interest, striving, and so on, it cannot lie in the subject as intrinsically valuable. For a relation to some good, thing, object, goal or other non-desire that fulfills or ends the desire is required or necessary. The subject lacks value, as lacking what will fulfill a desire. But of course many other intrinsically valuable agents live in the world: they do not derive their value from *my* desire. Nor can my desire establish their own intrinsic value. The theory again fails.

Moreover, the world is a condition of value, not the reverse; this would be true even if desire had something to do with value, that is, if the fallacious hypothesis that value were the object of desire were valid. Without a world, *no* desire could be fulfilled. Desire presumes value in the world: the value of the inventory of the many goods of the world, from which desires can be constituted and value derived. Otherwise desire would be in vain, even if it could be constituted. Since desires are constituted as effects of inherently valuable goods, the devaluation of the world in the reduction of value to desire should be critically evaluated as a fallacy. Further, inherent value of worthwhile goods in the world is the ground of the constitution of desire, for example, of water in quenching thirst, as I argued above. This value was inherent to water long before any one individual desired it and would remain with water whether that individual ever was born.

Creative actualization of the good as valuable is the condition of desiring. Goods constitute desires, that is, a desire for a computer presumes creative actualization of computers and their marketing as goods in order to creatively actualize "desires" for computers. Desires presume creative actualization of goods, past or present, for which a desire can "arise" or be instilled. Desire is a mere representation of what is good which presumes some criterion of good that can be represented. Creative actualization tells us if it is good and thus constitutes the ground that determines whether the good is wanted. Values are the ground of desires, not the objects of desires. Valuable goods constitute desires as different, as specific, and as intentional. (This may be a desire for the normative identity, chocolate bar, not an individual one—any chocolate bar would do. But I do not think that this affects the present point.) There can be no desire without creative actualization of goods, that which is desired. We could not desire a chocolate bar without creative actualization of chocolate. Desire is a response to the worthwhile, not its ground. What then of desires for new inventions, that is, for what has not yet been creatively actualized? As I noted above, desires that lead to novelties are grounded in the potential value of what must be creatively actualized to have value. Creative actualization of the novelty has value not the desire. Only as the effort of creative actualization can desire be satisfied: as desire for what has already been creatively actualized or the worth of what has potential value.

Nor are they desirable as such: desire in particular is something to be sated not pursued. Instead, creative actualization of the object of desire is good, not desire as such. We do not value desire as such, will as such or even liking as such. Desire is at best a representation of what might be good if it were to be creatively actualized (and for that matter, a representation of what is wanted). Only creative actualization can tell us if it is actually good and this is the ground that determines what was wanted.

Desire, then, is not value but the lack of value. Value is inherent in creative actualization of worthy ends, which sometimes end desires. Value is achieved not through desire but, where desire is a factor at all, by cessation of desire, whether because it passes, because the agent has moved on to other things or because the value of something is achieved through creative actualization. Desire is not enough for value; actualities are created apart from desire and may not reflect desire at all, but hard work, luck, trial and error, good judgment, imagination, and so on. I am not arguing that desire is never involved in any respect in the process of evaluation. Instead, the distinction of value and evaluation eliminates desire from value as such. Value is attached to actualities, as inherent in activities and goods while desire is at most one element in evaluation. Desire is not value but lack of value and pursuit of its satisfaction must be critically evaluated. Creative actualization of some objective ends desire: value is achieved by, at best, ending desire. Desire as such is never value.

As with affective theories, conative theories are egoistic. Like other subjective theories, conativism cannot generate obligation, since it is egoistic. Desires are my desires: they are subjective not altruistic. Despite arguments that we can have benevolent desires, interests, and so on, desires as such can never rule out bad desires or interests, for example, to torture someone. Good desires and likings are outweighed by bad. Duty is other-regarding: duty to someone toward whom we have responsibilities. Indeed, the egoism of conativism brings out its fallacies: what I want may do harm to others. Such desires must always be evaluated morally before we act. All of ethics teaches that the effect of our actions on others is a moral consideration, from the golden rule to Utilitarianism. The ethics of character adds that desires frequently must be limited for a good life. Even critical hedonists like Epicurus and John Stuart Mill argue for limiting desires to quality pleasures. Morals are a limit upon desires, not grounded in them.

Equalitarian norms justify giving equal moral weight to each person's desire as a general principle. But if desires conflict, as in the case of rapists and women, some desires are given preference, and are thereby not treated equally. Not all the desires even of one individual are equal. What of the desire for murder, sadism, theft, and other such acts? They are not valued by a society as wrong. But then moral values constitute a standard for desires that are not based on desire as such. The theory fails to account for moral values and cannot do so in principle, as I argued above, as egoistic. Indeed in both East and West, in Buddhism and Stoicism, the limitation of desires is the basis for the good life, not their enhancement. Moral considerations are similar: moral values form a limit on desires and are not derivative from desire.

Since whims, desires, and likings are not always in our power, they are sometimes not in the province of morals at all. No one would find someone who wants a million dollars but does not go about stealing it at fault. Having a desire is one thing; acting on it another. We are responsible for our actions, not our desires. The literature of intent, from Christ through Kant, argues for the moral culpability of even "coveting" our neighbor's goods or wife. But random desires may float through our consciousness at all times which we do not act upon. We cannot hold someone responsible for that over which they have no control.

Value, unlike desire is public. We may harbor a secret desire that we share with no one, but it can only be pursued in publicly observable actions. Secret or hidden actions are not private. We might sneak somewhere unobserved, but such actions are not unobservable in principle. Even in the cases where they are linked action must show the value of evaluations. Acts are guided by the worth of the projected value, but have their own implicit value as the general means of contact with the world. Even private sources of value in desires have public criteria, namely, that the value is pursued in action. Would an extremely thirsty person pass up a chance to drink water?

George E. Moore and John Dewey criticized Mill for attempting to justify the desirability of pleasure or happiness in terms of the fact that humans do desire it. Similarly, the conative-interest theory should be criticized for the reduction of the desirable to the desired. Desire must always be evaluated with respect to worth, subordinated to more important considerations. One of Dewey's criticisms of conativism emphasized the transformative capability of intelligent mediation on desires (Dewey 1939, Hickman 2007). Primitive or uncritical desires should be considered in relation to a host of other factors in any situation, by means of the methods of intelligence. We should critically evaluate desires in terms of long-term goals, dealing with present problems and situations, cultural and moral principles, and other factors. Desires emerge from this process transformed into what is desirable, which has a normative element. What is worthwhile is normative or worthy and evaluated in terms of standards beyond desire, including moral standards. Acting on desires without deliberation is unintelligent.

However, Moore and Dewey conceded too much to conativism by even using a term such as "desirable." Such a term implies the ultimate validity of subjectivity as the foundation of value and only raises a quibble with regard to whether whim or considered subjective states will be the foundation. We would argue that this entire model is axiologically false.

Desires themselves are distinct from evaluation since they require evaluation. Are the desires worthwhile? Are they superior to other desires, more worthy of pursuit? Are they moral? We never pursue "whatever we want"; desires must be evaluated. Some desires may be bad, for example, for tobacco. Desires are placed in a scale along with other options and evaluated or rated; the most highly valued are pursued. This decision involves weighing a number of factors other than desire, such as feasibility, overall goals, the place of the particular value of the situation in one's life, and so on. Since reference to the value of the objective is required, value determines whether an action is desirable, and subjectivity as a cause of action is revalued as an effect. Desire is determined by the value of the particular project. As I noted above, the object of desire may be unattainable or unfeasible and thereby without value. Such an objective cannot under any circumstances be creatively actualized, for example, a faster than light starship. Such an objective is without value: the desire cannot be satisfied with an objective. The value of what is desired must also be evaluated, that is, the value of its actualization: is it worthy of bringing about? Is it justified? Is it an improvement, an amelioration? Is it compossible with other values? Evaluation of the value of what is desired is distinct from desire as such, then. Desires may be evaluated, but desires are not equivalent to evaluations, as critical evaluation raises the issue of the value of the desire (cf. Moore's question: "but is it good?"). Thus the worth of the desire is distinct from the desire even for evaluation.

Acting on desires as a model is impulsiveness, not wisdom. Dewey argued that acting on whims and desires that have not met the test of critical reflection is the mark of an immature and spoiled person (Dewey, 1939). No evaluation, no axiological critique has been undertaken to determine the worth of satisfying the impulse, namely, should we desire x or even have a desire at time y: is it good? Value cannot reside in desire: what if my desire is harmful? Why is a desire of value, what if it becomes a source of anguish if fulfilled (pregnancy)? An impulse requires evaluation to determine if it is worth pursuing. A whim may not, upon critical evaluation, prove worthwhile. The value of desire must be established prior to action. Action reveals values instead of desires or whims, since we cannot act upon all desires, whims, likings, and so on, at once, but must evaluate these and choose which is worthy of pursuit.

2. Will

"Will" is perhaps the vaguest concept in Western philosophy. Aurelius Augustine is perhaps the first to differentiate will from desire, but did not do so consistently. In some cases Augustine writes as if will were a special form of desire, in others as if it were opposed to desire, and in still others as if will is equivalent to choosing or choice (Augustine, 396 in Burleigh 1953). This relation to desire has remained ever since, whether the "will" is distinct from desire, based on it, as in Kant's synthetic model of rational desire; or made into a metaphysic in conjunction with force, as in Arthur Schopenhauer and Friedrich Nietzsche. Either will is seen as a development out of desire or is confused with desire. In no case is the distinction of will from desire clear. Inevitably this has consequences for ethics, especially in terms of free will. In the context of value it is not clear in what sense value as what is willed is distinct from value as what is desired. If the medieval British philosophers defined good as whatever God wills to be good, is this distinct from what God desires or identical with it? Since the "good will" in Kant includes desire is the "good will" also "good desire?" If "instincts" in Nietzsche are reduced to forms of the "will to power" is this then a philosophy of Platonic "appetite" extended to the cosmos, the world as "will to power?" Such issues need to be addressed before a value theory based on will could even be intelligible, much less credible.

Voluntarism shares the defects of conativism, since value is independent of will, sometimes constitutes will, and should frequently regulate it. Value is the ground of willing even if we concede the validity of the actuality of will. Will does not create value: value creates will. For if will is refined desire, then it requires evaluation of desires in which creative actualization of diverse goods or options enables an agent to rank them and "choose" them. No option no choice. Thus even if will is choice not desire, value grounds it. Value may be creatively actualized willy-nilly: goods produced against my will. A plane is in good condition to fly not because I choose it but because it meets mechanical standards. I can will that my inventions work all I like but willing them alone will not make them work. A candidate may be elected whom I do not choose. I may choose to try to sell real estate in an undesirable locale, but this will not attract buyers.

Again, the theory is subjective and relational. Thus it has all the problems of subjective or psychological theories. These include: the value of willings themselves (the critical question), the devaluation of world in terms of a *pros hen* to a small part of the world (reduction), and the resulting worldlessness of such theories, the location of the relation of will and that which the willing intends, the confusion of such willing with other values, that is, the category mistake, the confusion of evaluation and value, of the language of willing and that of value, of willing and action, the egoism of such theories, and so on. Actions are public, but willing is not. It is difficult to see what value the notion of "will" adds to value theory even if it could be determined what the value of willings themselves are or the legitimacy of this norm. If steel is superior to paper for construction this has to do with the inherent character of steel, not will, "good" or not. Steel does not lose value when our whimsical attitudes change. Again, beauty has little to do with will. The Grand Canyon is beautiful independently of our will.

Will could be bad or evil; will alone, whether desire or choice, cannot make a bad action a good one or an immoral one, moral. Moral principles may be the contrary of what some people will, for example, the choice of gas-guzzlers despite pollution, global warming, and obligations to future generations. We may be obligated to change our will in view of moral principles, rules, and values. Changing such willfulness is often the purpose of moral exhortation: instilled as a result, not a basis. The social consequences of actions are more of an issue for morals regardless of personal choices or whim. We may have duties to others irrespective of whether they are what we would have willed. Such social obligations precede any specific willing. Again, who is to decide which wills, of contending ones, are correct? Could the will alone be used as the criterion, the will of a third party? But then the will of another party comes into play, and still another will would be required to choose between the three and so on ad infinitum. Above all, even if wills have value, their creative actualization as correct has actual value. The creative actualization of new art works is also required in order to will or choose them. At best objects willed are a species of value, and at worst a devaluation of values.

Proponents of this theory would no doubt argue that the epithet or qualification "good" is needed to complete the theory. Will alone is insufficient: only "good wills" have value. This qualification defeats the theory however. Will must be axiologically or normatively corrected. Thus a standard of value or of right other than and beyond willing itself is required to give will value. "Good will" cannot come from the will itself since it needs correcting. But then will is superfluous for determining value, since any psychological state, actuality or action can be corrected according to the non-voluntarist standard.[4]

3. "Demand" Value

I will next consider a group of theories interrelated in terms of demand. These include various theories used by economists in modeling but have implications for theory of value, society, and philosophical anthropology. The models make assumptions about humans that are self-fulfilling, since they affect public policy. Demand is a form of desire, but also has a stronger connotation, namely, that some object is not only wanted but "demanded." Economic models that posit "needs" are a form of this type but were covered separately. Talk by economists about "utility" comes under this type, as does any model of the life-style of or for consumers. Economists have frequently embraced utilitarianism as a model of humans for economics. "Utility" has two meanings in this context, the philosophical one of "the greatest good of the greatest number," and the more economic notion of "net aggregate utility." Insofar as utilitarianism is a theory of value, it is dubious if it can be encompassed by demand, since happiness or satisfaction is a feeling, not a desire. We may desire the feeling, but may as much lack it, which is why we desire it. Generally, economists are referring to means, not ends, but conflate the two in their models. Nevertheless, demand, self-interest, and the like are species of desire.

The amorality of this model is astounding. Economists argue that economics is a "science" and therefore value-neutral, but given the value of truth and knowledge this claim is preposterous. The refinements made by later generations of economists on earlier work, for example, that of Feldman, Galbraith, Samuelson, Arrow, and Friedman on that of Keynes, Marshall, and Adam Smith, is an improvement, that is, theories or refinements of superior value have been creatively actualized. The utilitarian value of their work is also obvious in the use made of it by governmental, financial, and corporate agencies and institutions. An example is the governmental control of the money supply in the "free market" by the Federal Reserve Bank, following models created by Milton Friedman. Thus we may infer from the evidence that the work of economists is not without value; more, they explicitly aim for valuable models. To wrap themselves in the mantle of science is propaganda, since predicting the market can never be an exact science. Economics is a quasi-science at this point and probably will never obtain the precision of physics or chemistry. Moreover, economists generally argue within the context of the "market," that is, they presume the superior value of one model of economic life over others.

The "demand" theory of value would be better described as the "brothel" theory since the argument is that anything for which there is a demand has value. There is a demand for whores, ergo, whores have value. Apparently most economists learned value theory at the Mustang Ranch. Another version argues that anything to which a price can be attached has value. Since there is a price for addictive drugs like heroin, heroin has value. Pederasts form a "consumer market" for children, therefore there is a "market" for "kiddie porn," kidnapping, and sex with children. Organized crime supplies a "demand," therefore they exemplify the model of "rational" economics. Examples like these, which could be easily multiplied, reveal what is wrong with this and all the theories surrounding conativism. They are indiscriminate and describe the life of a moral homunculus, freeloader, hypocrite, or outlaw. Consumer capitalism, based as it is on greed, has cultivated just such creations, the corporate managers of Enron, Tyco, World Communications, and other companies whose management put greed ahead of morality. Economics has indeed spawned "value free" managers, or rather those with purely economic values, Randian-Marxist values. The life style is one fit for a master with a concubine, that is, a model suitable for "managerial society."

The "demand" theory is self-defeating as it stands, since if demand alone were supreme, the market and the society utilizing such a market would collapse internally or be destroyed from without. Unrestrained demand for drugs would end with a society of addicts. Demand for atomic or biological weapons by terrorists to blackmail the society would bankrupt it or result in McVeigh-style acts of destruction but at a much larger scale, for example, the destruction of large cities and their markets. Demand is not desirable, as these two are not coextensive: a "demand" for ransom money by a group that threatens to blow up the world. A "value-free" model cannot critically evaluate such "demand values" since moral values have been deliberately excluded from any "rational" model for economics.

The economic model of humans is false in several respects, even as it helps create the society of *homo oeconomicus* that it models, and is thereby self-fulfilling. The first is that the value it posits as the *summum bonum* is axiologically false, apart from the issue of the validity of a *summum bonum*.

Few hold economic goods supreme, that is, as the highest good or end (misers); for subjects value happiness or other extrinsic goals like health that goods produce, not the wealth itself, and these are not for sale. Health and happiness cannot be purchased, although certain means that may aid it can. Economic goods are rightly valued as means to an end. Making wealth an end is fallacious. Economics is a distinct category of value, but should and could never be supreme (King Midas).

Economic materialism could also be accused of pure quantification of value: that human life has no value without pecuniary means (cost/benefit analysis). The implication is that poor people have no worth or intrinsic value by themselves. I am hardly arguing here for the perpetuation of poverty or against social justice, but social justice requires another norm of value than quantity or "willingness to pay" (cf. the phrase "human resources"). There are higher values than money alone; and quantification is a perversion of value. Indeed, utility in the sense of "useful for…x" implies good for something else, that is, it is incomplete as it stands, since that which it is useful for is not specified and cannot be determined by utility alone. Nor are all problems economic, contrary to economism, a philosophy that unites Marxist and capitalist.

Another closely related model is the purchaser as motivated by "self-interest." I am speaking of economic self-interest or self-interest in the wider sense, for example, Joel Feinberg's "realist" model for morals (Feinberg, 1978). These theories are distinct from the Perry-Prall interest theory, but are related as subjective. Self-interest, despite the unwise attempts of several philosophers to incorporate this very economic model into moral philosophy, is self-centered. The "self-interest" theory, whether narrowly economic or as a more psychological model for all human behavior, is actually an ethical egoism theory. As such it devalues the whole for the extremely reduced part. All the thousands of species, unique landscapes, environments, indeed the entire cosmos are reduced to their value for one paltry ego! This view belongs to comedy, not ethics, to hubris taken seriously by tragic-comic actors, blind to their relation to the world around them.

The "self-interest" theory of value shares the defects of other species of conativism, namely, that value is independent of self-interest, sometimes constitutes self-interest, and even regulates it. Value is the ground of self-interest even if we concede the validity or actuality of self-interest. Self-interest does not create value: value creates self-interest. For if self-interest is refined desire it is formed from evaluation of desires in which creative actualization of diverse goods or options enables an agent to rank and choose them. No option no choice. Value may be creatively actualized despite my self-interest: goods produced against my self-interest, as Detroit realized in the seventies. A car is in good condition not because of my self-interest but because it meets mechanical standards. I may choose to try to sell cars that are inferior in value in accord with my self-interest, but this will ultimately alienate, not attract, buyers.

Again, it is evident that the theory is subjective and relational. Thus it has all the problems of subjective or psychological theories. These include: the value of self-interest itself (the critical question); the devaluation of world in terms of a *pros hen* to a small part of the world (reduction) and the resulting worldlessness of such theories; the location of the relation of self-interest and that which the self-interest intends; the confusion of such self-interest with other values, that is, the category mistake; the confusion of evaluation and value, of the language of self-interest and that of value, and of self-interest and action; the egoism of such theories, and so on. Actions are public, but self-interest is not. I have difficulty seeing what value the notion of "self-interest" adds to value theory even if it could be determined what the value of self-interest itself is or the legitimacy of this norm. If water is superior to poison for quenching thirst this has to do with the inherent character of water, not self-interest. Water does not lose value when our self-interest changes. It may indeed be in our self-interest to use water for quenching thirst based on its inherent value, but then self-interest cannot be the ground of value. Again, beauty has little to do with self-interest. The Grand Canyon is beautiful independently of our self-interest.

Self-interest could be bad or evil; self-interest alone cannot make a bad action a good one or an immoral one a moral one. Moral principles may be the contrary of what some people perceive as in their self-interest, for example, the choice of embezzling money from a corporation despite obligations to stockholders. We may be obligated to reign in our self-interest in view of moral principles, rules, and values. Changing such self-interest is often the purpose of moral exhortation: instilled as a result, not a basis. The social consequences of actions are more of an issue for morals regardless of self-interest. We may have duties to others irrespective of whether they are in our self-interest. Such social obligations precede any specific self-interest.

Who is to decide what self-interest, of contending ones, should be served? Self-interest cannot give us a "decidability" principle, a principle by which to decide moral conflicts, since if my self-interest conflicts with your self-interest there is no neutral principle by which a moral interest can be decided based on self-interest alone. If either party is favored, the decision is arbitrary from the point of view of self-interest; even assuming we could somehow derive "moral" interest from "self" interest. Could the self-interest alone be used as the criterion, the self-interest of a third party? But then it constitutes the self-interest of another party and still another self-interest would be required to choose between the three, and so on ad infinitum. Above all, even if self-interest has value, it requires the creative actualization of that which fulfills or meets our interest. The creative actualization of economic goods and arrangements is also required in order for them to be in our self-interest. At best obtaining objects deemed in our self-interest is a species of value, and at worst a devaluation of values. Creative actualization has value, for without bringing about objects that satisfy our self-interest, our life would be one of perpetual frustration. In other words, the theory is axiologically deficient and requires creative actualization for completion.

An "ethic" based on self-interest is an oxymoron. If everyone consistently pursued their self-interest, the result would be the grossest sorts of immorality, as Hobbes argued: a *bellum omnium contra omnes*, a point Hobbes made forcefully enough that I do not have to expand upon it. Hobbes is often misread to mean that we can establish a society on self-interest. His point is just the opposite: the society founded on self-interest alone is impossible and life in it would not be worth living. But more, the pursuit of self-interest has resulted in the environmental crisis we now find ourselves in: global warming, changes in the weather that affect the food supply, deforestation, and so on. Pursuit of desire, self-interest or happiness as an object of desire may interfere with duties and virtues, especially duties to the environment. The consumer society is the culture of death, ironically, self-imposed. The assertion that self-interest can be the basis for any ethic much less an environmental ethic is wishful thinking. Competition as an ethic is similar, since this model is taken from relations of business concerns and "business is war."

4. The Interest Theory

The interest theory is a vague theory as presented by David Prall, and several elements that may be incompatible are conjoined (Prall, 1921, 179–290, Ralph B. Perry, 1926, 1954).[5] Prall has presented a conative theory based in the relation of a subject to an object that is liked. This allows him to argue for the identity of value despite different kinds, and the overall value of life. Prall inclines toward an esthetic subjective theory in which value is independent of judgment and action in the experiences of liking. Prall argues that desire, interest, and liking form one theory. In some cases this makes sense, for example, value as "satisfaction" of desire and thus an object of "liking," if we grant we like to have our desires satisfied. However, it is not at all clear that they are one theory. If value is the object of desire, it may not be something we like or take an interest in. We may desire water because we are thirsty but then take no further interest in it. We may not even like water. Desires are often connected to biological necessities but interests rarely are. Likings are intermediate: we may like our food, because it is tasty, but still not be interested in it. "Satisfaction" of desire is a feeling, but is taking an interest in something a feeling? More, is every object of interest something one likes? What of necessary chores? Even granting the relation of liking, desiring, and interest, we may question whether interest and desire have any necessary relation to either feeling or a "motor" response. Sentiments and satisfaction are affective, not conative. The satisfaction of an interest may involve a feeling. Liking may follow satisfaction, but may also be equivalent to it. In either case it is so close to a feeling that it would be difficult to separate it from an affective good. Prall may have chosen it for this reason, that liking something may cause us to strive toward it and thus motivate desire. Thus he may have viewed it as a term that combines conation and feeling in one. However, it is different than desire. In any case, if we took this as the character of value then value would be located in the satisfaction of the interest not the object of interest. The relation would then be entirely internal to the subject, and not involve a relation to an object. Nor do all desires result in action, behavior or striving, as they are evaluated for priority and desirability.

I have treated many of the elements thrown together by Prall above, namely, desire and self-interest. In this section I will treat interest and liking. The theory is uncritical: as Dewey noted, it would justify an interest in burglary as it stands. In terms of quality, we could like or take an interest in bad wine. Interest does not reveal worth, much less value, since water or a plane may have value regardless of our liking or interest. The theory also confuses evaluation and value, on the same ground; and confuses ground and consequent. What arouses interests is not itself an interest, since if we take an interest in astronomy, we do not take an interest in our interest but in the stars, planets and other celestial bodies. Prall's argument that life is an intrinsic, overall value containing subjective interests, but that it only a good life has value, is circular. If life has intrinsic value simply by virtue of living, its value cannot be contingent on interests.

The interest theory is ethical egoism with solipsistic overtones: egoistic as deriving value from subjective interests; solipsistic in its assumption that value could proceed from a subject. The vastness of the whole world is devalued for the extremely small part. The interest theory basically reduces value to value for a subject—it "makes value subjective"—as the object only receives value in relation to a subject. As Harold Osborne has argued, this position is basically relativistic (Osborne, 1933). If the subject confers value, and truth is a value, which is "not independent of the subject," then truth is ultimately a "subjective" value. This position would jeopardize the "objectivity" of truth: subjective biases and agendas could replace the distinctive value of impersonal knowledge (Fox News). Obligation would also be relative to subjective biases. Further, subjectivity of value confuses evaluation and value: the subject evaluates or ranks but if something has intrinsic or inherent value, which Prall argues, this value belongs to it independently of any subjective ranking. The hierarchy could be relative without affecting any intrinsic value.

The reduction of value to liking or interest is fallacious. We can like something but not value it (candy if we are on a diet; ice cream flavors which we like only a little); and value something we do not like (the dentist's drill; the diet itself). Liking or interest is an identity combining all that we value that does not differentiate kinds or modes, that is, the many different kinds of value. Liking and interest are personal: my value. They combine all the different kinds in a relation to self, based on what is liked by a self. But this does not reveal what another self likes, and thus is limited to one person. But value is ascribed not in terms of the identity of what is of interest to me but in terms of what differentiates the value as such. We cannot even know if what is liked or holds personal interest is valuable, since its justification is entirely private.

Inherent value is the value proper to each instance of value. Paper is better for writing; metal for cooking. Paper is no good for use in cooking utensils; metal is hard to write on. The value is proper to the differentiation of what is valued, since otherwise there would be no ground for differentiation of what is liked. Liking and interest may also leave out quality. If a skilled person performs a creative activity hundreds of times, s/he may derive little satisfaction from it. Yet the value of the product may even be superior due to the great skill with which the craft is practiced after enough time, for example, quilting or cabinetmaking. The value of the product is independent of liking: the product has inherent quality and comparative quality in relation to the work of beginners. If it did not it could not be exchanged, for others would not value it.

"Interest" resembles a term common in economics and there may be a subtle economizing of value in Prall. The origin of the interest theory may lie in economics, an origin that belies Prall's arguments for different kinds of value and would constitute a reduction of value. There is at least a hint that everything has a price in Prall: price as value, not just one measure of comparative value. This measure would devalue the person. The many elements in a person's life are devalued as merely instrumental to the meager one of liking or taking an interest in things, a relation which points beyond the intrinsic value of the person in intentionality, and even constitutes a denial of the value of the person. The person is not of value only their likings of other things are. Value as interest is far from the whole of life; we might have no direct interest in our health, and dislike exercise, medicine, a healthy diet and other means to a healthy life, but health might be an important element in a good life apart from any interest taken in it.

Finally, the interest theory is morally deficient. Should we take an interest in sadism? Why not on Prall's theory? If contemplation of another's displeasure arouses our liking what, according to the theory, is morally deficient in it? Without defining morals apart from interests, any interest would seem to be justified so long as it satisfies subjective likings. This would be the end of morals. The argument that obligations represent "future interests" is weak, as future interests are contingent but obligation has a sense of necessity. Obligation is also in the present. Moreover, what is in our future interests may conflict with our present interests: which has priority? Only a standard beyond interests entirely can decide and this defeats the theory. What if morality conflicts with self-interest? On the ground provided by the interest theory alone, following self-interest would be justified instead of morals.

Above all, the theory fails to locate value in general, as value is any object of interest or liking. Is value then in the *object* of interest, or the interest-liking itself? If value is any object of interest, it cannot be in the liking or interest itself, for it requires satisfaction by an object. But if the latter is the bearer of value, then the interest theory cannot locate intrinsic value in a subject. Unless interests themselves are valuable, they are not bearers of value. If they are, then they do not require a relation to a determinate object, and the theory is solipsistic. Objects of interest would then cease to have value, and the theory would fail. If interests do not have value then the theory fails as a theory of value, since interest is not the mark of value. But there is at least some notion of intentionality in the interest theory; we take an interest "in" something. Liking involves liking *something*, it involves a (necessary) relation to that which we like. The object of interest is required. But what is required is necessary and what is practically necessary is obligatory for value. The logic of the theory points toward the inherent value of the object of interest, then, and interest becomes a condition of value, like desire. However, it then has all the problems of such a condition: is the interest itself of value or the object of interest? But how could there be any interest or liking without an object of interest or liking? Thus there is a condition of the condition, and interest or liking is conditioned, and thus not the condition of all value.

Prall might argue that he does recognize the relation as necessary, but that such objects have value for the subject, not the objects. However, the states would depend upon an "other," a relation extrinsic to themselves, not intrinsic. Prall also attributes intrinsic value to esthetic objects and also to experience. Yet he also confines intrinsic value to subjects. True, subjects constitute the objects to which intrinsic values are attributed in part, as objects of experience. However, if the objects have inherent value, this must be independent of a subject, or value would not inhere. Moreover, Prall raised the question of why some objects are considered more valuable than other objects. This question is an argument against the subjectivity of value, since the comparison is of objects themselves.

The value of a good tool, to borrow an example from Dewey, is in many cases independent of the subject, or its interests. The tool has value as a tool whether we like it or not, whether we take an interest in it or not. Our liking the tool is irrelevant to whether it works and functions well: whether it is good tool. If liking/interest is irrelevant to the value of an instrument, why is intrinsic value confined to subjects, since the relation of instrumentality is to intrinsic value as intrinsic, for its own sake, not to a subject? If instrumental value is independent of subjective states, and the relation of instrumental value implies intrinsic value, as Prall argues, then intrinsic value is also independent of any subjective states, as the relation is, by definition, to an instrument, not a subject. The equation of intrinsic value with a subject is not required for the relation of instrumental and intrinsic value, by Prall's own reasoning, but logically independent of subjective states.

Affective theories do not fare better as relational theories than conative theories. As with conative theories, affective relational theories suffer from problems of location of value. Is the value in the feeling or the object of feeling? If in both, where is the value? It cannot be in the subject or the object, but there is no location of the relation as such in which value can inhere. If in the feeling, a relation to an object cannot be necessary, but by contrast, it is in a relational theory. If the object has an inherent property of causing the feeling, then value is located in the object as inherent. This undermines the theory, since value is not connected with feeling but objects. Affective theories cannot "extend" intrinsic value to "objects of" value, as only feelings have intrinsic value according to the theory. If objects have intrinsic value, it cannot be as an object of feeling, since this is an extrinsic relation to feeling *per se*. The *pros hen* relation to subjective feeling would deny that the object could have intrinsic value, since its value derives from subjective feelings. Despite the arguments that the theory upholds intrinsic value (Lepley, 1949), it undermines intrinsic value.

Prall's originality lies in his assertion that moral obligation is intrinsically valuable, instead of instrumental: moral obligation is not defined separately from value. With such a treatment, Prall's is the most axiological ethic of the theories that have been examined, for even morals are defined as intrinsic values. Value is as independent of ontology as it can be, for the existence of the object is not required to consider its value. However, like the theories of Nietzsche, Max Scheler, Dewey, Raymond Polin, and others, ethics is not entirely normative either, not confined to what ought to be. His lasting contribution to axiology lies in this point.

Philosophers have been stupefied by the psychological model for value. It appeared like an easy way to deal with problems of value given a so-called "scientific" worldview. Simply reduce them to feelings, desires, or other psychological states and the problem will disappear. There has been a certain amount of intellectual laziness or inertia, if not dishonesty about this tendency, which has ignored all the evaluations involved in science and in such models of philosophy themselves. But this is another story.[6] What is evident from our critical evaluation of psychological theories is how axiologically false any attempt to use price as a substitute for value must be, especially "willingness to pay" as a measure of inherent or intrinsic value. This entire project is erroneous, to speak nothing of immoral. Advertisers created this self-fulfilling model for a consumer society, financed by corporate interests. Psychologists in the pay of advertising agencies and economists in the pay of financial interests have added fuel to the fire. They have enhanced "demand," desire, "liking," and other trivialities to increase profits and consumption. They have fed the beast by cultivating the psychological capacities appropriate to acceptance of it.

In general, psychological states taken toward a valued object cannot be inherently valuable, as they are contingent upon such valued "objects," that is, goods. Feelings, desires, satisfaction, liking, and interest in the Grand Canyon require the actual Grand Canyon. If value lies in the "relation," it must be located in either the subject or the object: it cannot be in both. Yet the psychological state alone cannot have value, since the state is intentional. Intentional states have a teleological structure, that is, intention has a goal, attaining the object. Relation, however, is normative, not actual, since "relations" are not actualities in themselves but normative regularities between actualities, whether causal or other relations. It is not as if there is an additional actuality of the relation in the world; instead there is a normative identity, just as in relations of individuals of the same kind.

If value lay in the psychological state, and not the object, we could value the Grand Canyon without a Grand Canyon. The object would be irrelevant, since subjective states would be inherently valuable apart from the object. This is obviously absurd, for it contradicts the first premise, namely, that the states are intentional. If value is located in the "object" of desire, feeling, striving, satisfaction, willing, attitude, interest, or liking then it inheres and cannot be derivative from the relation to such subjective psychological states. Relational theories as such cannot account for or justify inherent value in either subject or object, since value is located in neither the subject nor the object: inherent in neither.

Psychologism as a model is also pre-social, that is, ignores the social aspect of human life and value. Advertisers do not ask if "creating desires" is moral, but if it is effective. Desire is used to manipulate; they use the blond on the hood to sell the automobile: not value but propaganda. Desires are not the basis of ends: the relevant question is, what is the good to be achieved by pursuing the end? This is an evaluative theory of goals, in distinction from a psychological theory. Psychologizing evaluation risks psychologizing all of the critical functions of philosophy, since evaluation of arguments is one of the tasks of logic, the primary tool of philosophy. No doubt some philosophers would welcome this development, but they do not love wisdom.[7]

This and the preceding chapter do not constitute an argument against the value of specific psychological or subjective states. Creative actualization of happiness within moral limits, which takes environmental impact seriously, can indeed have value. But value is not psychology, and the attempt at reduction is so patently fallacious, confusing two quite distinct realms, that the evidence points to an agenda driving the process.

Conclusion

VALUE HIERARCHIES AND VALUE AUTONOMY

The topic of this concluding section will include a summing up of the theory of creative actualization, but with an eye to its place in philosophy. The question for us is the place of values in philosophic structures: their origin and justification as a reflection of the structure of hierarchy in Plato's theory of the Form of the Good. Hierarchies of value, the model in previous philosophy, are based on the evaluation of knowledge as first in value in Platonic idealism. Knowledge of value was considered the condition of value, and thereby valued above other categories of value. Plato's structure was adopted by subsequent theories that were quite different in many respects from Plato's, from Aristotle's teleological *pros hen* relation, in which inherent value is treated as instrumental; to atomic foundationalisms and reductionisms in value theory. Value emanated from a *summum bonum*, a one of value, whether the content of the *summum bonum* was intellectualistic, voluntaristic, hedonistic, or what not, in a hierarchical, top down structure.

How far does creative actualization escape this specific structure? I will argue that creative actualization results in a very different relation of values: in accord with reciprocal relations of inherently valuable goods, and intrinsically valuable agents, good in themselves, a horizontal differentiation results in pluralism of values, not a monistic *summum bonum*.[1] Again, the relation of goods and values is to an inherently different "other," not a *pros hen*, a referring back of instrumental values to intrinsic values, from which their value derives. Above all, pluralism of value is a rejection of the reduction of value, whether to human good, subjective good or any other narrowly defined good considered as a *summum bonum*.

In the last section I will argue that the hierarchy of Plato's theory dominates not only value theory, but metaphysics as well. Not only value theory, but all philosophical systems contain such hierarchical relations, whether idealist (emanation from the "One"), participation in "degrees of "being"" (as the *pros hen* back to a justifying first), or atomistic, foundational models (in which the higher is built up out of the lower: a reverse hierarchy). Philosophy is indeed a "history of footnotes to Plato." The question then arises whether creative actualization can provide an alternative ultimate justification, an alternative to hierarchy. What is the relation of creative actualization to previous value structures as well as paradigms of philosophy that incorporated the Platonic structure, including "being," the subject, and language?

1.

Most value theories in Western thought have the structure of a value hierarchy, in the form of some *summum bonum*. It could even be claimed that value theory confuses all values with hierarchical structure, and that little or no distinction has been made between values and hierarchy. The central place that "standards" of value, a *summum bonum*, have taken in the literature of value is a reflection of hierarchical structure. A standard of value subordinates other values in a value relation of dependence. The worth of the subordinated value is not inherent in it, or intrinsic to it, but derives from the standard or the *summum bonum*. The direction of valuation is downward as in all hierarchies. Value is derivative from the standard downward so that inferior grades of value derive worth only from their place or rank. The further down they are in relation to the standard of value, the less worth they derive from it. Higher ranks derive greater value than lower, as they more nearly reflect or partake of the supreme standard.

If appeal to a "higher standard" carries weight in value considerations, it is not the only organizing principle of hierarchies. We can show fairly easily that not all hierarchies have value standards as their organizing principle. Thus value and hierarchy are not co-extensive, for some hierarchies are not organized around supreme values, a *summum bonum*, but around another first. Value hierarchies are but a "species" of hierarchy. Alternatively, hierarchies are the attempt to use the value model as a model for other fields.

Another hierarchy is formed by logical terms. An instance is subordinated under a kind or species, a species under a genus, one genus under another of greater scope, and so on. The "higher" terms have greater generality of scope, but not necessarily greater value. Term logic involves subordination under unifying generals, and thus exhibits the same structure. In turn, intellectualist hierarchies are created, and subsequently by systematic extension, hierarchies in other areas of philosophy. Indeed the importance of logic to philosophy might be evidence for the origin and pervasiveness of hierarchical models in philosophy as due to logical hierarchies. I will argue below, their origin is even earlier. It could still be argued that language has such hierarchies and these have influenced earlier thought, however.

Military, religious, political, and bureaucratic hierarchies are based on degree and extent of power. A general officer may issue commands to several aides of high rank, who in turn instruct junior officers, who carry out the orders with lower grade regulars. Power is proportional to rank within the hierarchy, for it extends to more subordinates the higher the rank. An analysis of a power hierarchy may be broached in terms of a logical hierarchy such that greater generality of scope roughly corresponds to greater power. A regional or national official may have several local subordinates who are inferior in both power and scope of operations.

Among the principles that characterize all hierarchies, two stand out in importance for value hierarchies in particular. Firstly, what forms an identity or a whole for subordinate parts, and is a unifying rubric as a higher element, is itself an individual aspect or part as a lower element, as subordinate. Organs are wholes with a distinctive function individuating them. They unify tissue and bio-chemicals into a distinct organ. But in turn the organ forms a part of an individual living thing. Corps and divisions form armies. This tendency of hierarchies is toward greater unity the higher the level. The vertical hierarchy proceeds from the authority of a first, a one in many, which is the supreme organizing principle, the highest authority. Hierarchies must culminate in such a "one," for their character is vertical and general: the higher the position in the hierarchy, the greater the unity of scope, forming the relation of levels in the hierarchy.

The significance of this structure for values is that supreme value standards are monistic, and tend toward intellectualism. Reason, the specifically human *differentia*, is the supreme value in the Platonic psychological hierarchy, the source of the hierarchical model of the *summum bonum*, for as a universal human essence, it alone has sufficient generality of scope to bring all other values into a relation. Other values derive value from their degree of rationality. The other values are subordinated under the single standard, reason. The tendency to subordinate all value under this intellectualist standard is based upon its generality of scope. The supreme value must be all encompassing, that is, general and universal. By contrast with the particulars of appetite and passion, reason deals with the universal or general. Virtue is measured against the virtue associated with reason, wisdom.

A later consequence of this subordination is the conglomeration of different levels of value in a reductive move. Ultimately, the positing of a one in value tends to reduce the finer distinctions in value to equal status as subordinate to the One. Thus all goods can be treated as equally increasing or decreasing happiness interpreted as pleasure, hedonism. This makes for ease of evaluation: is the good instrumental to pleasure or not? Differences in levels become irrelevant. However, the reduction follows upon and is determined by the problematic of hierarchy.

A second feature of hierarchies is that they are normative. "Normative" means both a norm or principle uniting all instances, that is, a universal; and a model or standard regulating those for which it provides the norm. Value standards exemplify this feature almost paradigmatically, for as I indicated above, reason as a standard is both universal and a regulative model. In Plato's myth,[2] it controls the appetites, or at least guides them and selects among them, and this model is held to be valid for all humans. The "irrational" must be either brought under the standard, that is, subordinated to reason; or excluded and disvalued: adjudged bad (Foucault, 1973, Preface, x-xi).

In hierarchies, value is not intrinsic but derivative from a "first," participating in "the Good" as the source of all goods. Plotinus serves as a clear example of such a structure of value: all good is an emanation from "the One" as The Good. Value is relative to and derives value from the source of value, the "highest" perfection. The relation is vertical and extrinsic: vertical because value forms a descending hierarchy from the standard; extrinsic because value derives from an exterior source, the highest, supreme value.

Surprisingly, non-rationalists have not questioned or criticized this intellectualist value structure. Instead, they only change the content of value. Atomistic theories build hierarchies as well, but substitute a non-intellectualist value theory for intellectualism. The hierarchical structure remains, only its source is changed. Hierarchies are generally "pyramidal" in structure, with a one at the top and the greatest number at the base. While foundationalists tend to reduce value to subjective states, these also have a hierarchical structure. The non-intellectualist standard is substituted for reason as the *summum bonum*. Feeling, desire or other psychological states regulate good from above as the source and justification of good. Despite their rejection of intellectualism, then, they have adopted the hierarchical model, in which a supreme good regulates good from "above." Another example is the "order of rank" of values in Nietzsche, who is otherwise an acute critic of intellectualism. The level and rank of values is distinct from the theory of values in these structural relations, for it can follow from either.

Hedonism exhibits a monistic value structure to which all other value is subordinated. Pleasure, not reason, is the one supreme standard, but it regulates as a norm just as surely. Pain, the *fons mali* of hedonism, is either "disciplined," such that it *ultimately* aims at pleasure; or is ruled out as bad or evil. Hedonists accept the value of reason, but subordinate it on their scale: its sole function is to aid in obtaining pleasure or avoiding pain. The value of reason is derivative from its relation to pleasure. Pleasure is deemed of sufficient generality to encompass all value in its scope. The generality is at the "base" of the pyramid, in the form of the individual pleasures, not the apex, but the apex reflects the base. A good life is constituted by the many pleasures within the life. Pleasure as a standard derives from the numerous possible pleasures, but still constitutes a value monism. While atomists would insist that pleasure as a standard is constituted by individual pleasures, they still borrow the hierarchical structure of intellectualism in value theory. Pleasure is still a standard of good, ruling over all other goods. Thus their value structure constitutes a reversal of their metaphysical structure.

The sole non-hierarchical theory of values of which I am aware in Western philosophy is relativism. Relativism avoids hierarchy only at the cost of absurdity, however. All values are devalued by relativism, for everything is as good as everything else. Nietzsche was right is his estimation that relativism ends in nihilism. All other theories of value subordinate value under a supreme good, the one supreme standard, whether this standard is eudaemonism (happiness in the qualified Aristotelian sense), self-realization, a moral will, life, or the diverse subjective theories: feeling, proper attitudes, desire, interest, and what not. This monistic standard is the standpoint from which subordinate differences can be judged.

Whether such hierarchies proceed from "above," as in intellectualism, or are constituted from "below," as in hedonism, such structures are authoritarian. For both are hierarchies of value, containing a structure in which worth is transmitted "vertically." Value is derived extrinsically from the standard of value, the "one, best." Subordinate values refer back (*pros hen*) to a first in value, the supreme standard. The value of subordinate values reflects the source of value, without which they would not have value. Since their value is extrinsic to them, they are actually devalued by *summum bonum* theories. If reason is the standard, it constitutes the authority by which different values are judged, and in whose light their value can be viewed. Reason is normative for the others, as reason judges their comparative worth. Other values are judged for their reasonableness under the authority of reason, which is the sole court of appeal as the standard. Reason alone prescribes subordinate worth, and actions are performed in accord with its authority. Hierarchy is the value structure appropriate to authoritarian societies, and, more subtly, to the transmission of power.

The divine voluntarism of Scotus and Ockham provides an explicit version of this authoritarian value structure. The law or commands of God should be obeyed, in this view, simply because of the indisputable goodness of transcendent authority, divine will as reflected in its fiats. The law of God is good because God has willed it so, rather than because it is intrinsically right or good. Actions are subordinated to these laws or commandments, because their ultimate authority lies in the one, infinite (and thereby universal), divine (and thereby highest) will. Thus hierarchical forms of Christian value theory incorporate Aristotle's contradictory account of values, and his use of the *pros hen* relation in value theory. Christianity argues both for the infinite value of the individual soul and also the emanation of value from God to creation. An instance cannot have inherent or intrinsic value if value is only granted through divine grace, an extrinsic relation. Thus the individual cannot have intrinsic value if the value of the individual derives from his or her relation to the divine will. Again, other theories of value just substitute a different content for the same structure.

All value hierarchies, including *summum bonum* theories, have in common a monism of value. One value determines the value of the rest, whether as participating in the supreme value (God, the Good), as means to the supreme value (pleasure, life), or as conditional upon the supreme value (reason, desire). However, no single good can either justify or ground all the rest. Firstly, the supreme good would have to be infinitely good. For it must provide an unlimited justification for all the different kinds of goods, all quantities of good, and all novel goods. But with the possible exception of God, the other aspirants to supreme goodness are not infinite goods, but limited. Pleasure and desire come and go, while life is finite in duration. Reason and the Good are defined by contrast with non-reason and the Bad or Evil, and thus are limited by their opposites or contraries. As for God, even divine goodness has a limit in Evil. Indeed, one problem for theology is to explain evil in a world created by a God of infinite power and knowledge. Why was Auschwitz allowed to occur?

Secondly, goods are compossible. For all supreme goods require means to them and these must have a good of their own, an inherent good, for them to achieve their end: a means to pleasure, to life, and so on. Such goods must be compossible with each other as well as distinctly good for them to become actualized. They are co-conditional.

Thirdly, the value of one good is not equivalent to that of another. The good of food is not the same as that of art. If all value derived from one principle, its own principle, its inherent good, would be subordinated to an extrinsic good. Its inherent value would not then be available to us or manifest. If all value were identical, we could not utilize the different plural values of inherent goods.

Fourthly, no one kind or category of value is inherently higher than any other, since they are distinct as kinds. The distinction of different kinds of categories is an insubordinate classification: separating distinctive ends. Knowledge does not serve faith, power or money, but is good in itself. Prosperity, bourgeois society notwithstanding, is not inherently primary; those cultures that pursue it to the exclusion or subordination of all else, Midas cultures, represent a cultural choice, not a model for categories of value. Categorial values only exhibit polarity, not hierarchy. Categories of value require differentiation in structure, not hierarchy. Thus different types are justified, a reciprocal structure: an argument against their derivation from a monistic first.

Subordination of the distinct kinds of value to a justifying first, a supreme category or kind of value, follows the *pros hen* model. The other kinds and categories are evaluated instruments to the achievement of the first, for example, knowledge as "power" or as an instrument in the achievement of wealth. The other, subordinate categories of value would not be of inherent value; their value would derive from the sole, supreme category, whether religion, economics, politics or any other category taken as a first. The only art that would then be evaluated worth producing would be art that is saleable, contains a religious theme, or a political message (propaganda).

Finally, personal evaluations cannot ground a plurality of goods, whether desire, will, attitude, feelings or any other subjective basis. Different persons have different priorities: which is correct? The same person may have different priorities at different times: health, when ill; feeling, when at leisure;

and desire when in need. Nor do such subjective ends indicate anything about the actual value of such ends. A desire for suicide, excessive food or drink, or any other personal end may be bad for the person's health, overall goals, or life. Hierarchy derives from personal *evaluations*, and these do not tell us the *value* of the end.

Indeed, several fallacies in value can be derived from these distinctions in value. One value fallacy involves making or positing one category of value as supreme over all the others, its source and justification. This includes all kinds of scientism and positivism (knowledge as the supreme standard), economism (Marx and some "neo-conservatives"), religion (Soren Kierkegaard's religious life, fundamentalisms of all types), or any other category of value taken as supreme, as the *summum bonum*. Another value fallacy is any personal value taken as valuable to or for all people, for example, pleasure, money, and the like.

The confusion of personal and categorial values is also a fallacy. Achieving the end may require procedures distinctive to the category, regardless of personal whims and subjective wishes. The institution of specialists, in health care, nutrition, investment, and other fields, constitutes evidence for the fallibility of persons. Although hierarchies derive from personal evaluations, a critical evaluation of such judgments can and should be made. While categories of value serve persons, it does not follow that personal whims are the best judge of what is good for them. Moreover, just because the achievement of some value is possible only through specific activities and rules does not indicate whether it should be achieved; or, indeed, whether it is worth achieving.

2.

By contrast with the "vertical" structure of hierarchies of value stands the "horizontal," that is non-structural relation of value autonomy.[3] This relation involves external relations of reciprocal, mutual difference in which values can form distinctive categories, inherent value and agents good in themselves. The "form" is not essential as in hierarchy, but extrinsic and incidental. To explicate what this means we may take knowledge as an example. In a hierarchy, knowledge would receive its value by the authority of reason, or from some other standard such as pleasure. It would be grounded in a higher value. If knowledge is autonomous, however, it stands by itself, not subordinated under a hierarchy. Knowledge has inherent value, not of derivative value, for there is no *pros hen* to a first value: knowledge has value by itself. Knowledge is self-authoritative, auto-nomic, and not in need of a censorious standard with which to decide what can or cannot be known. Knowledge is more truly valuable as it has value by itself. Inherent values are worth pursuing for their own sake, ends in themselves. Knowledge is properly sought for itself, not for wealth or any other value to which it might become subordinated. Its pursuit involves procedures requisite to it alone, for example, experiments and laboratories, not banks and factories. Its goals are autonomic, self-legislating of the means and actions proper to this distinctive goal. An end-in-itself, worthwhile by itself, involves means peculiar to itself, for the most part. As an end in itself it involves autonomy, with its distinctive goal, and requisite means are involved in its realization. The means are proper to the goal, evolved to achieve it, and regulated by it.

Authority in a reciprocal structure is autonomic not hierocratic. Subordination under a hierarchy ends autonomy. Yet autonomy is requisite for any value, as a value must be independent to reveal what is distinctively valuable about it. Standards of value, by comparison, are more standards than values, for they subvert value. They undermine intrinsic worth by subordinating all worth. Worth is reduced: to the worth of the one standard. Only one worthwhile value remains from such a reductive enterprise.

The differentiation of autonomous values increases value, while subordination decreases value. Values in relation to a first are not values in themselves, not distinctive. But what of the possible alternative of a "loose" hierarchy, in which values were ranked in order to distinguish higher and lower values, but in which the values would have no relation of derivation, avoids this reduction of value? (Scheler, 1916) We will discuss ranking in the next section; it is enough to state here that this movement away from monism of value is toward autonomy. Thus it implicitly supports a theory of fuller, greater value differences, even if it does not go far enough. The indication is that subordination is of less value than the differentiation of distinctive values that is the mark of autonomy. For the pluralization attending differentiation increases the movement away from monism, and toward greater value. Differentiation of value ends in a "horizontal," non-authoritarian, and pluralist theory of values.

Values not value: no one ruling authority gives exclusive worth (cf. "the" good, *agathon*). Instead, pluralities of values define spheres of distinctive worth autonomously. The relation of these values does not form a hierarchy, but is reciprocal. The reciprocal structure involves distinct, plural values, not a hierarchy, which tends to subordinate distinct values to one value. In principle, no one of the plurality of values can emerge as a "first among equals," for this would subvert autonomy and tend toward hierarchy. Autonomy involves reciprocity, for the autonomy of any value requires avoiding subordination, and the reciprocity must be mutually extended to any other value. It should avoid subordination under all the other values, just as they need insubordination to each other.

Autonomous values that have a reciprocal structure are valuable in themselves and by themselves. The logic of intrinsic and inherent value is anti-hierarchical.[4] Inherent value cannot derive its value from an extrinsic source and remain inherently good. As good in itself, its value must be inherent and thus free from subordination: autonomous. This principle or norm applies *a fortiori* to intrinsically valuable agents. The reciprocal structure is extrinsic to their distinctive value, though it safeguards their autonomy. The distinctiveness of an autonomous value shares nothing with any different value but inherent worth; worth is plural, for differences of value are revealed. Their exclusiveness suggests a category principle in the sense of a mutual, extrinsic, and systematic division of spheres. Different values imply multiple, plural values, distinct from one another. Plural values involve differentiated categories of value, as their mutual exclusiveness implies a number of different spheres of value. Were their number to be reduced, meliorism and autonomy would be threatened. A theory of value categories is a part of value autonomy. The structure of the categorial values is, again, reciprocal, for subordination would undermine categorial distinction.

Unlike Immanuel Kant's theory of categories, for example, this categorial principle is not superordinate. The principle does not constitute instances "under" a category. Categories of value serve to distinguish only. The "reciprocal imperative" demands reciprocity with factors intrinsic to the realization of their distinctive value. Their worth would remain unactualized were they to remain mere principles. The categories are extrinsically differentiated from one another but involve an internal relation to ends and means. To be creatively actualized, they must be enacted: the categories are reciprocal with the action they delimit or they would be practically worthless. Again, unless they are adopted as a goal or end, their value is restricted to speculation. While they differentiate goals, they must be internally related to their own goal to demonstrate their true value. Consistency demands that a multiplicity of internal relations within a category (goal, act, value) should be reciprocal not hierocratic.

While ends do regulate means to some degree, there can be no "teleocracy" within a category in any strict sense, for no *telos* or goal can be creatively actualized without action. Ends depend upon means for realization. They are reciprocal, for without means, ends are mere whims, while without ends, means are end-less: futile. Ends cannot justify means without evaluating them; the choice of means is not value neutral. Reciprocity may replace the normative structure of ends and means with a horizontal, temporal structure.

The reciprocal relation of differences in value autonomy replaces form and content with a pragmatic structure of differentiation. "Form" is the "being" of values, since the form is what endures as the "being" of the kind of value. Essentialism in value relations is superceded by creative actualization, since duration is replaced by meliorism: open-ended improvement. What is worth differentiating is self-grounded as autonomous and valuable in itself. There is no extrinsic relation to an alien ground that "justifies" subordinate values in a hierarchy. For example, value categories differentiate the ends of a culture. They are not the determinate, specific idiosyncrasies of a culture, that is, cultural values, but the differentiation of its goals. They are neither form nor content, as their structure is extrinsic to their worth as autonomous. They are evident through their adaptation by a culture; their value has been revealed to a culture in some manner.[5]

Given that a culture devotes some energy to a category of value, its worth is forcefully apparent to the culture. The value has been differentiated as a goal from the other goals and activities of the culture. They have emerged over the course of its history, by trial and error, or perhaps by sudden flashes of insight. The value categories are not obvious or "given," but have been differentiated by chance or insight over the course of time.

The "horizontal" structure of values in relation is echoed by the horizontal structure of value actualization over time. Action transforms the world, and creates novelties and gradual improvements. (The changes brought about by weathering and geological "actions" are of course actions only in a metaphorical sense. Nevertheless, we speak of the "action" of a river in "carving" a valley. The difference is that these changes are not deliberate.) This temporalization of value has several theoretical consequences. One is that differentiation of values is meliorist. Whatever judgment is made about the value of life as a whole, differentiation involves a relative improvement. The emergence of new values means an increase in the quantity of value, for more values have become differentiated. The quality is increased as well, for a distinct category of value involves the distinction of a new value quality from other categories. Another consequence is the priority of change over ""being.""

Horizontal power relations, as mutually limiting and reciprocal, are consistent with and embody justice. Reciprocal relations involve the relation of value equality; compossible and plural values call for collective judgment, not a "higher authority." Power is dispersed in proportion to value equality: equal power means equal freedom. Indeed, value autonomy invites insubordination of inherent values: knowledge for its own sake, not for the sake of faith, money, power or other extraneous ends. An end in itself is justified inherently, as valuable in itself.

Traditionalists could argue that only a supreme standard can truly measure improvement. In turn, this makes reference to a fixed hierarchy, by which improvement can be gauged. The improvement is in terms of the hierarchy. I would respond that value hierarchies have tended to resist changes in cultural values in practice because they cannot encompass the emergence of new values in theory. Value is a fixed relation to the highest standard in a hierarchy. Anything that cannot be subordinated to this standard is "bad." Novel values, in respect of their novelty, do not fit into this fixed relation. Their newness is a double threat: to the fixity of the relations, and to the supremacy of the standard. Because the novelties have not emerged from the standard, its position as the hierocratic source of value is threatened (Galileo, Darwin). Defenders of traditional, high culture originally viewed jazz as trash, because its relation to their standards of aesthetic value had not been worked out.[6] Its novelty was perceived as a threat. Now that jazz has become established, it has turned tables and become a standard itself (compare rock music).

A different lesson is provided by the contemporary differentiation of entertainment value from aesthetics. One function art used to have was entertainment. Entertainment was subordinate, however, for it was but a part of art; its value was derived from higher aesthetic standards. The development of film and television along lines of "popular culture" was condemned by this standard, but no one ever claimed that television is art. Movies and TV were the means by which entertainment was liberated from subordination to aesthetics. This has not "split" culture so much as increased value. Good art need no longer cater or pander to popular taste, while in turn, people have a choice between aesthetic enrichment and entertainment. Yet entertainment is still criticized by the extraneous standards of "high culture," as if sit-coms ought to resemble Shakespeare; or rock music, Wagner. The extreme reluctance with which novelty is accommodated in terms of entrenched value hierarchies should be evident here. Consider the Church's opposition to novelty from entrenched power hierarchies, for example, the Church's opposition to Thomas Aquinas, who eventually became its "angelic" doctor; and to the mysticism of Teresa of Avila, who became a saint. The novel value usually emerges in spite of the hierarchy, not because of it.

Such a diversity of values is the condition of a good life, since such a life includes distinct activities: plural goods. However, life is not the supreme standard for value, since values are autonomous from any standard. For the problematic of value is not identical with the problematic denoted by "The Good." Philosophy of Value was formulated within the context of differences in value, economic, religious, social, esthetic, and other kinds; as well as cultural differences and personal differences. The problematic of the Form of the Good was a one in many, an identity of all goods. The problematic of value is plural values, that is, differences in value. However, such a problematic does not imply that there can be no identity or principle of values, as opposed to a One in value. For creative actualization is a theory of values, that is, the identity of all differences. The categories and levels of value differ, but all require creative actualization.

3.

One potential problem for a value autonomous theory is rank. Is there no better and worse? Are all values undifferentiated in rank? Does plurality mean that any value is possible, that all values are equal in value? Does value as creative actualization avoid hierarchy? The tendency in the history of value theory has been to reduce values to rank: Plato (reason as the judge of value), Aurelius Augustine (faith), positivism (knowledge), Friedrich Nietzsche ("order of rank"), and so on. I will argue in this section that ranking as a model for value destroys the autonomy of value, and thereby reduces or eliminates the options available for choice. (Historical revaluations are sometimes subordinating, for example, in politicization of non-political values (the Nazis' approach to art). However, they can also be insubordinate: the freeing of religion from the state at the beginning of the Middle Ages; and of intellect during the Renaissance.) Ranking is also unjust to differences in value, plural values, reducing value by subordination. I will distinguish value and good from evaluation as ranking.

One way to approach such questions is to see if any historical theory combined autonomy with subordination. Kant's moral theory, which is the pioneer in the use of autonomy, is also intellectualist (Kant, 1785, 1788). Reason is the standard by which moral worth is judged, even as the autonomous individual legislates according to it. Autonomy means self-legislation, making laws for oneself (Paton, 1947, esp. ch. 4). In a sense we place ourselves under law by bringing will in line with it: willing universal law. Norms in Kant's morals are not necessarily standards. Obligation is universal in form, but autonomic: valid for one individual will. Reason is used to give universal form, but this form does not give content. Content is derived from maxims of the will which contain ends and rules governing types of situations and circumstances (*ibid.*). These maxims are particular, as governing distinct types of situations (if in all such cases) and therefore of limited applicability. More importantly, they are not derived from reason even if they must conform to reason, in the sense of universal law, to achieve moral worth.[6] Reason is not a value standard, as worth is not derived from it. My maxims are not necessarily your maxims, although each must conform to reason as a norm governing moral worth.[7]

There is a definite ranking in Kant's morals, but not a hierarchy. In short there can be rankings without a hierocratic principle that assigns rank, and from which value is derived. What other factor can create rank? It will be recalled that autonomous values stand in an extrinsic relation of mutual autonomy. Because the relation is extrinsic, it does not threaten the autonomy of any value category. Nor does it prevent comparative, extrinsic rankings. The comparison of equal values can result in an extrinsic ranking, for example, the quality of two different wines or the paintings of two artists. Such a ranking would be without an organizing principle, or standard, as rank is not based upon a relation to a "first." Instead, the intrinsic value of one is compared in relation to the intrinsic value of another. The comparison does not start from a standard but begins in time. The rank is based on quality of the values themselves in comparison. Again, competition may result in a ranking which cannot be based on any standard. There is no outside limit to how fast runners can run, for records can always be broken. Ranks may emerge over time and then be "revalued" later.

My argument here is that worth grounds rank: that no value rankings, hierarchies included, are possible without value autonomy.[8] As this flies in the face of most value theory, some justification must be given. Most value theories confuse all values with hierarchy. (Perhaps they have been misled by rank, which we tried to show above is not equivalent to hierarchy). Value is in proportion to rank within a hierarchy for such theories. We would argue, however, that there can be no ranking without difference. One glass of wine cannot be superior to another taken from the same bottle; differences are requisite for comparison. And there can be no differences without autonomy. For a hierarchy does not have differences in value, only in degree of the one value, the standard. No differences in values are available for ranking in a hierarchy. True difference requires autonomy from subordination to another value. Difference requires an extrinsic relation of independence; dependence would mean that worth would be shared not different. Value hierarchy requires value difference, that is, autonomous values and kinds or categories of value. Since worth grounds rank, reductionism is undermined. Hierarchies are a reduction of values to value. Autonomy of value, the condition of ranking, requires plural values. Both, not either/or. (Similarly, pluralism and autonomy are supportive of freedom of taste, and undermine any elite standard in taste.)

What of differences in degree? I would argue that differences in degree are not sufficient to rank different values.[9] Ranking and judgments of quality are distinct types of evaluations, also distinct from hierarchy. Both of the former involve different values arranged in a ranking, but the degree of quality of a value, or its means of fulfillment are distinct from other values. Hierarchy is rule of lower offices by a higher; as well as classification by higher and lower types, steps or concepts. Compare scales in science, for example, of hardness, of solubility, and so on. To say that different things are rational in different degrees is not sufficient to compare them, only these in relation to reason (or in relation to pleasure, happiness, and so on for other theories of value.) Their own worth cannot be ranked, that is, their differences in worth, or what is uniquely valuable about them. To argue that they have no worth without the hierocratic standard would support our argument that they cannot be ranked without autonomous value. For it denies both autonomous worth and any intrinsic principle of ranking. To reduce all ranks to differences in degree would be a denial of different kinds of value. But without such kinds how could differences of inferior rank be established? If pleasure and *thymos* are good only insofar as they participate in reason, what is the difference in value between pleasure and *thymos* themselves? Since there would be no difference in kind of value, they are of indeterminate value. Their degree of participation cannot be established without reference to kind. Again, rank cannot be established.

What of different kinds of values? Can prosperity or beauty be justified by a supreme value, for example, knowledge? The kinds or categories of value argue qua categories for autonomy. While any value can be evaluated for its instrumental value, including kinds, such evaluations are grounded in the inherent value of the kind: the distinct norms that separate kinds as such. What of values subordinated under the kind? Are individual actions "brought under" kinds, thereby creating a hierarchy? Are the value categories a case of subordination "under" a distinct value? In a sense this raises the problem of identities: cultural and categorial values are general, that is, collective terms for individual practices and actions.

If the hierarchy of value kinds is not actual, not in the structure of the world, how are they creatively actualized? The answer lies in action, which individuates norms as instances. The kind is "embodied" in individual acts and their consequences. Here the degrees of value achieved by creative actualization argue against any ranking, since they are instantiated as the same instance. An economic or esthetic good may instantiate both an individual instance of quality and the standard of quality: both instance and norm. There is no actual difference and thereby no hierarchy in actuality. Only in evaluation are inherent goods subordinated as instruments of other goods, and evaluation is grounded in this inherent value. The same considerations apply to cultural choices and personal evaluations of quality. Both require inherent value as their condition.

Such a hierarchy is found in Platonism and, derivatively, any foundationalism of values or *summum bonum*. Only Platonism demands hierarchy, as the higher is more truly real. However, forms embodied in instances could be imagined as reciprocal. The species and genus both identify the individual as a kind of individual equally. It is equally a species difference and a generic identity. The genus is quantitatively larger only: universals are a problem of involving quantification. Species and genus as normative identities are reciprocal with their instance in the sense that the instance creatively actualizes the normative identity in the world. Such normative identities are the relational element in the actual.

An important point to consider is that with value autonomy basic, values retain value regardless of derivative rankings. Their structure is the ground, the condition without which no ranking can be made. To emphasize this point the process of ranking should be entitled "evaluation" to distinguish it from the reciprocal structure of value autonomy. The rankings are not evaluated out of the autonomous values in a bottom-up hierarchy, for this might involve subordination, and thus violate the autonomy of distinct values. The autonomous values retain value apart from any evaluation; they are reciprocal in structure. For without this autonomy no ranking can be evaluated, as we have argued. Ranking retains the external relation of values without reference back to a first.

The emphasis on the autonomy of distinct values may suggest that what is unique may be more valuable, since it is less identical with other values, rarer. In turn a hierarchy or ranking is suggested with degrees of uniqueness as the standard. For example, a work of art is more unique than a product of agriculture (cf. Hannah Arendt on labor vs. works in 1958). But "unique" is more of a principle of difference than of identity, that is of difference from other values and thus an extrinsic relation of autonomy. Be this as it may, what is "unique" would be futile as a theory of value, since almost anything can be copied. It may be that the sweetness of tone of a Stradivarius or the peculiar tints of a masterpiece of painting can never quite be duplicated. But these are exceptions.

In sum, rankings must be derivative from and consequent upon reciprocity, that is, distinct autonomous values: inherent worth grounds all ranking. There can be no *a priori* rank structure, for rank is derivative and subject to re-evaluation, not fixed. Rank is earned, that is, emergent from historical changes. Just as elected officials in government have power in virtue of their office, but are subject to recall, and "being" voted out of office, so a value can appear more important to one age, and less so to another. Nor does ranking constitute a surreptitious insertion of a *pros hen* relation into value theory. Rank is not grounded in autonomous values, for (1) value remains the same, whether the value is autonomous or evaluated. Rank is not equivalent to value; value is distinct from rank. (2) Rank can vary with historical circumstances, notably in historical revaluations (Nietzsche). Thus there is no fixed relation to a justifying first, a timeless hierarchy. (3) Differences in value are distinct from evaluation of competing values. Values retain value in evaluation, but not necessarily rank. The value remains the same whether it is autonomous or ranked, or it could not be ranked. Values, then, can have different roles in different modes, autonomously.

Plural values *per se* undermine hierarchy with their extrinsic, reciprocal relation. Such a relation implies a type of axiological equality between autonomous values. All values equally are creatively actualized. Creative actualization differs from hierarchical and *summum bonum* theories in that the source of value is not some justifying first, but inherent in the goal achieved by the action and activity of creative actualization. Value is proven by the test of creative actualization, whether a kind of value, a personal value such as happiness, or a novelty. All values and goods are a function of creative actualization, and this characterization argues against inherent ranking. Other modes of value do not "refer back" to creative actualization for justification, since if they have not been creatively actualized they cannot have value. Potential happiness does not brighten our lives, but actual happiness. Creative actualization is axiologically required for happiness; it does not stand in a the relation of justifying happiness, but of bringing happiness. Certainly, the evaluation is made that creative actualization of the instance would be worthwhile: it is not as if evaluation and value are in different universes. However, the instance has inherent value, and once creatively actualized in the world is autonomous from evaluation qua value. Values have a reciprocal relation with other values in the world. As I argued in a previous chapter, values and goods that do not attain reciprocity are inferior, for they do not increase the good in the world (atomic war).

Similarly, personal values require creative actualization. We act to actualize pleasures that we evaluate worthwhile, for example enhancing recipes to accord with personal evaluations of quality. With such creative actualizations we may attempt to follow a particular mood and thereby particular, unique circumstances. Thus the actions or activity may be creative in a contextual sense. The good or value may meet standards based on personal evaluations of quality, but this is an evaluation. The personal value itself must be creatively actualized or it remains a mere potentiality.

How does evaluation proceed, then, if it involves procedures distinct from the values it ranks? Ranking cannot begin by building up from these values, in an atomist manner. Beginning is a problem for evaluation, then. But to begin on any project is to aim at some goal. The project is justified in terms of its projected, consequent value. Among competing possible rankings, that is, different orders of the values, a choice must be made of which is superior given some end in view. Choice presumes different values, and thereby, reciprocity. Comparing values in evaluation takes place in the context of inherent value. In practice, the values must compete for which will be chosen as a goal. The choice is made in view of the future of as yet unactualized possibilities. Evaluation includes choice, for competing rankings may be creatively actualized through action. As only one, or at most a few can be creatively actualized by any one action, some are ruled out in evaluation. Others will lose out or be subordinated as a mere means to a goal, for example, if knowledge is used as a means to assist health (medical science) and not pursued for its own sake (pure knowledge). Instrumental and utilitarian values require some autonomy for values, in order to evaluate differences between goals and means.

We may choose future goals, but we believe that they can be creatively actualized. We would not choose impossible values or goals. Some of these choices may be difficult, of course, since the value of each alternative possibility may be worthwhile. Our choice may involve a ranking of different possibilities in terms of worth or quality. However, the reciprocal structure of values compossible with each other in the world should be independent of our choices. Value autonomy means that the value of a distinct value or kind of value is independent of our choice. Choice is our own affair, but choices involve selecting between autonomous, inherent values. Choice follows upon evaluation, putting values into a ranking for selection, but the values must exhibit value to be chosen. Their independence or autonomy from choice as the source of value is reciprocal with choice, or they could not be chosen. We may also choose them for their own sake, but the ground of such a choice consists in evidence of their inherent value. Indeed, we prize their inherent value: that they have independent, autonomous worth. Their value does not proceed from the authority of the subject over the object, as in the Cartesian model, even if our pursuit of such a value or good aids in its creative actualization. On the contrary, choice is a reciprocal model. Creative actualization of the value or good in the world as a choice requires that the goods are grounded in the world: actual not potential. The value of a value or good is not subjective. Without differences in value, no choices among them could be made. Without mutual autonomy, an extrinsic relation, no differences in value would be possible. The reciprocal structure of values in the world is independent of choice and the ranking or evaluation that precedes choice.

Moreover, the self is also creatively actualized in acts of autonomous choice. We create ourselves through our choices, in self-actualization. The self is not an underlying "substance" but a series of historical choices that constitute a life (Jean-Paul Sartre, 1947 and Raymond Polin 1972). The self can thereby change over time. Partly these choices are culturally determined; partly they are autonomous. A choice of career, of life-style, or any other major choice requires creative actualization and changes the self in its enactment.

The ranking that results from evaluation is the outcome of comparisons, for example, in the evaluation that a hammer is more suitable as a tool in a specific situation, or that one hammer is better than another. Comparative value, then, is "a posteriori." Ranking emerges out of competing claims, theories, conceptual schemes, opinions, beliefs, and other candidates that must be evaluated. (Insofar as conceptual status is a species of rank, as degree of abstraction, for example in genus and species, it is grounded in fruitfulness for communication, identification/differentiation, knowledge, and so on.) The value question emerges as a consequence but at a deeper level. The question of rank arises in practical or theoretical situations in which mere relativism is unsatisfactory, for example: will this bridge collapse if we cross it? The value of preserving life is here ranked superior to concessions to tolerance for different opinions about the best way to construct the bridge. A corollary for evaluation is that comparative value is not eternal. Evaluation is historically situated. Indeed, evaluation of goals involves reference to the future: it is through and through temporal. Different ranks might result under different circumstances. If we may loosely say that religion was the highest value for the medieval epoch, .it can be fairly stated that it no longer holds the pre-eminent position. Philosophy is no longer the handmaiden of theology.

Second, justification can no longer be in terms of a "first," a standard to which appeal can be made. For there is no eternal hierarchy under which values are subordinated. Autonomic values do not form an intrinsic hierarchy. How then are rankings "values" given this theory? Any rank that is to be evaluated must be historically justified. Its value must become evident or be made apparent. The subordination is not justified by reference to a standard: fitting into or under a hierarchy. The subordination is justified only under the condition that it is to an intrinsically different value. For if no good end were served, how could subordination be justified? Thus a ranking of theories may be justified by the goal of knowledge. In turn, the value of knowledge is intrinsically different from, for example, health or wealth. If two competing versions of "truth" must be evaluated, for example, in court, the value of truth remains superordinate. Lies subordinate truth to alien ends, for example, gain or ambition. Such untruthful evaluations would be bad evaluations. Further, competing means within a category may result in historical selection of one over another (surgery over bleeding), but the distinctive worth of the category justifies this evaluation. Worth again grounds rank.

Evaluation rankings are justified historically in terms of pragmatic worth. We must pragmatically evaluate the worth of the goal prior to any "subjective" evaluation. Such evaluations have resulted in relatively superior practice. The practices need not be absolutely superior to be adopted. Surgery is risky, but nevertheless more efficacious than magical incantations in achieving the goal of health. Practice is a criterion of value, for what is practiced exhibits the evaluation in which, at one point in history, one practice was rated superior over another. Such practices are above all practiced, for regardless of their shortcomings, there is little else on the horizon that could replace them, or the alternative might itself be adopted (subject to qualifications undertaken in the next section.) The values they are intended to creatively actualize do not entail their own means; evaluation of competing means is distinct. It is not always obvious what will produce health or result in knowledge. The means must be worked out over time.

The end too must be justified pragmatically. The distinct value that has been adopted as a goal justifies the practice, but the practice justifies the goal as well. Autonomic values have proven their worth, thus they differentiate their distinctive goals. This has taken place through practice: they have been practically distinguished as worth pursuit. Practice is reciprocal with value, in an internal relation, for the value differentiates the practice, while the practice creatively actualizes the value. The practice is worthwhile, as distinctive from other practice: justified by the distinct value it creatively actualizes.

4.

Reciprocity of structure is not simply a consequence of value autonomy, but bound up with it more intimately. Despite the externality of the structure of reciprocity, it is not incidental. For what principle or norm can guarantee autonomy? How can subordination of one value by another be checked? What is to prevent aggrandizement of one category of value on the sphere of another? The answer lies in the reciprocal structure itself. The spheres have worth divided between them such that they collectively encompass all practical worth. Their relation is not absolutely autonomous then, as the relation forms a whole, the whole of value.

Within this whole, justice must be done to each kind of value. None can be subordinated unjustly without doing violence to value. Thus the reciprocal structure is good for autonomy, as it guarantees that each distinctive value shall have its proper sphere. Reciprocity has the relation of justice, a proper balance between different spheres. Reciprocal justice involves an equal relation of inherent values, as inherent. What of other candidates for justice, for example, justice in distribution by merit, involving hierarchy? Distributive justice presumes reciprocal justice, since each agent is equally a candidate for such merit. The inherent worth of each grounds their rank.

Autonomy is not structureless, but has a moral structure. Each category is mutually limited to a particular sphere. To exceed such limits would threaten autonomy. The limits are provided by the just structure of reciprocity. Violations of these limits cannot be justified in terms of any one category of value, for example, wealth, as they threaten categorical structure. If all were subordinated to wealth, no knowledge, health, beauty, or any other value that did not immediately result in remunerative gain would or could be of any value. The pragmatic value of categories becomes creatively actualized through reciprocal structure. Distinctive values are possible only within moral limits.

Moral *arguments* are frequently in a hierarchical form: pleasures and inclinations are evaluated against moral limits. Many pleasures and inclinations are within such limits and thereby can be evaluated as morally permissible. Others are ruled out, subordinated to duties. The ground of such evaluations is the superior worth of what is morally right. Are the values subordinated under moral values in a hierarchy, then? Do moral values constitute a standard for value categories? With extrinsic autonomy, values achieve a reciprocal structure, that is, justice to each distinct value. They exhibit and creatively actualize a moral structure of value. Moral worth is in turn creatively actualized by the realization of categorical autonomy, in a sense. Though pragmatic value, the worth creatively actualized by the autonomous value categories, requires a reciprocal, moral relation, the reverse is not true. They are co-conditional in one sense, but not another. Pragmatic value justifies historical and other differentiation. Yet creating new differences for their own sake, that is, quantity, is not requisite for reciprocity. Whether two or two hundred values were realizable, these would require a moral structure. In a sense, reciprocity involves differentiation, so is not its condition. For without differences there can be no relation of reciprocity. They are coeval, so some differentiation of pragmatic worth is requisite for realization of morals. Yet moral values do not determine how many or what kind of value categories will be practiced.

This answers the question on the relation of moral and category values. They are coeval. Morals do not subordinate the categories of value—their autonomy is preserved—as their autonomy demands a moral structure *per se*. Subordination of one by another would be unjust, therefore not moral. By maintaining a reciprocal structure, they creatively actualize moral worth.

The categories are formally equal if they stand in a just relation. In practice, that is, in a culture, one or another may predominate but none is, or should be, of higher rank *per se*. Reciprocal justice involves equality in an internal relation, for a higher or lower would involve subordination. Moral values are creatively actualized by an external relation of autonomous values, but themselves have an internal relation to each other. Autonomy means freedom from extrinsic values; freedom is involved as a moral value in reciprocity. Subordination of all categories by one category or another (hierarchy) is non-reciprocal, which means injustice to their differences, inequality, non-autonomy, and so on. Truth is also a moral value involved in reciprocal structure. For the true value of distinct categories cannot be revealed when subordinate; subordination ends autonomy. Autonomy is requisite for fully creatively actualized value. The value must be different to reveal what is truly distinctive about it. The true value does not derive from an authority to which it is subordinated, but is revealed reciprocally with distinct values. What is distinctly valuable must stand by itself for its truth, autonomously. Autonomy involves reciprocity. Thus truth stands in an internal relation to other moral values, justice (reciprocal structure), freedom (autonomy), and equality (insubordination). From these follow compossibility, pluralism, and other moral norms.

Moral values are not a standard for other values. The moral structure of autonomous values means only that a specific relation among values has been achieved. Moral structure is a relation of autonomous values, not a standard for them. What matters is the proper relation, not some absolute standard to which all is subordinated. Although autonomy is achieved with reciprocal structure, the autonomy is not absolute. For its distinctiveness is reciprocal and coeval with other values, and thus forms an open system of values. The categories of value divide up spheres of worth among themselves, forming a mutual system of all worth. Insubordination, however, means that it cannot be timeless, as absolute values might be. The "system" is an open one, in which greater worth is achieved through differentiation and amelioration, the axiological principle in the relation. In moral relations of duty, there is no hierarchy, since agents evaluate others as having equal moral worth. The subordination of actions to principles requires the reciprocal creative actualization of the principle through action. Reciprocity is also creatively actualized in the temporal relation of actions to the future. The future goal in one sense regulates, but in another is contingent upon the present action.

A moral structure for values makes no "ontological" claims or commitments. Firstly, its superior worth does not rest upon its realization. Value autonomy would have greater value than hierarchy whether or not it was fully creatively actualized. Secondly, realization involves goals, means, and especially action. These are analytically distinct from value structure, although they are not value neutral. Finally, structure or reciprocal form is only a relation, not an "entity" per se. The relation is not an entity or ""being."" Two or more values can stand in a reciprocal relation, without changing the structure. This is not to say that it is ideal, however. The distinction of is and ought is only abstract, not real. For were categorical values to become pragmatically creatively actualized in reciprocal balance, and manifest in institutions, structure would be as it ought to be. "Is" and "ought" would coincide. Nevertheless, the priority of change implied in creative actualization marks the end of subordination to the ontological of all values. "Ontology" is superceded by creative actualization.

What is the significance of this moral relation of values to personal values, the traditional province of morals? Autonomy implies duty, that is, mutual regard for others, Indeed, we would argue that all duty has a reciprocal structure. Without general obligation to others, my autonomy is threatened. However, value reciprocity does not rule out self-realization, happiness or pleasure. (For that matter, neither does Kantian ethics. Contrary to many prevailing interpretations, Kant viewed development of talent as a duty, and was far from forbidding pleasures that did not conflict with duty.) Different personal talents and favored pleasures may be deemed personal values. That these may be creatively actualized is part of autonomy, of personal choice. In accord with meliorism, it is better that as many personal values as possible be developed, which may improve the future, yet not interfere with the autonomy of others (reciprocal duty, compossibility). Personal autonomy is less a harmony—a subordination of "lower" notes to "higher"—than the reciprocal development of different personal possibilities. Just as the autonomy of the categories means self-grounding, or the development of their full worth for its own sake, so is personal development pragmatic. Personal values are within the bounds of duty, but not prescribed by duty as auto-nomic: self-regulating. In practice this would mean that personal differences would not be subordinated to intellectualist specialization. Personal capacities would be chosen for their own sake, although these cannot be characterized, given autonomy.

Duty would prevail only in value conflicts. Value conflicts make reference to evaluation, however, as such conflicts involve autonomous values which can potentially conflict. Personal values would remain autonomous, but not actualized in particular situations in which duty should prevail. Moral values would be evaluated higher in such situations, but only in them. Personal autonomy would remain or ensue at other times. Moral authority would be limited and temporary. The justification is that autonomy involves reciprocity. The alternative to moral evaluation is subordination and an end to value autonomy, whether personal or categorical.

5.

Autonomy of values means that a new value relation is basic to ranking. Rank is consequent upon reciprocity. Differentiation and not hierarchy is the relation of values. Morals are not hierarchical, as a reciprocal relation of values involves the creative actualization of a moral structure. The Greek conception of values, which can only be hierocratic, is upset.

The authoritarian structure of value hierarchies subordinates reciprocity to a first, a standard. The authoritarian standard resembles a monarchial form of government, in which authority flows from a single person, a highest authority. Value autonomy is more akin to a federation or a republic of values, not a monarchy. It forms a republic of equals but with differences, of emergent pluralism. Plural values reciprocally emerge in historical time. They become valued and revalued according to evident worth.

However, value autonomy cannot be confined to value theory. For worth grounds rank: rank has a ground in difference, that is, autonomous values. As values are insubordinate, autonomy constitutes a challenge to metaphysics, which creates general hierarchical structures. All metaphysical structures require ranking, whether hierarchical, foundational or composite-developmental. (Rank cannot be a ground, since the separation of ground and consequent itself constitutes a ranking.) All metaphysics is hierocratic, for the "first principle" or "first philosophy" of all reality is the subject matter of metaphysics. All reality and all knowledge is brought into a relation to one cause or unifying principle, be this "being," atoms, reason, sense data, language, "the Form of the Good," process, and so on. The hierarchy may proceed from a first, as in Plotinus' emanation, or be built up from a uniform principle at the base, as in atomism and phenomenalism: metaphysical foundationalism. Both form a hierarchical structure in which relations are in terms of a first principle. Hierarchy philosophies have some one systematic, organizing principle, be this emanation, participation, development, *pros hen*, composition, reduction, or what not. This principle regulates the formation of the hierarchy, the relation of levels. Intellectualist norms involve increasing generality of scope as one ascends, namely, instance, species, transcendental, unifying standard. Yet intellectualism is not the only hierocratic principle. Compare atoms, molecules, organs, individuals, kinds, laws, "unified field theory"; with sense data, "association of ideas," compound ideas, "objects of sense," forms, principles, ideas; and again (phonemes), words, concepts, phrases, sentences, paragraphs, grammar, meanings, and language as a whole.

The "form of the good" is the metaphysical basis for later hierarchies of value, including *summum bonum* theories. Lower values derive their value from the *summum bonum*, just as copies derive their reality from the form of the good. Other paradigms of metaphysics are based upon this hierarchical model, despite their rejection of the first principle of the Form of the Good. Plato's hierarchy of Good becomes a hierarchy of "beings" in Aristotle and subsequently the One or God in Neo-Platonism,[10] and in medieval philosophy for which God is at the apex as the most perfect "being" and the source of all other "beings". For Descartes it becomes a hierarchy of degrees of certainty of knowledge for the knowing subject, to which all knowledge refers back. For logic it becomes the priority of propositions utilizing the copula as models for language. This originated in Aristotle, but was adopted by Moore in his analysis of the meaning of value judgments. Bertrand Russell called for a "hierarchy of meta-languages" in his Introduction to Ludwig Wittgenstein's *Tractatus*. Thus hierarchy permeates all subsequent paradigms of philosophy at their root (McDonald, 2004). In turn, the hierarchical model grounds ranking as the model for value theory, *summum bonum* models, originally based on participation in the Form of the Good. Thus subsequent metaphysics is based on a hierarchical relation of values to the supreme good in Plato.

Such vertical hierarchies proceed from the authority of a first, whether from above (for example, emanation), or below (as in atomism). Systematic philosophy is often a vertical ranking, a closed hierarchy. "First" philosophy is superordinate, while other areas of philosophic purport are subordinated, either logically, in order of knowledge, in order of "being," or another ordering principle. Subordination of matter to form, of attributes to substance, of existence to essence are all different sorts of hierarchies. The *pros hen* relation also forms such a hierarchy: of the changing to the unchanging. These reflect evaluations that one reality or concept is "higher" than the other. Reduction in philosophy, for example reductive materialism, reduces phenomena, objects or other candidate actualities to the first, whether matter, spirit (Berkeley), or another monistic principle. Metaphysical foundationalism is the reverse of idealism, but replacing the One with matter does not eliminate a hierarchy of values based on internal states such as pleasure. All systems contain such hierarchies, whether hierarchical, emanation from the "One"; *pros hen*, or referring back to "being" as degrees of "being"; or (metaphysical) foundationalism: the higher are conditioned by and a composite of the lower. The hierarchy relation, then, involves a participation, *pros hen* or reductive relation: referring back to a first, a one-in-many. The Heideggerian-Derridaian project of "Deconstruction" confuses hierarchy with polarity. To overturn a hierarchy, for example, logocentrism, is not to reverse a polarity—not to take the contrary of the first. It does not reverse truth and falsity, for example, since the truth of deconstruction is presumed.

Any actualities/bodies/objects/percepts[11] can be juxtaposed such that they lie next to one another, insubordinate to any one. We would argue, therefore, that neither spatial location nor perception can provide a ground for hierarchy. Hierarchies are revealed to be evaluative-normative: things are subordinated and classified under a first principle (for example, reason: "logical" subordination) which is less descriptive of "reality" than prescriptive. The first principle of the hierarchy is a model or norm for that which is subordinated under it.

If worth grounds rank this hierarchical structure has a challenger. For hierarchies attempt to rank, to subordinate under a norm or organizing first principle. We have argued that ranking presumes value autonomy, that is, that which is to provide differences for rank. Hierarchies cannot escape this requirement, for they are normative. Indeed, it might be argued that metaphysics is a barely disguised value hierarchy of higher and lower in which all is subordinated to a normative value standard. Be this as it may, hierarchies are clearly not reciprocal in structure, nor grounded in difference.

The metaphysics of substance and attributes divides an instance into what is "unchanging" and what changes, and projects this as real. The "unchanging" or "being" of the instance is then made into a first, a ground. This judgment constitutes an evaluation of some features as subordinate to others, since the instance cannot be ontically separated but only logically: the distinction is "ontological" not actual: a normative distinction. Further, "being" cannot ground universally except with a structure of hierarchy: as the legitimating first in a hierarchical relation of referring. The subordination of other attributes to substance, especially value to "being" and of "accidents" to "essence" is a hierarchical evaluation, the evaluation that the prior of the contrasted pairs is of greater importance and first in order: an ontological evaluation. The vertical authority of "being" emanates from the top of a hierarchy as the first philosophy in a hierarchy based on relations of dissimilar "attributes." The substance and attribute metaphysics, started by Aristotle but echoed in Hegel and Moore, presumes hierarchy in its hierarchy of degrees of "beings", that is, a relation that was evaluative in origin. Indeed, Hegel's entire system is evaluative, the placement of concepts and categories as higher or lower in the system a hierarchy. The "*pros hen*" relation is an evaluation disguised as onto-logos analysis.

If I am correct, a significant corollary of autonomic values is that the "intentionality" of value categories is avoided. There can be no *pros hen* of the categories to an underlying substance which unites them. For such a substance is a "first" or unifying principle, which subordinates its categories in the order of "being." If value autonomy is the ground of rank, however, value categories cannot be subordinated to substance. The substance-attribute relation is a hierarchy relation that is ultimately derivative from idealist-transcendent systems, based on evaluation, namely, hierarchies of Goods in Plato. I would include value as an attribute or "character" of action or behavior (Dewey) as the same relation. By contrast with Aristotle's reduction or subordination of the good to a transcendent attribute of "being," value autonomy involves their independence from ""being."" Value categories are not categories "of," for they would not then have autonomy, but metaphysical subordination to ""being.""

Similarly, there can be no privileged meaning in conceptual analysis: no *pros hen* of equivocal terms. Analysis of terms in the model of referring back to a primary meaning is replaced by historical norms of equal validity qua norm, but of relative importance historically as some meanings come into fashion and others fade. In this view, concepts, as universals, are norms.

Similarly, autonomous values cannot enter the division of the world, Cartesian in origin, into subjects and objects. Their value is autonomous from any individual subject, and is neither an attribute posited by a subject to an underlying object, nor a "real" attribute of any object ("objective values" in Hartmann). Objectivity of values would subordinate categories of value to "being," just as in Aristotelian philosophy. Subjectivity would subordinate value to reason, feeling, will, or some other psychological or transcendental principle.

Finally, "objective values" or values as "ideal objects" (Scheler) presumes a Cartesian universe of subjects and objects. Values are not "objective," since inherent worth is the ground of the evaluation that the distinction of subjects and objects in the world has value. As Martin Heidegger correctly argued, the subject is in the same world as the object and, moreover, is in-formed with values. Values are not "phenomena," qua creative actualization, nor even objects of experience in all cases: goods are completed actions or actualized values that have achieved their effect, that is, they make or create "objects." The order of rank that Scheler assigns "objectively" to values is derivative, since worth grounds rank.

Autonomous value categories are radically axiological: values at the root. The categories do not reveal value only by virtue of the non-axiological realm, "being" or consciousness. They are worthwhile in themselves, autonomically, and not in relation to some higher standard, whether subjective or objective. Self-worth means self-grounding, which follows from autonomy. The values cannot be grounded in a higher principle if they are the condition of higher and lower. A radical axiology replaces the vertical authority of a hierarchy of fixed "beings", including the "being" of atoms, with autonomous, non-coercive, and thereby pragmatically evident, inherent worthiness. In sum, radical axiology is the rejection of subordination of value to other first principles and first philosophies, whether as attributes of "being"-as-substance, will (the subject as substance) or "concepts." To this we counterpoise reciprocity of values and goods in which inherent worth grounds evaluative rank, and thereby hierarchies are derivative. Values are not instruments of the "one." The moral relation of reciprocal difference replaces the political relation of hierarchy: justice should ground power. Each element of the whole has its own value.

Moreover, creative actualization reveals a world of change, not of "being." Values cannot be attributes of "being," since novel creative actualizations "were not." A world of "being" as first would be static: "being" must have enduring attributes. Values are not fixed but can expand. Creative actualization of something new is beyond "being," genuine change in the world. Value cannot be a *pros hen* to substance, a mere instrument. On the contrary, value is its own ground: inherent value. Creative actualization as an act cannot have its origin in a substance as a *telos*, bringing about an end that "is not." On the contrary, changes are creatively actualized if they are evaluated superior to actual or potential alternatives. New "substances" are thereby created in a comparative relation of values: aluminum, synthetic chemicals, and so on. "Substance" is not fixed but by accident, the result of evaluations grounded by creative actualization. (Nor is "form" an object of attribution: the instance is since form is attributed to the instance.) Actions come to an end, a limit when value is achieved. Such limits of action make actions complete in themselves: finite and finished. Action is not a mode of "being," since action is not a becoming, not a coming to be. Action takes us out of the problematic of "being-becoming." Action is distinguished from becoming by values that limit it to completion and thereby give it meaning and change the world.

A more "radical" thesis is that only what is self-grounded can be a first ground for philosophy. For its value is intrinsic and not derivative. As rank is grounded in autonomy, the attempt to construct hierarchies must be consequent upon self-grounded values. Value autonomy provides a new structure for first philosophy, of autonomy, involving a radical change in structure and meaning of first philosophy. First philosophy has been equated with metaphysics, but the latter attempts to build hierarchies (Aristotle, *Meta.*, VI, 1). This attempt is secondary to autonomic reciprocity, an open system of values in which values emerge coevally. If metaphysics claims independence from history, value categories reflect it. The question is how can a ground for rank be first? Is being "first" a rank? Can the reciprocal relation constitute an ultimate grounding?

First philosophy as a ground for knowledge is a legacy of the Delphic problematic for Philosophy. The oracle's advice to Socrates, "Know thyself," became Plato's quest for knowledge, and resulted in the problematic of knowledge as the premier problem of philosophy in Aristotle's grounding of knowledge in "being," that is, metaphysics. (See *ibid.*, I; McDonald, 2004, ch. 1). An intellectualist hierarchy is the model of most philosophy, in which all is subordinated to the goal of knowledge. (For Aristotle even first philosophy itself is a field of knowledge, despite his additional claim that it must ground knowledge.) A first philosophy of values means that intrinsically different values are brought into a reciprocal relation. There can be no intellectualist first: knowledge is dethroned as the supreme value to which others must pay obeisance (compare the role of science in positivism, early and late). I would not claim that this is the first revaluation of the intellectualist hierarchy in the history of Western philosophy. On the contrary, this history may be read as a series of revaluations by different ages. If Plato promoted reason, the skeptics of late antiquity demoted it. The early Christian writers (Tertullian, Augustine in *Civitas Dei*) subordinated reason to faith, a ranking which was maintained for one thousand years, up to the time of Rene Descartes. In the modern age reason also has its rivals.

These revaluations all remain within the framework of hierarchy, however, although skepticism may form an exception. Our project goes beyond hierarchy, that is, beyond the supremacy of identity. Hierarchy means subordination of all difference under the one most general idea, or concept, or standard, whether "being," experience, "the One," language, or others. The one-in-many identifies what all differences have in common. The many are instances of the one, their regulative principle or norm. If, as we claim, such rankings under a one must be secondary, how can one term, "value," apply to all differences among reciprocally autonomous categories? How can value differentiate?

Identity and difference are correlative. To differentiate a species is to identify it as such and such a species, its species difference. Again, to identify something differentiates it from what it is not. Thus it might appear that neither could be basic, for they imply each other. In one sense this is true, for the most exclusive difference must be identifiable to be spoken of. Yet this may not reveal what is distinctly different about it. It does reveal that difference cannot stand alone. What can justify the insubordination of any difference? While the source of the differentiation may lie in intuition, accident, imagination, and so on, these cannot justify the difference. Why differentiate? these sources may be arbitrary. The value of the distinction must still be justified. The difference cannot be the nature of the case, for it is itself a differentiation whose value must be justified, in terms of its distinctive value. Thus "value" is the most general differentiation, as its value justifies each difference. The differentiation is justified because it is inherently worthwhile, that is, autonomously valuable. Differentiation must be justified in terms of its value.

All differences can be deemed "values" as they are differentiated from each other. The "one-in-many" is value, but this is not their identity. Value is their differentiation. The values are alike only in revealing different value. They are "values" as intrinsically different: equally different values. A first philosophy of values brings differences into a relation, namely, a value relation of mutual difference. Mutual difference is not absolute, as morally structured into just spheres. Thus there is a common aspect of worth, if in intrinsically different ways. Value can be the term of greatest generality as it brings not only identities but even differences into relation, without injustice to their intrinsic merits. Hierarchy is not needed for generality, as in metaphysical standards and norms. Values have the greatest scope for no difference is excluded given value autonomy.

Value autonomy implies radical axiology, that is, ultimate grounding in value. At the root of philosophy no difference may be excluded which has proven worthwhile. Each difference is entitled to an equal say at the bar of justice. Autonomy differentiates at the root, which involves a radical project and program for philosophy, to use the term in its other sense. The form-content program of metaphysics is superceded, through the just relation of equal difference which autonomic values involve. The distinction of form and content does not arise within this relation; its problematic can be of derivative importance only. First in importance requires grounds to reveal what is important, values. For the value of other projects may be questioned, while the value of a radically axiological philosophy—of an inquiry into values in themselves—can never be at issue.

Ranking, hierarchy or evaluations are not an ultimate ground, since they have to do with potentiality, not actuality. Since they are grounded in potential or actual worth, difference can be evaluated prior to identity in one respect. Differences in value, not their underlying identical substratum, are basic to evaluation into hierarchies. Thus substance is historicized as revaluations. Similarly, different, incompossible hypotheses that must be ranked.

Values identify with general concepts or categories only as differences, thereby as correlations of identity and difference, norm and instance. The identity of worth is grounded in distinct worth. Subordination into a hierarchy of such distinct worth—evaluation, ranking, a *summum bonum*—must be based on differences in worth. Because "being" cannot encompass differences, a hierarchy is necessary for it to reduce differences and subordinate them: attribute to substance, species, genus, transcendental; change to becoming to coming to be. "Being" is the most general term of identity.[12] Values, then, are plural instead of general: not an identity of all values but the different values revealed by creative actualization. Values as creative actualizations reveal the potentialities of the world as distinctive, novel, and thereby different goods. The difference of the novel is its inherent value: the value that belongs to itself alone. Relations in the world are not fixed, but indeterminate; there is room for creativity. What unites differentiations as "value" is creative actualization. Pleasures, feelings, objects of desire, novelties, needs, and other differences over time have been creatively actualized. Value ultimately differentiates over time, creating meaning in a better world.[13]

Radical axiology dissolves the dialogue between "idealists" and "realists" that has dominated Western philosophy. The idea that worth grounds rank is close to realism; while evaluation of competing realities is non-realist. Realism has been a check on idealism, since worth grounds rank: no idea, however valuable it may seem, has value unless creatively actualized in practice. Ideas must be "brought about" or "realized." But idealism has equally been a check upon realism since no reality is ultimate: reality can be transformed by novelty and is thereby indeterminate and plastic. "Reality" can be improved and is thereby only provisionally or temporarily "real." Realism and idealism are reciprocal, but superceded by creative actualization, the improvement of the world. Ideas are creatively actualized in the world and transform it: an inferior "reality" left behind for a creatively actualized reality that is no more "real" than the next transformation. The ideas that were the inspiration for the superceded reality are themselves superceded with the next creative actualization, transforming the world.

NOTES

Preface

1. For Thomas Hobbes, see *Leviathan*, P. I. Also, cf. the Greek word *axios*, denoting what is worthy.
2. Plotinus, *Enneads*, Augustine, Aurelius, *De Natura Boni*, and Boethius, *The Consolation of Philosophy*. A modern form of this theory can be found in Rene Le Senne, "Ethics and Metaphysics," from *Traité de Moral Générale*, (Presses Universitaires de France, 1961), reprinted in *Contemporary European Ethics*, trans. and ed. by Joseph J. Kockelmans, (Garden City, New York: Anchor/Doubleday, 1972), 133-156.
3. *Will to Power*, trans. Kaufmann, New York: Vintage/Random House, 1967, sect. 1045. For the "naturalization of morality" see ibid. sect. 462.
4. This basis turns out to be the human good.
5. Everett's theory should not be confused with subjective vitalism, for example, that of A. Campbell Garnett (in Lepley, R., 1949, ch. 4). Another theory that argues for value in "relation to life processes," is that of C. E. Ayres, (ibid. ch. 2). Albert Camus treats this question as problematic.
6. Particularly in *Phys*. II, 7, but also in *Meta* I. Thomism follows Aristotelianism on this issue, that is, value as an attribute of being. Nicholas Hartmann's ethics is a modern version of Aristotelianism in grounding value in being. However, it has several features of its own. (Hartmann, 1932).
7. Lewis, 1947, P. III. Another relational affective theory is that of Bert Jessup in Lepley, 1949, ch. 7. For Meinong see Werkmeister, 1970.
8. There is also some notion of self-realization in several of the authors in *New Knowledge of Human Values* beside Maslow: Allport, Lee, Goldstein, Fromm and perhaps Dobzhansky.
9. Louis Lavelle defends a form of the theory that emphasizes the object of will: value as the *object* of willing. (Lavelle in Kockelmans, 1972, 6-30). Lavelle, however, like Spinoza, argues for an Ethics of Self-Realization or perfectionism. Jean Paul Sartre, by contrast, emphasizes the *choice* of the existential subject. Sartre did not write any treatise on value but argues that values are valid only insofar as we have chosen them, not because of any divine sanction ("Existentialism Is a Humanism"). If I may be permitted to identify choice and will, or at least make the connection between them, then the willing of the individual, subjective choice, is the source of value.
10. A purer example of desire as the locus of value is in Parker, DeWitt H., in Lepley, 1949, where he characterizes value as the "satisfaction of desire," and also Henry Aiken, ch. 1 of ibid. See also S. Alexander, 1933; and S. Pepper, in *ibid.* Lepley, and in his discussion of von Ehrenfels in Ferm, V., ed., 1950. Cf. Brandt, R.B., 1979, in which good is defined in terms of "rational desire" (*inter alia*, 126). This is a composite theory, however, as it combines desire with reason.
11. The main source on Cyrenaic theories as well as Epicureanism is the *Lives* of Diogenes Laertius (1915). However, one work of ancient Epicureanism

survives, namely, *De Rerum Natura* of Lucretius (1951). For Utilitarianism also see Jeremy Bentham, (1973), and Henry Sidgewick, 1966. Another similar theory may be Phillip B. Rice's theory that value is connected with "joy" (ch. 14 of Lepley, 1949).

12. For Hume's views see Hume, 1970). Another such theory may be Harold Lee's "caring for" an object as the property cared for is not "real" but "putative." However, this may border on a relational view. See ch. 8 of Lepley, 1949.
13. In Alfred J. Ayer's emotive theory of value, value judgments are the expression of emotions (Ayer 1952). A closely connected theory is that of Charles Stevenson, 1938.
14. I have argued this point in more detail in "Kant's Theory of Value," ined.
15. Cf. Raymond Polin's "creation of value" by a subject in ch. 7 of Kockelmans, 1972.
16. This issue is covered by most of the authors in Lepley, 1949. Dewey proposed it in ch. 3 of *ibid.*
17. The term "bearer" of value has been adapted from Scheler, 1916.
18. I am indebted to John Cooper, 1975 on this point.

Introduction

1. Just as the "Golden Rule" involves a reciprocal relation of self and other, so Immanuel Kant's Categorical Imperative (Kant 1785) requires universality as a condition of moral worth, and thereby equality, the structure of reciprocity.
2. For more discussion of evaluation and first philosophy see my McDonald, 2004.

Chapter One

1. Stack, George J., 1968.
2. A precis of Creative Actualization as a theory of value was presented in my article, "Toward a Deontological Environmental Ethic," (McDonald, 2001).
3. Dewey stressed the importance of consideration of circumstances and the situation.
4. I will examine the latter point in chapter three; here the point that creation requires the value of action as distinct should be sufficient.
5. This may be a ground for revaluation of means, against its subordination as merely valuable as a means in a hierarchy. I will discuss this point at greater length in a later chapter.
6. Cf. Dewey's distinction of end-in-view from consequence.
7. I will argue this point in more detail in chapter four. See also McDonald, 1998.
8. In this sense, as our own, the world is made part of one's self to some degree and loses its alien quality. Creativity may help overcome alienation. I am indebted to the late Prof. Albert Hofstadter on this point (Hofstadter, 1970).
9. I am indebted to the late Karin Stokkink for certain ideas on the imagination, although their exposition and development are largely mine.
10. George E. Moore has claimed both that everything is natural and that good is a non-natural predicate (1903, 1). This is not only an egregious contradiction, but impossible. Creative actualization is within nature taken as a "whole" and could not but be or would not be possible. Breathing is natural: it is also good for our health. This is to speak nothing of his uncritical use of "natural." If everything

is natural, then Auschwitz, lawrencium and synthetic chemicals are "natural." I have critically evaluated nature and naturalism in my paper "Does Nature Exist?" (McDonald, Nov./Dec 2000).

11. There are also similarities between his notion of good as functioning and mine, although I would reject his distinction of natural and artificial good.
12. Evaluation of competing possibilities to create alternative "realities" was covered in my *Radical Axiology, a First Philosophy of Value* (McDonald, 2004).
13. Again, values are the most general term of difference, as they give limits and meaning to actions that differentiate them. However, identity and difference are correlative, as I will argue below. Thus the problematic of identity and difference is of limited value.
14. Cf. Hannah Arendt on the distinction of work and labor (Arendt 1958). Work creates goods while labor is tied to becoming, the cycles of nature, in her view.
15. I have argued elsewhere (see n. 2) that the distinction of environmental ethics from ethics is untenable. Environmental ethics is the ethics of the future, since our decisions as humans are both within the environment and affect that environment. The environment is a larger whole than the merely human and thus human ethics is only a subset of environmental ethics. I will expand these points in chapter four and in my book *Revaluation of Cosmopolitan Ethics*, ined., which I hope to publish shortly.
16. This topic is addressed below and in my book *Speculative Evaluations, Essays on a Pluralistic Universe*, which I hope to publish shortly. Change is not a variation on becoming for creative actualization, that is, within a paradigm of beings. Nor is change an "eternal recurrence of the same." Nothing can be eternal in principle that is subject to human actions. Indeed, genuine novelties and differences arose before humans: cf. Henri Bergson's book *Creative Evolution*.
17. I am following Aristotle, for whom the end is equated with the good (N.E., I, 1 et al.). Cf. also *Meta.* I and II and XII (partly quoted above); and *Physics* II, 7 for his account of the good as the end. *Telos* has several aspects that are disguised by the unity of the concept. Firstly, *telos* is both end or goal and also good. Secondly, Aristotle does not carefully distinguish end as goal and end as result or consequence.
18. John Stuart Mill recognized this paradox.
19. This is despite Aristotle's identification of the good and the end (*N.E.* I, 1 et al.). This equation correctly identifies achievement of the end with the good, as a quantitative unity, but is likely to confuse the issue by identifying good with goals. Thus value is more than an analytic element.
20. The relation of theory and practice will be examined in more detail in another work.
21. Radical axiology and creative actualization are the denial of "being."
22. However, they inform imagination with normative identities during acculturation and development. I will cover this topic in a projected future work on imagination.
23. I am following William James (1903) here who argued that the world could be improved: meliorism. The critical question, that is, what is a genuine improvement, will be addressed in more detail in a later chapter.
24. I examined how evaluation of competing ideas or models of reality must be evaluated in my book *Radical Axiology, a First Philosophy of Values*

(McDonald, 2003). The question might arise of non-living elements of the world. Differentiation of such elements is itself a normative evaluation, and involves other evaluations. However, these non-living elements too are transformed by creative actualization. Further, their "survival" from geological and other transformations implies that their actuality, though fortuitous and not planned, was superior to that of what has not persisted. In any case, their "reality" is contingent upon their not being transformed, as I will point out below. This question is examined more fully in my *Speculative Evaluations*.

25. Beauty also involves representations of the world by the imagination and private evaluations of quality. I will address this point in more detail in a projected future work on imagination.
26. Martin Heidegger's attempt to conflate being and value (in *Brief über Humanismus*) misses the significance of the novel. To view being in terms of Being is a *pros hen* to an enduring, despite Reiner Schurmann's arguments to the contrary. Heidegger's attempt to make Nietzsche into a philosopher of being and to ignore Nietzsche's concentration on issues of value is even more misplaced.
27. I will examine this point in more detail in a projected future on Teleological Grounding.
28. See "Values and Facts" in Clarence I. Lewis's *Values and Imperatives* (Lewis, 1969, 91 (cf. 95). I agree with Lewis that moral knowledge is not derived inductively in this manner.
29. I will discuss this point again below. I agree with Dewey that the ethics of good intent (Augustine and Kant *inter alia*) is an "attempt to avoid responsibility for consequences." Further, intending wrong is simply not as bad as doing wrong.
30. I have adopted this term from Risieri Frondisi (Frondisi, 1971). "Polarity" has a connotation of a logical "contrary," but without the logical implications.
31. Cf. James O. Urmson's "On Grading," (Urmson, 1950).
32. See Davidson, Donald, 2004, 20, as an example of and also confusion of evaluation with "attitudes" and "desires." (cf. Dewey's critique: values require critical evaluation).
33. Cf. Simon de Beauvoir on the "Ethics of Ambiguity" with Dewey's approach to finite value: that new problems may emerge from solutions to old problems, and that these constitute new challenges to the method of intelligence. Strictly speaking, this problem belongs to evaluation and justification, which will be addressed in chapter three.

Chapter Two

1. See the Preface for a definition of these terms.
2. As I noted in the preface, Clarence Irving Lewis (1946) first distinguished inherent from intrinsic value, although my use of the distinction differs slightly from his.
3. Wilhelm Windelband explicitly raised this problem (Windelband, 1980).
4. This point is the heart of the difference in values discussed by Windelband in the address cited above (ibid.).
5. This point is in response to P. Miller's equation of value and richness (Miller, P., 1982).

6. Such a relation is implied by Robert Hartman, in "The Axiometric Structure of Intrinsic Value," (Hartmann, 1974), where value is treated as a "property" of a "thing" (86). This mode of relation is virtually identical to the substance-attribute relation in Aristotle.
7. I will cover perfectionism in more detail in ch. 5.
8. *Cat*. I. Aristotle would never have analyzed the concept substance ("*ousia*") unless the *hyle* had been transformed by creative actualization into another good, for example, a statue. Thus substance was discovered due to creative actualization.
9. Self-actualization is consistent with habitat changes. Plant succession is from pioneer to climax species. Amelioration *as* change does not mean stasis. But amelioration as bettering includes flourishing. I address this point in more detail in my *Revaluation of Cosmopolitan Ethics*, ined.
10. Predators also use other self-actualizing agents for food, that is, as instruments. This creates problems for vitalistic theories of value such as Albert Schweitzer's (Schweitzer, 1976, sect. 10). I will address this point in more detail in the next section.
11. I am indebted to Manfred Frings and Roger Funk for the notion of "bearer" of value, based on their translation of Max Scheler's *Formalism in Ethics*.
12. I am indebted to J. Baird Callicott (1989) for the distinction of locus from instance of value, which forms an example. The unique value placed on individuals is part of the Christian value schema. For this view see Windelband (1894/1980).
13. Like most terms "value" is equivocal, having more than one meaning, as well as several modes. I examined these in the preface and will again below.
14. I will examine the psychology of the imagination in a projected future volume. I am indebted to Karin Stokkink for her originary notion of imagination.
15. Life as the bearer of intrinsic value and survival were discussed in the last section. Intrinsic value is not the same as life, since this would be impossible. I will return to this point in chapter four. On the intrinsic value of the individual see above, sect. 1.
16. However, this issue is too large to be examined here; I will address freedom in the metaphysical sense in a projected future volume.
17. Cf. Talcott Parson's *The Structure of Social Action* (Parsons, 1937). The underlying principle of the book is that action can be social, or socially determined.
18. Quality may also admit of degrees, as Kant implied in his table of categories (from zero to full actuality) (Kant 1929). For Kant quality is an *a priori* category of the mind, while for our theory it is inherent. "Absolute" quality would be unconditional; "relative" by comparison with that degree in other instances of the quality.
19. Harmony of the parts working together in a functioning whole was emphasized by Plato in the Republic, I-II (cf. Neville, Robert, 1981, and his other works on value. For a critique see Callicott, 1989). Symbiosis is from biology and part of a healthy environment. I will return to symbiosis in ch. 4.
20. I have covered evaluation as a first philosophy in my previous work, (McDonald, 2004). I will examine both evaluation and norms as a method in a projected future volume, *Evaluation.* Some standards for evaluation have already been covered in this work and others will be covered *passim* below. For example, practicability, feasibility, and greater efficiency in functioning well are

practical standards for evaluation of improvements. Moral evaluation will be covered in chapter three. Evaluation may also involve contraries ("polarity" in Frondisi, 1971) as well as comparisons: one possibility is good another, bad.

21. Theories of standards in the history of philosophy can roughly be divided into natural and non-natural. The Sophists, Aristotle, the Stoics, the materialists, Jean-Jacques Rousseau, and Friedrich Nietzsche upheld nature as a standard. Nature was regarded as something to be overcome by Plato, Thomas Hobbes, the Idealists, and Karl Marx. Naturalistic standards include life, health, strength, and the human species nature. Non-natural standards include order, higher ends, and human realization. Some movements have been split on this issue, namely, the pragmatists, who split into idealist (Charles Peirce) and naturalist (John Dewey) camps; and the positivists, who are naturalistic but include some members who uphold a higher calling for humans and human science.
22. Thus judging one potential good better than another according to some standard is an evaluation, not a good. Platonic value theory confused value and evaluation, since it thought of value hierarchically, that is, in terms of better and worse, or evaluation. This confusion was then transmitted into all later value theory as the thesis of the *summum bonum*. I will cover this point in more detail in chapter five.
23. For an example of confusing values and norms see Davidson, Donald, 2004 (even though the latter are required for language!). The Heidelberg Neo-Kantians and Max Scheler also appear to have made this mistake.
24. Dewey has made an implicit critique of the reverse case of "one can't derive an ought from an is." A problematic situation is one which, as problematic, ought to be improved. Cf. McDonald 2004, ch. 5-6.
25. This applies particularly to the Frankfurt School and its neo-Hegelian, rationalist approach to value.
26. Cf. Georg Hegel's famous critique of Schelling in the Introduction to the *Phenomenology of Spirit*.
27. Thus as I noted in the first section, William of Ockham's razor is an evaluation. Cf. Peirce's statement of pragmatism.

Chapter Three

1. Immanuel Kant recognized this problem in his treatment of the "courage of a brigand," in the beginning of his *Groundwork of the Metaphysics of Morals* (Kant, 1785 (1964). John Stuart Mill (1863, (1973) recognized this problem in his defense of Utilitarianism.
2. This division follows Frederick Copleston (1946-74), probably in turn following W. D. Ross 1930 (1961) and William Frankena, 1963. The division has been disputed more recently, however.
3. Although Kant's ethics differs from Aristotle's in structure, they agree that ethics generates practical philosophy, since Aristotle argued for a moral classification of states and for ends of scientific study. John Cooper argued that they are closer than is usually supposed in other respects as well (Cooper, 1975).
4. I hope to write a volume on evaluation in general in the future.

5. Although this is John Dewey's formulation of obligation, it is consistent with Kant's. See Dewey, Tufts, 1908 (1932 rev.).
6. As Dewey noted, the ethics of intent seeks to avoid responsibility for consequences (ibid.)
7. Another issue is, if the relation of intrinsic value and obligation is logical, is logic required for its discernment? That is, is the specific difference of humans, rationality, required? But is this morally relevant? or is recognition of obligation distinct from obligation to intrinsically valuable bearers *per se*? I will examine anthropology in ch. 4.
8. Clarence I. Lewis, who incorporated transcendental arguments into a pragmatic ethic, anticipated this project (Lewis, 1969). I hope this establishes my pedigree with the pragmatic tradition.
9. Indeed, Brogan argued that judgments of value are comparative, running together value and evaluation, which evaluates or compares different possible values. I will address this point in more detail in chapters five and six.
10. I am referring to Aristotle's *pros hen* relation, in which equivocal terms refer back to a legitimating first or foundational value. I will examine this point also in chapter five. See below n 18.
11. The pernicious model of Utilitarianism, for which duty is defined in terms of non-moral values, does not weaken our argument, despite its widespread use in our own times as a model for consequentialism. Utilitarianism is only one form of consequentialism and a weak one. I will examine Utilitarianism in ch. 6. See also my critical evaluation in *Revaluation of Cosmopolitan Ethics*, ch. 6, ined.
12. Cf. the "golden rule" and a similar formulation in Confucius' *Analects* (II, 5 and XII, 2). The rule is not culture-bound; when I was in South Korea, at the 2008 World Congress of Philosophy, a Nigerian professor presented a paper on similar norms in Africa.
13. Cf. McDonald 2001, 411-430. In that essay I argued that the principle of reciprocity is derived from compossibility of values. However, it is a condition that constitutes such compossibility, and thus not a consequent. Cf. the third statement of the Categorical Imperative in Kant: the kingdom of ends formulation.
14. The rules are "abductive" hypotheses that allow us to face the future with as much wisdom as we can. I have kidnapped this term from Charles S. Peirce See (1898), 1992, esp. ch. 2.
15. I will deal with these perversions of ethics in chapter six. See also my *Revaluation of Cosmopolitan Ethics*, ch. 6, ined.
16. Georg W.F. Hegel argued that "it is only in the Christian religion that the principle of the individual's infinite worth makes its appearance." (Hegel, 1956; cf. Windelband, 1894).
17. For his view see Rachels, J., "The Challenge of Cultural Relativism," repr. in Cahn and Markie, eds., *Ethics—History, Theory and Contemporary Issues*, P. II, ch. 36. I will discuss cultural relativism subsequently.
18. The *pros hen* relation is the relation of equivocal terms in which the different terms refer back to a justifying first. See Owens, Joseph, 1957 (1978), for an extended discussion of *pros hen* equivocals.
19. I will discuss this relation in more detail in a projected future volume, *Teleological Grounding*.

20. To this extent I follow Kant, who justified principles in his Critiques, especially the Deductions. However, there is the paradox that moral standards are principles applying to temporally and spatially different situations. I will examine this point in more detail in a future work.
21.The Neo-Kantians spoke of the "validity of values." Although I reject this characterization of value in general, moral principles that regulate actions that creatively actualize non-moral goods are not goods themselves, but are "valid" nevertheless, that is, have a valid role to play in limiting creative actualization of other goods.
22. Are moral views "true," as Paul Taylor argues (Ladd, ed., 1973, 96-7) or 'right'? Or rather is it true that they are right? Asking if they are true or false seems like a category mistake, but not the truth or falsity of their rightness.
23. This phrase was adopted from the article "Choosing, Deciding and Doing," in the *Encyclopedia of Philosophy*, Edwards, Paul, ed., (New York: Macmillan, 1967) near the end.
24. I can trace my pedigree to pragmatism on this point to Clarence I. Lewis, who discusses the other in ethics, and the relation of morality and the other to value (Lewis, 1958, 34 ff.)Thus he anticipated Levinas by a generation, just as Ferdinand C. S. Schiller first proposed the idea of ethics as first philosophy.
25. Cf. cosmic justice in some of the pre-Socratics, perhaps the origin of "natural law," another metaphor.
26. Feinberg has distinguished psychological and ethical egoism, but I will defer consideration of these to another chapter. For Feinberg's view see Feinberg, Joel, 1978 (in Cahn and Markie, 1998 ed.) 557 ff.
27. I will discuss these relations in more detail in chapters five and six.
28. The ethic of future times and the notion of applying meliorism to ethics were inspired by William James.
29. The idea of moral progress is taken from Lecky's *History of European Morals*. I am indebted on this point to Roderick. F. Nash (Nash, 1989).
30. However, as Callicott has convincingly argued, talk of "rights" in the wild is an illicit extension. As I argued above, consideration of the wildness of the wild rules out extension of political and social rights from civilization to the wild. Otherwise, the wild is domesticated and ceases to be wild.

Chapter Four

1. Cf. "love and strife" in the philosophy of Empedocles, who also hinted at cosmic justice as a part of the world processes. Cf. the "balance of nature."
2. In turn, the "Protestant Ethic" may have been the cause of capitalism, or a major factor in its development, with its instrumentalization of the environment for economic accumulation. See Max Weber, *The Protestant Ethic and the Spirit of Capitalism* (trans. by Stephen Kalberg, Englewood Cliffs: Prentice Hall, 1977) for this thesis.
3. This view is implicit in Friedrich Nietzsche's view that "becoming an organ" is to become an instrument of a more powerful will, an active or dominant will to power. See his *Will to Power*, passim
4. Callicott, Rolston and the Ecocentric School of environmental philosophy also uphold the latter thesis.

5. Similarly, the relation of meaning to the actual world in George E. Moore is unclear. Is meaning in one universe and that which is meant in another? Clearly not, which argues against any strong analytic-synthetic distinction.
6. I will argue elsewhere (*Speculative Evaluations*, forthcoming) that facts are not actualities in the world but judgments of knowledge about the actualities in the world. Facts are epistemological, not actual.
7. This view unites Nietzsche, the classical pragmatists (Ferdinand C. S. Schiller, John Dewey, and Clarence I. Lewis), Albert Schweitzer, and some eastern philosophies and religions.
8. Ecocentric holists such as Callicott and Rolston argue for the value of species, instead of individuals, and thus for the locus of value in species, not individual bearers of specific identity. Locus of value instead of theory of value is thus the point of contention between such holists and those who uphold individual rights such as Tom Regan. I have addressed this point in more detail in McDonald 2003, ch. 1.
9. Kant (1785, 1788) implies, and Hugo Münsterburg (1909, 28) and Roubiczek (1969) argue that intrinsic value is absolute. However, life comes to an end and the good or moral will also comes to an end with it.
10. I am indebted to Callicott, 1989 on this point.
11. I have examined these social categories at greater length in McDonald 1997, ch. 4.
12. This is the "Apollonian" element in *The Birth of Tragedy*. Nietzsche later changed his view; see *Twilight of the Idols*, passim. Cf. Camus and also cf. the "Second Coming" of Christ in some sects, for example, Jehovah's Witnesses.
13. In this view, suicide is not wrong, only bad. The value of life is not apparent to the agent contemplating suicide. If circumstances were to blame, a change in them would be a valuable goal: meliorism. For a defense of the view that morality concerns relations to others from another perspective see William James's "The Moral Philosopher and the Moral Life," in James 1910, (1975); and Peter Caws, 1967, ch. 9.
14. This is in answer to specific Marxist and neo-Marxist criticisms of contemporary society. The actor creatively actualizing possibilities provides the unity of a life. It is not a function of society or the state.
15. Dewey, John and Tufts, James, Ethics, New York: Henry Holt, 1908 (1932 rev.). The influence of the environment on ethics was already treated to some degree by Herbert Spencer (Social Darwinism); cf. Charles Darwin on the effect of climate on the *Origin of Species*, 106 ff., cf. 158: the environment as the condition of life (also ch. 11-13). Cf. also William Graham Sumner on "folkways" as responses to an environment, in Ladd, 1973, ch. 3. Cf. also the chapters on culture and environment, by Karl Dunker, (Ladd , 45 ff.), Clyde Kluckhohn, (Ladd, 78 ff.), and Robert Redfield, (Ladd, 129 ff., esp. 133 and 139-40.)
16. Virginia Held (2002) has wisely argued (implicitly) that the scenario of the "state of nature" in Hobbes is unlikely, to say the least, in that nurturing is left out of his account. Most infants of higher species require nurturing.
17. Cf. Arnold Toynbee's (1946) reading of history as "challenge and response."
18. Protagoras remarked that "man is the measure of all things, of things that are that they are, of things that are not that they are not." Heidegger tried to fit this within the problematic of being. But Heidegger has missed the basic dismissal

of the problem by some ancient figures, just as in George Berkeley's reduction of being to being perceived. To wit: "Gorgias of Leontini...in his work on "Not-Being and Nature"...maintains that nothing exists.... even if something does exist it is incomprehensible to man.... even if it is comprehensible it cannot be communicated or explained to anybody else." Quoted from Sextus Empiricus, in Smith, T.V. ed., 1934 (1965 ed.), 66.

19. According to Prof. Reiner Schürmann, Heidegger read Aristotle's metaphysics through the physics as dealing with the problem of technology. The reverse is more likely: Aristotle tried to encompass the problem of technology within the problematic of being and becoming. However, he also "bracketed" the problem by assigning *techne* to practice.

20. For a parallel but different reading of this debate see Pirsig, Robert M., *Zen and the Art of Motorcycle Maintenance: An Inquiry into Values*, New York: William Morrow and Co. Publ., 1974 (Bantam ed. 1975)*passim.*

21. Each individual has absolute value despite the inconsistencies in the notion of granting Divine Grace, and thereby that value derives from God alone. The latter represents the subtle importation of Platonic metaphysics into the Judeo-Christian value schema, and may be incompatible with it. I discuss this point in chapter six under "agapism."

Chapter Five

1. For a discussion of the *pros hen* relation see Owens, 1957. As Dr. Owens has noted, Aristotle's *Metaphysics* is a study of such *pros hen* equivocals, which are used as evidence for the substance-attribute metaphysics delineated by Aristotle. I am indebted to an anonymous critic for my clarification of this point.

2. For a critical evaluation of the form of such theories, that is, the metaphysics of value, see McDonald, 2004, ch. 3-4.

3. Max Scheler (1916) argued that his theory "belongs to modern thought" since it is inseparable from the metaphysics of a subject confronting objects of value. However, value is in an "ideal" beyond for Scheler: he combines Cartesian and Platonic notions. Similarly, the Neo-Kantians argued for a separate realm of value, a kind of normative "validity."

4. That is, Robert Hartman's "axiometrics." I discuss the "ontology" of form above and in the section on *pros hen.*

5. This view is in contrast to philosophies such as Clarence I. Lewis's (1946, P. I) and others in which analytic statements are definable but involve logically necessary essences, as opposed to the contingent and empirical properties of synthetic judgments. Thus analytic judgments state definitions in this view. However, Moore's theory does not totally exclude such judgments; only "simple" predicates are indefinable for Moore, as I will describe below.

6. Moore later modified his view; he was no longer totally convinced that good was indefinable (1959, 97) although he saw no good reason to think it was definable.

7. Moore concedes that good as a non-natural predicate of pleasure as opposed to being defined as pleasure may be simply a "verbal" dispute (1903, 12). Nevertheless, he thinks that the naturalistic fallacy is widespread and can legitimately be called a fallacy. For a contrary view, see Osborne, H., 1933, ch. 9.

8. *Ibid.* (Moore) ch. 1, 27. Thus a thing cannot lose certain qualities, according to Moore and still retain its value. Cf. Moore's critique of hedonism in which he argues against the notion that we can never add to the value of any whole except by adding pleasure (1912, 105).
9. For a critique of facts as entities, see McDonald, 2004, ch. 4, near the beginning. A summary of this argument is that facts cannot be both entities and knowledge of entities without circularity. As there is no class of entities in the world "facts," along with elephants, lilies, and gypsum rocks, facts must be knowledge of these. I will address this point in more detail in the essay "Critique of Facts" in *Speculative Evaluations*, forthcoming.
10. Moore also does not consider that the analytic-synthetic distinction may be fallacious, as Willard V. O. Quine argues in "Two Dogmas of Empiricism." However, Quine's article appeared fifty years later, so Moore cannot be accused of ignoring the issues raised by Quine. Lack of rigor is not equivalent to lack of foresight.
11. If Polin had gotten past the Cartesian metaphysics, he might have been even closer to our theory. Another forerunner is Caws, Peter, *Science and the Theory of Value,* New York: Random House, 1967, who makes similar mistakes by arguing for creativity and bringing about but basing value in desire and the subject. (John Dewey also anticipated our theory in outline, in his notion that value is connected with "bringing about.")
12. While such a view is allied with idealism, there are materialist versions, for example, in behaviorism in which the "correct" conditioned response is sought. See, for example, Burrhus Skinner 1972.
13. Ancient Chinese philosophy debated this point as well, since Kong fu-tse thought of virtue as a "striving" toward perfection, while Lao-tse thought we should go along with our natural tendencies.
14. The use of and sacrifice of one generation for another also runs the danger of totalitarianism, in which persons are molded according to an ideological formula. I have discussed this point in more detail in McDonald, 1997.
15. Cf. *Rep.* II-III with *N.E.*, X. The good state needs classes corresponding to other parts of the soul than the rational according to Plato; the life of wisdom is not for all according to Aristotle.
16. This was my early view and I am indebted to J. Baird Callicott for pointing out the limitations of this view (Callicott 1989).
17. He was, however, inconsistent on this point. See his "The Ethics of Animal Experimentation," *Atlantic Monthly,* 138, September 1926, 343-6; reprinted in *John Dewey, The Later Works*, ed. by Boydston, J., Carbondale: So. Illinois Univ. Press, 1984, 98-103.
18. I have covered Dewey's view in more detail McDonald 2004, 189-208. This paper is also in McDonald 2003, ch. 1. I have criticized the concept "nature" as well as naturalism in McDonald, 2000. Other "naturalistic" theories of value, for example, Epicurus', will be covered below.
19. The absolute value of all of life is taught by some eastern religions; of human life by some Christians. The equation of intrinsic value with absolute value is in Münsterberg (1907), and perhaps Kant. Intuitively this makes sense: what has intrinsic value does not derive its value from any extrinsic source or relation and thus its value is not relative to time, place, culture or any other value. What

is not relative is absolute. However, intrinsic value need not be absolute to be intrinsic.

20. Certain economists, to whom I am indebted, have made this point before. Unfortunately, I forget in which text I first read it, but I claim no originality for it.
21. Strictly speaking, the theory was replaced by marginalism in the late nineteenth century, according to Robert Heilbroner, in *The Worldly Philosophers: The Lives, Times and Ideas of the Great Economic Thinkers,* (Simon and Schuster, 1999). However, there were critics of it prior to the formulation of the marginalist theory. I am also indebted to Lee McDonald's *History of Western Political Thought* on the criticisms of the labor theory.
22. For a counterargument see Ernst F. Schumacher, (1973) where the value of "right livelihood" is upheld. If Schumacher is referring to crafts, he may have a point: the artist indeed may value the process of creation (cf. Thorstein Veblen's *The Instinct of Workmanship*). But hard labor was meted out as a *punishment*.
23. Dewey has been read as arguing against intrinsic value as an instrumentalist. However, Dewey did not argue against ends as such but against the complete separation of ends and means (McDonald, 2003).
24. The Cartesian category of "objective" is superceded by creative actualization.

Chapter Six

1. John Cooper has argued that this is not the best translation of *eudaimonia* (Cooper, 1975. However, this does not affect the present point.
2. I will cover relational theories in ch. seven, including desire, will, and so on.
3. I have argued against happiness as only a subjective state in McDonald, 1997, ch. 6. I will also argue against its identification with pleasure below.
4. Cf. the "Creation of Values" in Raymond Polin, which is not grounded in the actual as a transformation of it (Polin in Kockelmans, 1972).
5. As Nietzsche remarked, "a thought comes when 'it' wishes, not when "I" wish.
6. Cf. the "last man" argument in Environmental Ethics.
7. By needs I mean values except in the case of biological needs; I covered this point in ch. 5. Cf. above on self-sufficiency.
8. Again, value is in no way an "underlying substance," nor phenomenal. Such distinctions belong to being. I have examined this point in more detail in McDonald, 2004, ch. 3.
9. Cf. William James who thought that subjective and objective were stages in the larger experience.
10. Due to the strong affective strain in value theory in hedonism and more recently in the Ayer-Stevenson emotive theory, Held's argument (1990) that this side of the human psyche has been ignored in favor of reason by Western Ethics is debatable. True, Held interprets Utilitarianism as a closet form of rationalism, for which she has much justification. Economistic forms of the theory have indeed stressed rational decision making ("net aggregate utility," and so on). But the foundation has always been feelings of some sort as the basis for desires, whether feelings of pleasure, happiness, or more irrational forms as in Ayer.

11. Behaviorists, such as Dewey, Burrhus Skinner and Ludwig Wittgenstein would dispute this view. However, behaviorism, whatever its merits, is not the view of affectivists, who generally argue for feelings of a *subject*.
12. I argued for a normative understanding of concepts in McDonald, 2004, ch. 7. Definitions of concepts are also normative. I am indebted on this point to the thoughts expressed by Edward Schiappa (in Katz, E., and Light, A. 1996, ch. 10), but he should not be blamed for my failings.
13. Thus James Rachels argues that morality consists in socially necessary rules in his critique of cultural relativism, (Rachels, in Cahn and Markie, ch. 26, 555). Peter Singer, who takes a utilitarian view (in *ibid.* ch. 45, 803) has a footnote about Henry Sedgwick in which the same point is stressed. Kant's different formulations of the Categorical Imperative involving universality and other persons can be read as justifying this point.
14. See ibid. Singer, where Utility is tied to rules. Cf. also the work of R. M. Hare, 1963 (1972), and 1952 (1967).
15. Richard M. Hare, who analyzed morals in terms of prescriptivism in *ibid.* but I would argue that his prescriptivism ties him more to the Kant of the Categorical Imperative, which is both a moral rule in the form of a command and also rational, than to Ayer and Stevenson. For a similar view, see Saydah, J. Roger, 1969. However, Hare does accept the separation of the moral and factual realms, like many analysts.
16. It could be argued that Ayer's and Stevenson's theories are not as different as this separate treatment implies. However, I am not alone in reading Ayer as the more radical of the two and in his denial of any relation. See, e.g., S. Pepper (Ferm, V., ch. 39, 498). Stevenson, however, believed that feelings could be about objects, and thus expressed a less subjective view. In this reading I am in concurrence with Abraham Edel, in Marvin Farber, 1950.
17. This theory has also been questioned. If logic is a normative science that has been derived from studies of language, Aristotle's term logic, norms were derived from facts.
18. Non-hedonistic affective theories are held by Hume and Adam Smith, Meinong, Ayer and Stevenson, some feminists, and several authors in Lepley, 1949, viz. Bertram Jessup and Phillip B. Rice.
19. Jeremy Bentham argued for the theory under the rubric of the theory of "utility," but utility implies utility for something (Bentham, 1973). That is, utility is measure of means, not ends, of function and efficacy, not what they function for. But in speaking of a highest good, we do not mean a function but a standard. How then can means give us ends, much less a highest good? Use can never be a measure of value as such.
20. I argue this point in more detail in my forthcoming *Revaluation of Cosmopolitan Ethics*.
21. I am indebted to an old lecture on this point, but cannot remember the lecturer.
22. Cf. Martin Buber's "I and thou" relationship.
23. Augustine, *De Natura Boni* and *De Libero Arbitrio*. These works were the prelude to voluntaristic theories in Duns Scotus and others, that goods are such because of divine will, not inherently.

Chapter Seven

1. The theory also occurs in Thomas Hobbes' *Leviathan* and can be traced back to the Sophists, for example, in Plato's *Gorgias*, Callicles argues that the good is "doing what you want." Thus the theory is pre-modern. It was also adopted by Baruch Spinoza as a conativist theory, by Stephen Pepper (in Lepley, R., 1949) and by Samuel Alexander.
2. Cf. the Copernican revolution in astronomy, in which our perspective to the heavens changed. The stars themselves did not move. Nicholas Copernicus realized that appearances, namely, the apparent motion of the sun across the sky, are deceiving…
3. John Dewey and Virginia Held emphasize this point. Prof. Held notes that the model of the autonomous individual ignores the process of socialization in the family and household.
4. Kantians will note that Kant would agree that moral good is not simply what is willed but determines what ought to be willed. But then Kant is not a voluntarist in value theory, a point I have argued in my paper, "Kant's Theory of Value" (ined.) The question becomes why the psychological element is brought in to Kantian ethics: the answer is that Kant was a subjectivist and had to both argue for the subjective elements of agency and how the moral law determines the agency to action. Once we purge ethics of subjectivity we can jettison the "will" as well as a superfluity.
5. I am indebted to Stephen Pepper for the characterization of Prall's theory as relational (Pepper in Ferm, Vergilius, ed., 1950).
6. See McDonald, 2004 for a critique of such models.
7. Perhaps they should be classed as philomyxomycetes.

Conclusion

1. This value relation is consistent, I believe, with the "Pluralistic Universe" of William James.
2. *Phaedrus*, 246-257 and *Rep*. passim. Cf. the divided line as an epistemological hierarchy in *Rep*. VI.
3. "Autonomy" has been adopted from Immanuel Kant's ethics, even though Kant also created a hierarchy of values and thereby undermined autonomy. I will discuss the relation of Kant's views to mine below.
4. Aristotle recognized intrinsic or inherent value in the form of ends in themselves, notably virtue (*Nich. Ethics*). However, in a reductive move, he also argued that ends could also be used as means and thus superimposed the relation on value theory. But if a good is valuable in itself, it cannot derive its *value* from an extrinsic source. This reduction involves confusion of value and evaluation, or of the function of goods with their value.
5. Their abstraction from culture is hypothetical; it is justified if it describes cultural practice better than competing hypotheses. I have discussed value categories in relation to culture in some detail in McDonald, 1997, ch. 4.
6. Thus Georg W.F. Hegel's well-known criticism of the "empty" formality of Kant's ethics misses the point of placing our maxims under universal law. The content of morals is autonomously derived, not rationally derived (Hegel, 1952/1978

Preface). Another characterization of Kant's ethics, as "intuitive," has no basis in Kant's texts. Intuition in the first critique is tied to sensibility, not reason.

7. The formalist program in ethics of William D. Ross and others claims to be non-teleological. Be this as it may, Kant's formalism is teleological. Duty is done for the sake of morals (cf. 1785, I with 1797 a, I (*Rechtslehre*) 239 & 354 with 230). For moral worth, the will must aim at morality for its own sake, and not at any extrinsic incentive.
8. John Dewey used the term "esteem" instead of "worth." However, the spirit of his philosophy was a movement away from the Cartesian subject. "Esteem" seems too subjective; "worth" alludes to the character of the good or the value.
9. I will discuss these topics more fully in a projected future work: *Evaluation and Norms*.
10. Cf. Plotinus' *Enneads* in which the One is equated with the Good to Augustine's *De Natura Boni*, in which the one Good is identified with God. (Both of these models are justified readings of Plato, who argues that God is fully good in *Rep.* II.) The substitution of humans, instead of God at the apex of the hierarchy of creation, the "great chain of "being,"" began with the anthropocentric adaptation of Plato by Renaissance Neo-Platonists such as Pico della Mirandola and reached its culmination in atheistic humanism.
11. We hope the use of several terms will avoid ontological or other metaphysical commitments.
12. The copula identifies A=A or A is itself. The copula signifies an abstract relation of equality, that is, identity. Thus "Ontological difference" is a contradiction. Despite Heidegger's attempt to creatively actualize a value-free ontology ("Brief über Humanismus") his work is loaded with evaluations. Heidegger is a seeker looking for a comprehensive value philosophy.
13. Gorner has argued that whatever arises historically cannot have "ultimate foundations." (Gorner, Paul, 2000, 192) Is first philosophy than a-historical as *a priori*? But evaluation is historically situated – what would (result in) improvements given the historical situation? The result is melioristic, a change for the better. But is the form of radical axiology *a priori*: principle of principles, value, *telos*, etc.? Thus first philosophy provides the framework and justification not "ultimate foundations."

BIBLIOGRAPHY

Aiken, Henry. "Reflections on Dewey's Questions about Value," in *Value: A Cooperative Inquiry*. Edited by Lepley, Ray. New York: Columbia Univ. Press, 1949, ch. 1.

Alexander, Samuel. *Beauty and Other Forms of Value*. New York: Thos. Crowell Co., 1933/1968 ed.

Arendt, Hannah. *The Human Condition*. University of Chicago Press, 1958/1974.

Aristotle. *Metaphysics, Nichomachean Ethics, Politics, Physics, Poetics* reprinted in *The Basic Works of Aristotle.* Edited by McKeon, Richard. NewYork.: Random House, 1941/1970.

Augustine, Aurelius. *Civitas Dei.* Tanslated as *City of God*, by M. Dods et al., (abridgment edited by Oates, W.H.). New York: Random House, 1948.

———. *Concerning the Nature of the Good* (*De Natura Boni),* in *Augustine: Earlier Writings*. Edited by Burleigh, J.H.S. Philadelphia: The Westminster Press, 1953.

———. *De Libero Arbitrio*. Trans. by Benjamin and Hackstaff as *On Free Choice of the Will*. Indianapolis: Bobbs-Merrill.). Orig. 396.

Ayer, Alfred J. *Language Truth and Logic*. New York: Dover, 1952.

Ayres, C. E. "The Value Economy" in *Value, a Cooperative Inquiry*. Edited by Lepley, Ray. New York: Columbia Univ. Press, 1949, ch. 2.

Beauvoir, Simone de. *The Ethics of Ambiguity*. Kensington Publishing Co., 2000.

Bentham, Jeremy. *Principles of Morals and Legislation.* Repr. in *The Utilitarians*, Garden City: Anchor/Doubleday, (1973).

Boethius. *The Consolation of Philosophy.* Trans. by Green, R. Indianapolis, In: Bobbs Merrill, Library of Liberal Arts, 1962

Brandt, R.B. *A Theory of the Good and the Right*. Oxford, Clarendon Press, 1979.

Callicott, J. Baird. *In Defense of the Land Ethic, Essays in Environmental Philosophy.* Albany: State University of New York Press, 1989.

Camus, Albert. *The Myth of Sisyphus and Other Essays by Albert Camus*, Trans. by Justin O'Brien, New York: Knopf Publishing/Vintage International, 1991.

Caws, Peter. *Science and the Theory of Value,* New York: Random House, 1967.

Confucius. *Analects*. Dawson translation, Oxford University Press, 2008 ed.

Cooper, John. *Reason and Human Good in Aristotle*. Cambridge: Harvard Univ. Press, 1975.

Copleston, Frederick. *A History of Philosophy*. In Nine Volumes, Garden City: Doubleday/Image, 1946-74.

Darwin, Charles. *The Origin of Species, by Means of Natural Selection*. Amherst, New York: Prometheus Books, 1859 (1991 ed., based on the sixth, final edition).

Davidson, Donald. *Problems of Rationality*. Oxford: Clarendon Press, 2004.

Dewey, John, and Tufts, J. *Ethics*. New York: Henry Holt, 1908 (1932 rev.).

———. *Reconstruction in Philosophy*. Boston: Beacon, 1920, (1972).

———. *Human Nature and Conduct*. New York: Modern Library, 1922, (1957).

———. "The Meaning of Value," *Journal of Philosophy*, 22, Feb. 1925, pp. 126-33. Reprinted in *John Dewey, The Later Works*. Edited by Boydston, J. Carbondale: So. Illinois Univ. Press, 1984, pp. 98-103.

———. "Value, Objective Reference and Criticism," *Philos. Review*, 34, July 1925, pp. 313-32. Reprinted in *John Dewey, The Later Works*. Edited by Boydston, J. Carbondale: So. Illinois Univ. Press, 1984, pp. 98-103.

———. "The Ethics of Animal Experimentation," *Atlantic Monthly,* 138, Sept. 1926, pp. 343-6. Reprinted in *John Dewey, The Later Works*. Edited by Boydston. J. Carbondale: So. Illinois Univ. Press, 1984, pp. 98-103.

———. *Experience and Nature*. New York: Dover, 1929a, (1958).

———. *The Quest for Certainty, A Study of the Relation of Knowledge and Action*. New York: Capricorn/ Putnam, 1929b.

———. "Three Independent Factors in Morals," *Educational Theory*, 16 (July 1966), 198-209. Trans. by Jo Ann Boydston. Originally published in French.

———. *Freedom and Culture*. New York: Capricorn Books, 1939

———. "Theory of Valuation," from the *International Encyclopedia of Unified Science*. Chicago: Univ. of Chicago Press, Vol. II, #4, 1939.

———. "Experience, Knowledge and Value, A Rejoinder," in *The Philosophy of John Dewey*. Edited by Schilpp, P.A. New York: Tudor Publ., 1939 (1951), pp. 517-608.

———. "Ambiguity of Intrinsic Good," Journ. Of Philosophy, XXXIX, June 42, pp. 328-30. Reprinted in *John Dewey, The Later Works*. Edited by Boydston, J. Carbondale: So. Ill. Univ. Press, 1984.

———. "The Field of Value," in *Value, a Cooperative Inquiry*. Edited by Lepley, Ray. New York: Columbia Univ. Press, 1949, ch. 3, p. 66.

Diamond, Jared. *The Third Chimpanzee: the Evolution and Future Development of the Human Animal*. New York: Harper-Collins, 1992, p. 236.

———. *Guns, Germs and Steel: The Fates of Human Societies*. New York: W. W. Norton, 2005.

———. *Collapse: How Societies Choose to Fail or Succeed*. London, England: Penguin Books (Penguin Viking in USA), 2006.

Dunker, Karl. "Ethical Relativity," in *Ethical Relativism*. Edited by Ladd, John. Belmont: Wadsworth Publ., 1973, p. 45 ff.

Edel, Abraham. "Some Trends in American Naturalistic Ethics," in *Philosophic Thought In France and the United States.* Edited by Marvin Farber. Albany: State Univ. of New York Press, 1950.

Empiricus, Sextus. *Adversus Mathematicos*, Book VII, 65 ff. Trans. by Osborne, C.P. Reprinted in *Philosophers Speak for Themselves.* Edited by Smith, T.V. Chicago: Phoenix/Univ. of Chicago Press, 1934 (1965 ed.), p. 66 ff.

Epicurus. "Letter to Menoeceus" in Diogenes Laertius, *Lives of the Eminent Philosophers*, X. Trans. by Yonge, C.

Everett, Walter G. *Moral Values, A Study in the Principles of Conduct*. New York: Henry Holt and Co. 1918.

Feinberg, Joel, editor. *Moral Concepts*. Oxford Univ. Press, 1970.

———. "Psychological Egoism." from *Reason and Responsibility*. Belmont, CA: Wadsworth, 1978, 4th ed.

Foucault, Michel. *Madness and Civilization*. New York: Vintage, 1973.

Frankena, William. *Ethics*. Englewood Cliffs: Prentice-Hall, 1963.

Freud, Sigmund. *Civilization and its Discontents*, [*Unbehagen in der Kultur*]. Trans. by Strachey. James. New York: W. W. Norton, 1962.

Frondisi, Risieri. *What is Value-An Introduction to Axiology*. Translated by Lipp. La Salle: Open Court Publ., 1971.

Garnett, A. Campbell. "Intrinsic Good: its Definition and Referent" in *Value, a Cooperative Inquiry*. Edited by Lepley, Ray. New York: Columbia Univ. Press, 1949, P. I, ch. 4.

Hare, Richard M. *Freedom and Reason.* Oxford Univ. Press, 1963 (1972 ed.).

———. *The Language of Morals*. Oxford Univ. Press, 1952 (1967).

Hartman, Robert S. "The Axiometric Structure of Intrinsic Value," *Journal of Value Inquiry*, Vol. 8, #2, 1974.

Hartmann, Nicholai. *Ethics*. London: Allen and Unwin, 1932).

Hegel, Georg W.F. *Lectures on the Philosophy of History*. Trans. by Sibree, J, New York: Dover Reprints, 1956

———. *Phenomenology of Spirit*. Trans. by Miller, A.V., and Findlay, J.N., Oxford Univ. Press, 1979.

———. *Philosophy of Right*. Trans. by Knox, T.M. Oxford Univ. Press, 1952/1978.

Heidegger, Martin. "*Brief über Humanismus*," in *Basic Writings*. Edited by Krell, New York: Harper and Row, 1977, sections I and V.

———. *Sein und Zeit.* Translated as *Being and Time* by Macquarrie, J. and Robinson, E., New York: Harper and Row, 1962.

———. *Kant and the Problem of Metaphysics*. Trans. by Taft, R. Bloomington: Indiana Univ. Press, 1990

Heilbroner, Robert. *The Worldly Philosophers: The Lives, Times and Ideas of the Great Economic Thinkers.* New York: Simon and Schuster, 1999 ed.

Held, Virginia. "Feminist Transformations of Moral Theory," *Philosophy and Phenomenological Research*, Fall 1990, pp. 321-344.

———. "Reason, Gender and Moral Theory," in *Philosophy and Phenomenological Research*, Fall 1990, pp. 321-344.

Hickman, Larry. *Pragmatism as Post-Postmodernism: Lessons from John Dewey*. New York: Fordham Univ. Press, 2007.

Hobbes, Thomas. *Leviathan*. Indianapolis: Hackett Publishing Co., 1994.

Hofstadter, Prof. Albert. *Agony and Epitaph: Man, His Art and His Poetry*. George Braziller Inc. Publ., 1970

Hume, David. *An Enquiry Concerning the Principles of Morals.* Reprinted (with *Treatise of Human Nature*) in *Hume's Moral and Political Philosophy*. Edited by Aiken, G. Darien: Hafner, 1970).

James, William. *A Pluralistic Universe*. in one volume with *Radical Empiricism*. New York, E.P. Dutton, 1971.

———. *Pragmatism and Other Essays*. New York: Washington Square Press, 1910, (1975).

Jessup, Bertram. "On Value," in *Value: A Cooperative Inquiry*. Edited by Lepley, Ray. New York: Columbia Univ. Press, 1949, ch. 7.

Kant, Immanuel. *Groundwork of the Metaphysics of Morals*. Trans. by Paton H. J. New York: Harper and Row, 1785 (1964).

———. *Critique of Practical Reason*. Trans. by Beck, L., Indianapolis: Bobbs-Merrill, 1788, (1956 Trans. by, 1976 ed.).

———. *Critique of Pure Reason*. Trans. by Smith, N.K. New York: Macmillan/St. Martin's, 1929/1965.

———. *Metaphysics of Morals*. Trans. by Ladd. Indianapolis: Bobbs-Merrill, 1797a (1965).

———. The "Doctrine of Virtue", Part II of the *Metaphysics of Morals*. Trans. by Ellington, J. Indianapolis, Hackett, 1797b (1994).

———. *Lectures on Ethics*. Trans. by Infield, L. Indianapolis: Hackett, 1775-80 (1930, 1963 ed.)

Kluckhohn, Clyde. "Ethical Relativity, Sic et Non," in *Ethical Relativism*. Edited by Ladd, John. Belmont: Wadsworth Publ., 1973, p. 78 ff.

Kockelmans, Joseph J., editor. *Contemporary European Ethics*. Garden City, New York: Anchor/Doubleday, 1972.

Ladd, John, editor. *Ethical Relativism*. Belmont: Wadsworth Publ., 1973

Laertius, Diogenes. *Lives and Opinions of the Eminent Philosophers*. Trans. by C. D. Yonge. London: C. Bell and Sons, 1915

Laird, John. *The Idea of Value*. Cambridge University Press, 1929.

Lao tse. *Dao de Jing*

LaSenne, Rene. "Ethics and Metaphysics," from *Traité de Moral Générale*. (Presses Universitaires de France, 1961), reprinted in *Contemporary European Ethics*. Edited by Kockelmans, Joseph J. Garden City, New York: Anchor/Doubleday, 1972, pp. 133-156.

Lavelle, Louis. "The Problem of Evil," from *Evil and Suffering*, Trans. by B. Murchland, and repr. in *Contemporary European Ethics*. Edited by

Kockelmans, Joseph J. Garden City, New York: Anchor/Doubleday, 1972, pp. 6-30).

Lee, Harold. "Methodology of Value Theory," in *Value: A Cooperative Inquiry*. Edited by Lepley, Ray. New York: Columbia Univ. Press, 1949, ch. 8.

Lemos, Noah M. *Intrinsic Value: Concept and Warrant*. Cambridge UP, 1994.

Leopold, Aldo. *A Sand County Almanac and Sketches Here and There*. Oxford University Press, 1949/1968.

Lepley, Ray, editor. *Value: A Cooperative Inquiry*. New York: Columbia Univ. Press, 1949.

Lewis, Clarence I. *An Analysis of Knowledge and Valuation*. (The Paul Carus Lectures Series 7) La Salle, Ill.: Open Court, 1946, (1962 ed.)

———. *Values and Imperatives: Studies in Ethics*. Edited by Lange, J. Stanford: Stanford University Press, 1969.

Lotze, Rudolf Hermann. *Metaphysics*. Trans. by Green, T.H. and Bosanquet, B., Oxford, Clarendon Press, 1887-8.

Lucretius. *De Rerum Natura*. Trans. as *On the Nature of the Universe*, by Latham, R. New York: Penguin Books, 1951/1977.

Marcuse, Herbert. *One Dimensional Man*. Boston: Beacon, 1964 & 1967.

Maslow, Abraham, editor. *New Knowledge of Human Values*. Chicago: Henry Regnery Company, 1959/1970

McDonald, Hugh P. "Dewey's Naturalism," *Environmental Ethics* (24:2 (Summer 2004), 189-208.

———. "Does Nature Exist?" *Contemporary Philosophy,* Vol. XXI, #5 & 6, Sept./Oct and Nov./Dec 2000.

———. *John Dewey and Environmental Philosophy*. Albany, New York: SUNY Press, 2003.

———. *Radical Axiology, a First Philosophy of Values.* Amsterdam: Rodopi, Value Inquiry Book Series, 2004.

———. *Political Philosophy and Ideology, A Critique of Political Essentialism*. Lanham, MD: University Press of America, 1997.

———. "Principles: The Principle of Principles," *The Pluralist*, December 2009.

———. "Toward a Deontological Environmental Ethic," *Environmental Ethics,* 23:4, Winter 2001, pp. 411-430.

McDonald, Lee. *Western Political Theory, From its Origin to the Present*. New York: Harcourt, Brace & World, 1968.

Mead, George Herbert. *Selected Writings*. Edited by Reck, A.J. Indianapolis: Bobbs-Merrill, 1964.

Mill, John S. *Utilitarianism*, in *The Utilitarians*. Garden City: Anchor/Doubleday, 1863, (1973 edition).

Miller, P. "Value as Richness: Toward a Value Theory for an Expanded Naturalism in Environmental Ethics," *Environmental Ethics*, Vol. 4, #2, 1982.

Moore, George E. *Principia Ethica*. Cambridge: Cambridge Univ. Press, 1903 (1959 ed.).

———. *Ethics*. Oxford: Oxford Univ. Press, 1912 (1965).

———. "Nature of Moral Philosophy," from *Philosophical Studies*. London: Routledge, 1922 (1958) ch. 10.

Münsterburg, Hugo. *The Eternal Values*. Boston: Houghton-Mifflin, 1909.

Murdoch, Iris. *The Sovereignty of Good*. New York: Ark/Routledge, 1970/1985.

Nash, Roderick F. *The Rights of Nature*. Madison, University of Wisconsin Press, 1989.

Neville, Robert C. *Reconstruction of Thinking*. Albany: SUNY Press, 1981.

Nietzsche, Friedrich. *The Birth of Tragedy*. Trans. by F. Golffing. Garden City, New York: Doubleday/Anchor, 1956.

———. *On the Genealogy of Morals*. Trans. by Kaufmann, W., (with *Ecce Homo)*, New York: Vintage Press/Random House, 1969.

———. *Twilight of the Idols* in *The Portable Nietzsche*. Trans. by Kaufmann New York: Viking Press, 1954 (1972 ed.).

———. *Will to Power*. Trans. by Kaufmann. New York: Vintage/Random House, 1967.

Norton, Brian. *Why Preserve Natural Variety?* Princeton: Princeton Univ. Press, 1987.

Oldenquist, Andrew. "Choosing, Deciding and Doing," in *The Encyclopedia of Philosophy*. Edited by Edwards, Paul. New York: Macmillan, 1967, 96-104.

Osborne, Harold. *Foundations of the Philosophy of Value: an Examination of Value and Value Theories*. Cambridge University Press, 1933.

Owens, Joseph. *The Doctrine of Being in the Aristotelian Metaphysics; a Study in the Greek Background of Medieval Thought*. Toronto: Pontifical Institute for Medieval Studies, 1957 (1978).

Parker, DeWitt "Discussion of John Dewey's 'Some Questions about Value'," in *Value: A Cooperative Inquiry*. Edited by Lepley, Ray. New York: Columbia Univ. Press, 1949.

Parsons, Talcott. *The Structure of Social Action*. New York: The Free Press, 1937/1968.

Paton, Herbert J. *The Categorical Imperative, a Study in Kant's Moral Philosophy*. Philadelphia: The University of Pennsylvania Press (1947), 1971.

Peirce, Charles S. *Reasoning and the Logic of Things: The Cambridge Conference Lectures of 1898*. (Introduction by Kenneth Laine Ketner and Hilary Putnam). Cambridge, Mass: Harvard Univ. Press, 1992.

Pepper, Steven. "A Brief History of General Theory of Value," in *A History of Philosophical Systems*. Edited by Ferm, Vergilius. New York: Philosophical Library, 1950, ch. 39.

———. "Observations on Value From an Analysis of a Simple Appetition" in *Value, a Cooperative Inquiry*. Edited by Lepley, R. New York: Columbia Univ. Press, 1949

Perry, Ralph Barton. *General Theory of Value*. New York, 1926.

———. *Realms of Value, A Critique of Human Civilization*. Cambridge: Harvard University Press, 1954.

Pico della Mirandola, Giovanni. "Essay on the Dignity of Man" [*De Hominis Dignitate*] trans by Sears, Jane. New York: Peter Lang, 1984.

Pirsig, Robert M. *Zen and the Art of Motorcycle Maintenance, An Inquiry into Values*. New York: William Morrow and Co. Publ., 1974 (Bantam ed. 1975)

Plato. *Gorgias, Protagoras, Phaedrus*, and *Republic*. Repr. in *Collected Dialogues*. Edited by Cairns, H. and Hamilton, E. Princeton: Princeton University Press, 1973 ed.

Plotinus. *Enneads*. Reprinted in *The Essential Plotinus*. Translated and edited by O'Brien. Indianapolis: Hackett, 1964.

Polin, Raymond. "Values and Truths," ch. 7 of *Contemporary European Ethics*. Trans. by Joseph J. Kockelmans. Garden City: Anchor/Doubleday, 1972.

Prall, David. W. *A Study in the Theory of Value*. Berkeley: Univ. of Calif. Publications in Philosophy, Vol. 3, #2, pp. 179-290, 1921.

Quine, Willard van Ordman. "Two Dogmas of Empiricism," in *From a Logical Point of View*, ch. 2. Cambridge Ma: Harvard Univ. Press, 1953.

Rachels, James. "The Challenge of Cultural Relativism," repr. in *Ethics—History, Theory and Contemporary Issues*. Edited by Cahn and Markie. Oxford Univ. Press, 1998, P. II, ch. 36.

Redfield, Robert. "The Universally Human and the Culturally Variable," in *Ethical Relativism*. Edited by Ladd, John. Belmont: Wadsworth Publ., 1973, p. 129 ff.

Rice, Phillip B. "Science Humanism and the Good," in *Value: A Cooperative Inquiry*. Edited by Lepley, R. New York: Columbia Univ. Press, 1949, ch. 14.

Rolston, Holmes III. "Are Values in Nature Subjective or Objective?" in *Environmental Ethics*. 4/2, 1982, p. 134.

———. "Challenges in Environmental Ethics," from *Ecology, Economics, Ethics: the Broken Circle*. New Haven: Yale University Press, 1991. (reprinted in Zimmerman, et al. eds. *Environmental Philosophy, From Animal Rights to Radical Ecology*. Upper Saddle River, NJ: Prentice Hall, 1998, 2nd ed., p. 124 ff.

———. *Environmental Ethics, Duties to and Values in the Natural World*. Philadelphia: Temple Univ. Press, 1988.

———. "Feeding People versus Saving Nature," in *The Ecological Community*. Gottlieb, R.S., ed. New York: Routledge, 1997. p. 208 ff.

———. "Naturalizing Values: Organisms and Species," in *Environmental Ethics: Readings in Theory and Application*. Edited by Pojman, L. Belmont, CA: Wadsworth, 2001 3rd ed., 76 ff.

———. *Philosophy Gone Wild: Environmental Ethics*. Amherst: Prometheus Books, 1989.

Ross, William D. *The Right and the Good.* London: Oxford at the Clarendon Press, 1930 (1961).

Roubiczek, Paul. *Ethical Values in the Age of Science*. Cambridge Univ. Press, 1969.

Sartre, Jean-Paul. "Existentialism is a Humanism." (Paris: Nagel, 1946) Trans. by Frechtman, B. New York: Philosophical Library, 1947.

Saydah, J. Roger. *The Ethical Theory of Clarence Irving Lewis*. Athens: Ohio Univ. Press, 1969.

Scheler, Max. *Formalism in Ethics and a Non-formal Ethics of Values*. translated by Manfred S. Frings and Roger L. Funk. Evanston: Northwestern Univ. Press, 1916 (1973).

Schiappa, Edward. "Towards a Pragmatic Approach to Definition: 'Wetlands' and the Politics of Meaning," in *Environmental Pragmatism*. Katz, E., and Light, A. eds. New York: Routledge, 1996, ch. 10.

Schiller, Ferdinand Canning Scott. *Humanism, Philosophical Essays*. New York: Macmillan, 1903.

Schumacher, Ernst F. "Buddhist Economics," from *Small is Beautiful: Economics as if People Mattered.* New York: Harper-Collins Publ., 1973.

Schweitzer, Albert. *Civilization and Ethics*. Trans. by J. Naish, (reprinted in *Animal Rights and Human Obligations*. Edited by T. Regan and P. Singer. Englewood Cliffs: Prentice Hall, 1976, sect. 10.

Sidgewick, Henry. *Methods of Ethics.* NewYork: Dover reprints, 1966.

Singer, Peter. *Animal Liberation*. New York: New York Review Press, 1975.

———. "Famine, Affluence and Morality," reprinted in *Ethics, History, Theory and Contemporary Issues*. Edited by Cahn, S and Markie, P. Oxford Univ. Press, 1998, ch. 45, 803 ff.

Skinner, Burrhus F. *Walden II*. New York: Bantam, 1972.

Spinoza, Baruch. *Ethics Demonstrated in a Geometric Manner*. reprinted in *The Rationalists*. Garden City, New York: Anchor/Doubleday, 1974

Stack, George J. "Nietzsche and the Phenomenology of Value," in *The Personalist*. 49, 1968, 78-102.

Stevenson, Charles L. "The Emotive Meaning of Ethical Terms," *Mind*, XLVI, 1938, pp 10-31. repr. in Margolis, J. ed., *Contemporary Ethical Theory: A Book of Readings*. New York: Random House, 1966.

Sumner, William Graham. "Folkways," in *Ethical Relativism*. Edited by Ladd, John. Belmont: Wadsworth Publ., 1973, ch. 3.

Toynbee, Arnold, J. *A Study of History*. Somervell Abridgement. Oxford University Press, 1946.

Urmson, James Opie. "On Grading," in *Mind*, 59, 1950, pp., 145-159.

van den Haag, Ernest. "In Defense of the Death Penalty," reprinted in *Ethics, History, Theory and Contemporary Issues*. Cahn and Markie, eds. Oxford Univ. Press 1998.

Veblen, Thorstein. *The Instinct of Workmanship*. selections repr in *The Portable Veblen*. Edited by Max Lerner. New York: Viking, 1948/1969,

Weber, Max. *The Protestant Ethic and the Spirit of Capitalism*. Trans. by Stephen Kalberg. Englewood Cliffs: Prentice Hall, 1977.

Werkmeister, William H. *Historical Spectrum of Value Theories*. Lincoln: Johnsen Publ. Co., 1970.

———. *Man and His Values*. Lincoln: Univ. of Nebraska Press, 1967.

Windelband, Wilhelm. "History and Natural Science," (Rectorial Address of 1894) in *History and Theory*, xix, 1980, pp. 165-185. Trans. by G. Oakes.

Yoon, Carol K. "Scientists Say Orangutans Can Exhibit 'Culture'," in *New York Times,* 1/3/2003.

ABOUT THE AUTHOR

Hugh P. McDonald (Ph.D. New School for Social Research) is Associate Professor of Philosophy at the New York City College of Technology (CUNY). He has taught at several schools in the New York City area. He is the author of *Political Philosophy and Ideology* (University Press of America, 1996), *Radical Axiology, a First Philosophy of Values* (Rodopi, Value Inquiry Book Series, 2004), and *John Dewey and Environmental Ethics* (SUNY Press, 2004). He has also written several forthcoming books, including and *Speculative Evaluations: Essays on a Pluralistic Universe*. He is co-editor, with Dr. John Shook, of *F. C. S. Schiller on Pragmatism and Humanism: Selected Writings, 1891-1937,* (Prometheus Books, 2007). His numerous journal articles span ethics, value theory, pragmatism, and the history of philosophy. His interests are in historical pragmatism, where he is one of the few F.C.S. Schiller scholars, value theory, and applied ethics, particularly environmental ethics and biotechnology. He is the co-founder of the New York Pragmatist Forum, a group that meets monthly to discuss issues in American philosophy. In 2009 he was elected Scholar on Campus by his peers at City Tech.

Prof. McDonald is also the founder of the Calochortus Society, a group devoted to a beautiful family of native American wildflowers, and past editor of its newsletter. He has also written articles in botany and horticulture journals. An avid wildflower photographer, he maintains a web site of his published and unpublished photographs.

INDEX

VIBS

The **Value Inquiry Book Series** is co-sponsored by:

Adler School of Professional Psychology
American Indian Philosophy Association
American Maritain Association
American Society for Value Inquiry
Association for Process Philosophy of Education
Canadian Society for Philosophical Practice
Center for Bioethics, University of Turku
Center for Professional and Applied Ethics, University of North Carolina at Charlotte
Central European Pragmatist Forum
Centre for Applied Ethics, Hong Kong Baptist University
Centre for Cultural Research, Aarhus University
Centre for Professional Ethics, University of Central Lancashire
Centre for the Study of Philosophy and Religion, University College of Cape Breton
Centro de Estudos em Filosofia Americana, Brazil
College of Education and Allied Professions, Bowling Green State University
College of Liberal Arts, Rochester Institute of Technology
Concerned Philosophers for Peace
Conference of Philosophical Societies
Department of Moral and Social Philosophy, University of Helsinki
Gannon University
Gilson Society
Haitian Studies Association
Ikeda University
Institute of Philosophy of the High Council of Scientific Research, Spain
International Academy of Philosophy of the Principality of Liechtenstein
International Association of Bioethics
International Center for the Arts, Humanities, and Value Inquiry
International Society for Universal Dialogue
Natural Law Society
Philosophical Society of Finland
Philosophy Born of Struggle Association
Philosophy Seminar, University of Mainz
Pragmatism Archive at The Oklahoma State University
R.S. Hartman Institute for Formal and Applied Axiology
Research Institute, Lakeridge Health Corporation
Russian Philosophical Society
Society for Existential Analysis
Society for Iberian and Latin-American Thought
Society for the Philosophic Study of Genocide and the Holocaust
Unit for Research in Cognitive Neuroscience, Autonomous University of Barcelona
Whitehead Research Project
Yves R. Simon Institute

Titles Published

Volumes 1 - 190 see www.rodopi.nl

191. Eugenio Garin, *History of Italian Philosophy*. Translated from Italian and Edited by Giorgio Pinton. A volume in **Values in Italian Philosophy**

192. Michael Taylor, Helmut Schreier, and Paulo Ghiraldelli, Jr., Editors, *Pragmatism, Education, and Children: International Philosophical Perspectives*. A volume in **Pragmatism and Values**

193. Brendan Sweetman, *The Vision of Gabriel Marcel: Epistemology, Human Person, the Transcendent*. A volume in **Philosophy and Religion**

194. Danielle Poe and Eddy Souffrant, Editors, *Parceling the Globe: Philosophical Explorations in Globalization, Global Behavior, and Peace*. A volume in **Philosophy of Peace**

195. Josef Šmajs, *Evolutionary Ontology: Reclaiming the Value of Nature by Transforming Culture*. A volume in **Central-European Value Studies**

196. Giuseppe Vicari, *Beyond Conceptual Dualism: Ontology of Consciousness, Mental Causation, and Holism in John R. Searle's Philosophy of Mind*. A volume in **Cognitive Science**

197. Avi Sagi, *Tradition vs. Traditionalism: Contemporary Perspectives in Jewish Thought*. Translated from Hebrew by Batya Stein. A volume in **Philosophy and Religion**

198. Randall E. Osborne and Paul Kriese, Editors, *Global Community: Global Security*. A volume in **Studies in Jurisprudence**

199. Craig Clifford, *Learned Ignorance in the Medicine Bow Mountains: A Reflection on Intellectual Prejudice*. A volume in **Lived Values: Valued Lives**

200. Mark Letteri, *Heidegger and the Question of Psychology: Zollikon and Beyond*. A volume in **Philosophy and Psychology**

e201. Carmen R. Lugo-Lugo and Mary K. Bloodsworth-Lugo, Editors, *A New Kind of Containment: "The War on Terror," Race, and Sexuality*. A volume in **Philosophy of Peace**

202. Amihud Gilead, *Necessity and Truthful Fictions: Panenmentalist Observations*. A volume in **Philosophy and Psychology**

203. Fernand Vial, *The Unconscious in Philosophy, and French and European Literature: Nineteenth and Early Twentieth Century.* A volume in **Philosophy and Psychology**

204. Adam C. Scarfe, Editor, *The Adventure of Education: Process Philosophers on Learning, Teaching, and Research*. A volume in **Philosophy of Education**

205. King-Tak Ip, Editor, *Environmental Ethics: Intercultural Perspectives*. A volume in **Studies in Applied Ethics**

206. Evgenia Cherkasova, *Dostoevsky and Kant: Dialogues on Ethics*. A volume in **Social Philosophy**

207. Alexander Kremer and John Ryder, Editors, *Self and Society: Central European Pragmatist Forum*, Volume Four. A volume in **Central European Value Studies**

208. Terence O'Connell, *Dialogue on Grief and Consolation*. A volume in **Lived Values, Valued Lives**

209. Craig Hanson, *Thinking about Addiction: Hyperbolic Discounting and Responsible Agency*. A volume in **Social Philosophy**

210. Gary G. Gallopin, *Beyond Perestroika: Axiology and the New Russian Entrepreneurs*. A volume in **Hartman Institute Axiology Studies**

211. Tuija Takala, Peter Herissone-Kelly, and Søren Holm, Editors, *Cutting Through the Surface: Philosophical Approaches to Bioethics*. A volume in **Values in Bioethics**

212. Neena Schwartz: *A Lab of My Own*. A volume in **Lived Values, Valued Lives**

213. Krzysztof Piotr Skowroński, *Values and Powers: Re-reading the Philosophical Tradition of American Pragmatism*. A volume in **Central European Value Studies**

214. Matti Häyry, Tuija Takala, Peter Herissone-Kelly and Gardar Árnason, Editors, *Arguments and Analysis in Bioethics*. A volume in **Values in Bioethics**

215. Anders Nordgren, *For Our Children: The Ethics of Animal Experimentation in the Age of Genetic Engineering*. A volume in **Values in Bioethics**

216. James R. Watson, Editor, *Metacide: In the Pursuit of Excellence*. A volume in **Holocaust and Genocide Studies**

217. Andrew Fitz-Gibbon, Editor, *Positive Peace: Reflections on Peace Education, Nonviolence, and Social Change*. A volume in **Philosophy of Peace**

218. Christopher Berry Gray, *The Methodology of Maurice Hauriou: Legal, Sociological, Philosophical*. A volume in **Studies in Jurisprudence**

219. Mary K. Bloodsworth-Lugo and Carmen R. Lugo-Lugo, *Containing (Un)American Bodies: Race, Sexuality, and Post-9/11 Constructions of Citizenship*. A volume in **Philosophy of Peace**

220. Roland Faber, Brian G. Henning, Clinton Combs, Editors, *Beyond Metaphysics? Explorations in Alfred North Whitehead's Late Thought*. A volume in **Contemporary Whitehead Studies**

221. John G. McGraw, *Intimacy and Isolation (Intimacy and Aloneness: A Multi-Volume Study in Philosophical Psychology, Volume One)*, A volume in **Philosophy and Psychology**

222. Janice L. Schultz-Aldrich, Introduction and Edition, *"Truth" is a Divine Name, Hitherto Unpublished Papers of Edward A. Synan, 1918-1997*. A volume in **Gilson Studies**

223. Larry A. Hickman, Matthew Caleb Flamm, Krzysztof Piotr Skowroński and Jennifer A. Rea, Editors, *The Continuing Relevance of John Dewey: Reflections on Aesthetics, Morality, Science, and Society*. A volume in **Central European Value Studies**

224. Hugh P. McDonald, *Creative Actualization: A Meliorist Theory of Values*. A volume in **Studies in Pragmatism and Values**

CPSIA information can be obtained at www.ICGtesting.com
Printed in the USA
LVOW06*0747250214

375084LV00003B/28/P

9 789042 032538